weed/dated

W9-AYM-641

ELSEVIER'S
DICTIONARY OF
CYBERNYMS

ELSEVIER'S DICTIONARY OF CYBERNYMS

Abbreviations and Acronyms
used in Telecommunications,
Electronics and
Computer Science

in

English, French, Spanish, and **German**
with some **Italian, Portuguese, Swedish, Danish** and **Finnish**

by

T.R. PYPER and C.A.C. STOUT
Horncastle, Lincs, United Kingdom

2000
ELSEVIER
Amsterdam – Lausanne – New York – Oxford – Shannon – Singapore – Tokyo

ELSEVIER SCIENCE B.V.
Sara Burgerhartstraat 25
P.O. Box 211, 1000 AE Amsterdam, The Netherlands

First edition 2000

Library of Congress Cataloging in Publication Data
A catalog record from the Library of Congress has been applied for.

ISBN: 0-444-50478-8

⊚   The paper used in this publication meets the requirements of ANSI/NISO Z39.48-1992 (Permanence of Paper).
Printed in The Netherlands.

Preface

This dictionary grew out of a common vexation: the proliferation of quasi-industrial jargon in the field of information technology, compounded by the fact that these somewhat esoteric terms are often further reduced to acronyms and abbreviations which are seldom explained. Even when they are defined, individual interpretations continue to diverge.

The term "cybernym" was invented precisely to encompass all of these variants. It is derived from Norbert Wiener's coinage of the word *cybernetics* (1948), embracing the science of automatic control and communications systems — with specific reference to the field of telecommunications, which has progressed rapidly on the strength of recent advances in micro-electronics and computer science. The neologism might be defined as follows:

> An acronym, abbreviation, mnemonic, convenient contraction, cypher, signalling code, control message or vernacular initialism used in cybernetics applications (especially telecommunications).
>
> [Gr *kybernētes* a steersman, and *onyma/onuma = onoma* name]

If we were asked to name the wonder of the past century, few would immediately think of the telephone. Yet the single most astonishing development that has penetrated all our lives is the subsequent application of that very instrument, first patented in 1876, to virtually all aspects of our day-to-day existence.

The apparent simplicity of making a telephone call is something we take for granted. But the vast array of complex operations actually involved when a subscriber (or "user") lifts the telephone receiver is very cleverly concealed. First, a ringing current is generated by the exchange, then dial tone is heard (the steady thrum that confirms the closing of the line, or loop). If the user is hesitant — or, perhaps, premature — when dialling, a whole new sequence of operations can ensue. (Some countries still have an intermediate dialling pattern, where one or two digits are composed to await a second, different tone before proceeding with the number.)

If the call is directed to a local subscriber, the dialled digits provoke a particular sequence of actions in the nearby exchange. A national call generates an entirely different series of operations — as does an international call. If it is a data or facsimile call, yet further analyses and procedures (and therefore signalling codes) are implemented in the network. The call may be routed via a satellite link, it may be addressed to a cellular mobile network, or it may even pass through a microwave link. Each of these creates quite different patterns of response in the system.

On the other hand, the user might inadvertently drop the handset and leave it dangling off-hook: a further software routine immediately releases the howler tone as an alerting stratagem. Another user might vandalize a payphone or break into the cashbox, or mistakenly — perhaps fraudulently — enter a spurious personal identification number (as with charge and credit cards).

The point is that each of these operations generates a unique binary sequence (the activating zeroes and ones, or off/on switches) in the network and every occurrence is marked by a code, which may be a signal or a message. These codes are identified by abbreviations or mnemonics of between two and five (or sometimes more) letters. Clear-back (CBK) when the caller hangs up, or clear-forward (CLF) when the called party goes on-hook, for instance, are just two of thousands of such abbreviated codes.

Unfortunately (from the standpoint of the engineer, network planner, software designer, translator or research analyst, let us say), national preferences dictate that many — indeed, most — of these codes differ by name from country to country and the international standards organization has sought over the years to harmonize and rationalize the codes to establish a degree of compatibility between sovereign states. So, for the French, CBK becomes RAC (raccrochage), whilst in Spain it is rendered as COL (colgar); CLF, however, is interpreted as FIN by both the French and the Spanish.

Naturally enough, confusion persists. Let us suppose, for instance, that the national telephone company in Venezuela (CANTV) wishes to modernize, or simply upgrade, its telephone system in the capital Caracas (say half a million lines, or installations, for the business community alone); it will probably issue a public invitation to tender calling for bids to refurbish the system. Enquiries for the specifications are received from manufacturers worldwide — most of which, incidentally, require the documents and software to be translated into English. Flowcharts and tables, tariff scales, frequency allocations, specific national preferences and numerous exceptions will abound in the specifications. Even worse, those ineffable codes are often presented in combination (some English, some Spanish) and additional, non-standard codes (often related to call-charging and security) may be introduced for use in the Venezuelan network exclusively.

The problem is that until now the codes have been reproduced in separate (language) publications; there is no universal listing in alphabetical order that covers all three variants (to say nothing of the German, Italian, or Greek offerings). This dictionary sets out the English, French and Spanish alternatives as a single, merge-sorted whole. Even so, this is scarcely universal, since many language options are available. But it is a start — and an important one — which will assist the many specialists who have need of such information.

Today, of course, most of the codes have passed into the public domain, simply because they exist in most of the telecommunications systems installed throughout the developed (and developing) world and are largely known to most of those who work in that particular area. But the foreign variants often defy even the most astute observer. Where they are known, they are included here.

Our dictionary seeks to clarify this bewildering situation as much as possible. The 26 000 definitions set out here, drawn from some 16 000 individual cybernyms, cover computing, electronics, telecommunications (including intelligent networks and mobile telephony) together with satellite technology and Internet/Web terminology (oftentimes "cutting edge" or, to be more precise, facetious, as in PCMCIA — people can't memorize computer industry acronyms — and occasionally rather vulgar but since they exist and are in frequent use they are included here).

Annex I lists some of the innumerable file types encountered by those who peer into their computer's filing system using powerful desktop managers. These file extensions, or three-letter acronyms (TLA) are also found on the Properties tab of every folder and in the Windows "Open with" dialogue. An awareness of the company or organization that generated the software takes the user a long way towards understanding what to do with an unknown file type. Some of the extensions listed may well be uncommon but these "older" programs are still widely used (in the same way as morse code is still a preferred medium of communications — despite its formal banishment): obsolescence is a crass marketing ploy.

Annex II lists the abbreviations of country names found in universal resource locators (URL) — in other words, Internet/Web addresses. If nothing else, it shows how pervasive the World Wide Web actually is and how that name is truly justified. How else, for instance, would one have first heard of the place which is referred to by the letters *nu*? (In fact, Niue, an island territory to the east of Tonga.)

A list such as this is interminable, much the same as the painting of the great bridges of the world. The final addition to this dictionary was made in the first week of February 2000. There will be more tomorrow.

Editorial Annotations

Non-English cybernyms and their expansions are translated, whilst the English equivalent is indicated by the use of square brackets (*ie* [ADC]).

Lower-case letters in parentheses (round brackets) indicate the language concerned: (f) French; (s) Spanish; (g) German; (p) Portuguese; (i) Italian; (nl) Dutch; (d) Danish. The names of other languages are given in full.

Other notations

aka (also known as) offers an alternative name for a term or organization — usually a colloquial usage (e.g. OO/O2, or pixel/pel). It might also be a diversionary reduction.

cf (*confer* compare) indicates a comparison or contrasting definition.

qv (*quod vide* which see) suggests a cross-reference to a specific, related cybernym.

tn (trade name) indicates that the cybernym and its expansion are closely associated with the company or organization to which they are attached in this dictionary. This usage in no way implies or confers any copyright ownership, nor does it in any way suggest or affect the legal status of any registered trademark. Often they are simply not known to us.

v (*vide* see) refers to a similar (or the same) definition, but with a different cybernym (e.g. Netnews/Usenet). Otherwise an expansion.

Comments

Hyphens are inserted liberally to separate colliding vowels, or to assist pronunciation. Mid-Atlantic spelling is preferred in most cases (-ize, rather than -ise, but adaptor in preference to adapter; program for computer-related control systems, but programme for a radio or television broadcast, a schedule, timetable, plan or prospectus.

A

A administration, administración (s), gestion (f); attenuation, affaiblissement (f), atenuación (s); additional; ampere, ampère (f), amperio (s); availability

A# programming language (component of AXIOM v.2)

A&CP access and control point

A&SG accessories and supplies group

A(t) instantaneous availability, disponibilité (instantanée) (f), disponibilidad instantánea (s)

Å Ångström

A/B answerback; indicatif de réponse (f); distintivo (s)

A/D analogue-to-digital, conversion analogique/digitale (f), analógico-digital (s)

A/m³ ampere per cubic metre

A/N analogique-numérique (f) [A/D]

A/R alternate route

A/SYS access system, système d'accès (f), sistema de acceso (s)

A/UX Apple Mac version of UNIX

A3 authentication algorithm 3

A38 algorithm performing A3 and A8 functions

A5 stream cipher algorithm

A5/2 encryption algorithm A5/2

A8 ciphering key generating algorithm A8

AA atomic absorption, abort accept; acceso aleatorio (s) [RA]; acuse de recibo de datos acelerado (s) [EA]; signal d'acceptation d'appel (f) [CA]

A-A analogue-analogue [A/D]

AAA Aces for Ansi Art; adaptive antenna array

AAAI American Association for Artificial Intelligence

AAAS American Association for the Advancement of Science

AAB auto answerback

AAC automatic amplitude control

AACPC Apple communications interface adaptor *tn*

AACR Anglo-American Cataloguing Rules

AAD analogue alignment disk(-ette); appel à discussion (f) (conference call)

AADL axiomatic architecture description language

AAFE advanced applications flight experiments

AAIC accounting authority identification code, code d'identification de l'autorité chargée de la comptabilité (f), código de identificación de la autoridad encargada de la contabilidad (s)

AAL ATM adaptation layer

AAL(1/2) ATM adaptation layer (1/2)

AAL-x ATM adaptation layer type x

AAP applications access point (DEC); amplificador lineal de alta potencia (s) [HPA]

AAPM Associação dos Accionistas Privados da Marconi (p)

AAPTM Agrupamento de Accio para a Privatização Total da Marconi (p)

AAR automatic alternative/alternate routing

AARE A-ASSOCIATE RESPONSE application-protocol-data-unit

AARNET Australian Academic Research Network

AARP Apple address resolution protocol

AARQ A-ASSOCIATE REQUEST application-protocol-data-unit

AARTS automatic audio remote test

AAS active accessibility support (Microsoft Windows NT); atomic absorption spectroscopy; automatic announcement subsystem; automatic area segmentation (Epson)

AASP ASCII asynchronous support package

AAT average access time

AATRA Asociación Argentina de Telegrafistas Radiotelegrafistas y Afines (s)

AAU alarm adaptation unit

AAV appel à voter (f) (voting call)
AAX automated attendant exchange
AB abort; access burst; Anarchy Burger
abA abampere
ABA adaptive bandwidth adjuster; atenuación en bucle abierto (s) [OLL]
ABATS automated bit access test system
ABBS Apple bulletin board system *tn*
abC abcoulomb
ABC activity-based costing; adaptive bus controller; acuse de recibo de bucle (s) [LPA]; advanced broadband communications; answer-back code; American Broadcasting Company; Atanasoff-Berry computer (first electronic digital); Australian Broadcasting Company; automatic bill calling; automatic beam control; automatic brightness control
ABCD AltaVista Business Card Directory *tn*
ABCF adaptive break-in control function
ABCN Asia Broadcasting and Communications Network
ABDS adaptive break-in differential sensitivity
ABE agent building environment; mensaje de acuse de bloqueo de grupo de circuitos por fallo del equipo (s) [HBA]
abend abnormal end, fin anormale d'une tâche (f), final anormal (s)
ABET Asociación Boliviana de Empresas Telefónicas (s)
ABGM acuse de bloqueo de grupo de circuitos por fallo del equipo (s) [MBA]
ABGSF mensaje de acuse de bloqueo de grupo por fallo del soporte físico (s) [HBA]
ABGSL mensaje de acuse de bloqueo de grupo por fallo del soporte lógico (s) [SBA]
ABHC average busy hour calls
ABI application binary interface
ABIOS advanced BIOS
A-bis interface between base station controller and BTS

ABIST automatic built-in self-test (IBM)
ABL mensaje de acuse de bloqueo de grupo de circuitos generado por soporte lógico [SBA]
ABLE adaptive battery life extender
ABM activity-based management; asynchronous balanced mode; mensaje de acuse de bloqueo de grupo de circuitos para el mantenimiento (s) [MBA]
ABNF augmented Backus-Naur Form (RFC 2234)
ABO open-loop loss [OLL]; señal (eléctrica) de abonado ocupado (s) [SSB]
ABR answer-bid-ratio; available bit rate (ATM standard); automatic baud rate (detection)
ABRS automated book request system (British Library)
ABRT A-ABORT application-protocol-data-unit
ABS absolute function; address book synchronization (IBM); Alcatel Business System; alternative billing service; average busy stream
ABSBH average busy season busy hour
ABT abort
ABTS ASCII block terminal services
ABU Asia-Pacific Broadcasting Union
A-buffer visible surface information-holding algorithm
ABUL Association Bordelaise des Utilisateurs de Linux (f)
ABUT Association Belge des Utilisateurs de Télécom (f)
abV abvolt
abWb abweber (enhed for magnetisk mængde) (d)
Ac actinium
AC acceptation (f); accept; access; access channel; access control; activated carbon; assicurata convenzionale (i); Assistenzcomputer (g); acuerdo contractual (s) [CA]; acuse de recibo de conexión (s) [CA]; acuse de recibo de datos (s) [AK]; administration centre; advice of charge; alternating current;

answer complete; application context; approved correction; appel en cours (f) [CP]; application channel; application context; autocheck; automatic computer; awaiting connection; autoridad de certificación (s) [CA]; axiom of choice

AC PPDU alter context PPDU

AC&R American Cable and Radio System

ACA adaptive channel allocation; Application Control Architecture (DEC) *tn*; asynchronous communications adaptor; automatic circuit assurance; automatic conference arranger

ACA PPDU alter-context acknowledgement PPDU

ACAP application configuration access protocol

ACARB Australian Computer Abuse Research Bureau

ACARS Arinc communications addressing and reporting system, système d'adressage et de compte rendu type Arinc (f)

ACAST Advisory Committee on the Application of Science and Technology to Development

ACATS Advisory Committee on Advanced Television Services

ACB access-barred signal; annoyance call bureau

ACC accumulator, acumulador (s); asynchronous communications control, control de comunicaciones asincrónicas (s); access control class; Advanced Computer Communications, Alliance for Competitive Communications; arquitectura de contenido de caracteres (s) [CCA]; automatic call-back calling; automatic chrominance control; automatic clamp control; automatic congestion control

ACCA asynchronous communications control attachment, conexión de control de comunicaciones asincrónicas (s)

ACCE accessibility of signalling points (Nº 7)

Accent OS component of SPICE

ACCESS American Computerized Commodity Exchange System and Services; analysis computer for component engineering services support

ACGG arquitectura de contenido de gráficas geométricos (s) [GGCA]

ACCH associated control channel

ACCIS automatic command, control and information system

ACCOLC access overload class

ACCS automated calling card service

ACCT Ad hoc Committee for Competitive Telecommunications (EU), Comité en cada país para las telecomunicaciones competitivas (s)

ACCU Academisch Computercentrum Utrecht (nl); alternating current connection unit

Accunet switched (56 kbit/s) packet service (AT&T); réseau à commutation par paquets de AT&T (f)

ACD acheminement de débordement (overflow routing) (f); automated call delivery/distribution, distributeur automatique d'appels (f), distribución de llamadas automáticas (s), distribuidor automático de llamadas (s)

ACD-ESS automatic call distributor-electronic switching system

ACDI asynchronous communications device interface

ACE advanced communications enhancement; advanced computing environment (SCO, DEC, Compaq, Microsoft), entorno informático avanzado (s); adverse channel enhancements (Microcom); Ansi Creation Enterprise; asynchronous communication element; transmission et réception de données sous forme asynchrone (f), transmisión y recepción de datos en forma asíncrona (s), transmissão e recepção de dados sob a forma asíncrona (p), trasmissione e ricezione di dati sotto forma asincrona (i), Sendeempfangsbaustein für

4

asynchrone Daten (g), programmerbart asynkront kommunikationskredsløb (d); audio-connecting equipment; automatic calling equipment; automatic computing engine (Turing); Adaptive Contrast Enhancement (Bell & Howell *tn*)

ACELP algebraic code-excited linear prediction (Frame Relay Forum)

AceM Association of Contract Electronics Manufacturers (UK)

ACF access control field; advanced communications function/facility

ACF/NCP advanced communication function/network control program

ACF/VTAM advanced communication function/virtual terminal access method

ACG adjacent charging group

ACGG arquitectura de contenido de gráficos geométricos (s) [GGCA]

ACH acheminement (f) (routing); Association for Computers and the Humanities; attempts per circuit per hour; automated (automatic) clearing house

ACI after clean inspection; Automatic Channel Installation; señal de acuse de recibo de servicio de interfuncionamiento (s) [IACK]; accés interdit (f) [ACB]

ACIA asynchronous communications interface adaptor, interface d'adaptation pour communications asynchrones (f), interfaz de adaptación para comunicaciones asíncronas (s)

ACIAS automated calibration interval-analysis system

ACiD ANSI Creators in Demand

ACID atomic - consistent - isolated - durable

ACIS American Committee for Interoperable Systems; Andy, Charles, Ian's System (Spatial Technologies *tn*)

ACITT National Association for Coordinators and Teachers of IT (UK)

ACK acknowledge (character); caractère accusé de réception (positif) (f)

ACL access control list, liste de contrôle

(f); A Co-routine Language; advanced CMOS logic; appliance computer language; Association for Computational Linguistics; atenuación de conversión longitudinal (s) [LCL]

ACLL agente de control de llamada (s) [CCA]

ACM accumulated call meter; acheminement multiple (f) (multiple routing); address-complete message; addressed call mode; alarm control module; Association for Computing Machinery (US), Asociación para las Máquinas de Cálculo (s); audio compression manager (Microsoft)

ACME application of computers to manufacturing engineering

ACMOS advanced CMOS

ACMS application control management system

ACN accusé de réception négatif (f) [NACK]; advisory Committee on Networking (UK)

ACNA access carrier name abbreviation

ACNV automatically-controlled natural ventilation

ACO Aerosat Coordination Office; Bureau de coordination AEROSAT; message d'adresse complète (f) [ACM]

ACOC area communications operations center

ACOLI advance circuit order and layout information

A_COM combined link set; faisceau combiné de canaux sémaphore (f); atenuación combinada (s)

Acorn Access to Course Readings via Networks (UK)

ACOST Advisory Council on Science and Technology (UK)

ACP accept; algebra of communicating processes; ancillary control program/process; signal de numéro complet publiphone (f) [ADX]; airlines control package (IBM *tn*)

ACPE Association of European Private

Cable Operators, Association des câblo-opérateurs privés Européens (f)

ACPI advanced component power interface; advanced configuration and power interface

ACPM association control protocol machine

ACR absolute category rating (ITU test); allowed cell rate; Association Canadienne des Radiodiffuseurs (f); attenuation to crosstalk ratio; automatic character recognition; automatic control radar

ACRI Advanced Computer Research Institute

ACRONYM a convenient reduction of neologisms and yesteryear mnemonics

ACROSS automated cargo release and operations service system

ACRTF Association Canadienne de la radio et de la télévision française (f)

ACS access control set; alarm and control system; application connectivity services; asynchronous communication server; atelier de création de services (f) (service creation environment) [SCE]; attitude control system; Australian Computer Society; automated computer (time) service; automatic class selection; automatic computing system; address complete signal - no charge [ADN]

ACSA Allied Communications Security Agency, Bureau allié de Sécurité de Transmissions (f), Alliiertes Amt für Fernmeldesicherheit (g)

ACSE application common service elements; associated control service element/environment

ACSEPT automatic character-set exchange processing technique

ACSI Atari Computer Systems Interface

ACSO East Africa Common Services Organization

ACSR aluminium conductors - steel reinforced

ACSS aural cascading stylesheets

ACT activate, message d'activation (f); actual cycle time; alternative control techniques; annual change traffic, Application Center for Technology; Association of Commercial TV in Europe; automated credit transfer; (bloque de) aceptación de conexión de transporte (s) [TCA block]; atenuación de conversión transversal (relación de conversión transversal) (s) [TCL]; signal de numéro complet, avec taxation (f) [ADC]

Act1/2/3 actor language (from Plasma *tn*)

ACTA America's Carriers Telecommunications Associations

ACTAS Alliance of Computer-Based Telephony Application Suppliers (US)

ACTD advanced concept technology demonstration initiative (US)

ACTE Approvals Committee for Terminal Equipment (EU), Comité d'approbation des équipements de télécommunications (f), Zulassungsausschuß für Telekommunikationseinrichtungen (g)

ACTI active

ACTIUS Association of Computer Telephone Integration Users and Suppliers

ACTLU activate logical unit (SNA)

ACTOR object-oriented language (Whitewater Group Ltd)

ACTP Advanced Computer Technology Project

ACTPU activate physical unit (SNA)

ACTS Advanced Communications Technologies and Services, technologies et services avancés des communications (f), tecnologías y servicios avanzados de comunicación (s), Zukünftige Kommunikationstechnologien und -dienste (g); Advanced Communications Technology Satellite systems; algorithms and computational tools for complex system; automated coin toll service

ACTT Advanced Communication and Timekeeping Technology (Seiko *tn*)

ACTV advanced compatible TV

ACU acknowledgement signal unit, unité de signalisation d'accusé de réception (f), unidad de señalización de acuse de recibo (s); dispositivo de llamada automático (s); alarm collection unit; antenna combining unit; automatic call unit

ACUN acuse de recibo negativo (s) [NACK]

ACX Active Control Experts (US corp)

A-D analógico-digital (s) [A/D]

AD acceso digital (s) [DA]; activity discard service; address; administrative domain; application development; almacenamiento de documentos (s) [DS]; asignación en función de la demanda (s) [DA]; awaiting disconnection; message d'addresse, émis vers l'avant (f) [FAM]; tiempo de acuse de recibo distante (s) [AR]

AD/cycle application development cycle

Ada programming language named after Ada Lovelace (US DoD)

ADA activity discard acknowledgement; arquitectura de documento abierta (s) [ODA]; asynchronous device adaptor; automatic data acquisitions

ADABAS A Database System (Software AG *tn*)

Adaline adaptive linear neurons (architecture)

ADAM Animated Dissection of Anatomy for Medicine; All-Digital Answer Machine (Casio *tn*)

Adamicro Asociación para el Desarrollo de la tecnología y Aplicaciones de los Microprocesadores (s)

ADAPSO Association of Data Processing Services Organizations

Adaptor automatic data parallelism translator

ADAS advanced directory assistance system

ADB Apple Desktop Bus *tn*

ADC active desktop computer; adaptive data compression (Hayes); address complete signal - charge; signal de numéro complet avec taxation (f), señal de dirección completa con tasación (s); acuse de recibo de datos de capacidad (s) [CDA]; administration centre; air data computer, centrale aérodynamique (f), calculador anemométrico (s); American digital cellular (phone) system; analogue-to-digital converter (*also* A/D converter), conversor analógico-digital (s), convertidor analógico-digital/digitalizador (s)

ADCCP advanced data communication control procedures/protocol

ADCSP Advanced Defense Communications Satellite Project

ADCU alarm display and control unit

ADD addition; Applied Digital Devices (UK); automatic document detection (WordPerfect)

ADDD A Depository of Development Documents

ADDER automatic digital data error recorder

ADDS Applied Digital Data Systems Inc; automatic direct distance dialling system

ADE advanced development environment; ATS data exchange; attenuation distortion equivalent; Audible Doppler Enhancer *tn*; mensaje de acuse de desbloqueo de grupo de circuitos por fallo de equipo (s) [HUA]

ADELI Association pour le développement de la logique informatique (f)

AD-Employ project analysing impact on jobs of transition to IT

ADES automatic digital encoding system

ADF automatic direction finding; radio-compás automatique (f); radiogoniomètre de bord (f), radiogoniometría automática (s); automatically defined function; automatic document feed/feeder

adger to do something stupid without thought

ADGLS mensaje de acuse de desbloqueo de grupo generado por el soporte lógico (s) [SUA]

ADGM mensaje de acuse de desbloqueo de grupo para mantenimiento (s) [MUA]

ADGSF mensaje de acuse de desbloqueo de grupo por fallo de soporte físico (s) [HUA]

ADI address incomplete signal, signal de numéro incomplet (f); after develop inspection; Agence de l'Informatique; AutoCad/AutoDesk device interface *tn*

ADIBAN Association pour le développement de l'utilisation de l'informatique et de l'automatique en Basse Normandie (f)

ADIG application development interface guidelines

ADIJ Association pour le développement de l'informatique juridique (f)

ADILOR Association pour le développement de l'informatique en Lorraine (f)

ADIM Association pour le développement de l'informatique médicale (f)

ADIO analogue-digital input-output

ADIPC Association des informaticiens de Poitou-Charentes (f)

ADIRA Association pour le développement de l'informatique dans la région Rhône-Alpes (f)

ADIRC Association pour le développement de l'informatique dans la région Centre (f)

ADIREB Association pour le développement de l'informatique et de l'automatique dans la région Bourgogne (f)

ADIS automated data interchange system, système automatique d'échange de données (f), sistema automático de intercambio de datos (s)

ADISP automated data interchange systems panel (EU), groupe d'experts sur les systèmes automatiques d'échange de données (f)

ADIT Agence pour la diffusion de l'information technologique (f)

ADIZ air defense identification zone

ADL add lot; Ada development language; address data latch; Adventure Definition Language (Cunniff/Brengle); API definition language; associated data line; acuse de desbloqueo de grupo de circuitos generado por el soporte lógico (s) [SUA]

ADLAT adaptive lattice filter

ADLC adaptive lossless data compression (IBM); asynchronous data link control

AdLog language which adds a layer of Prolog to Ada

ADM adaptive delta modulation; add-drop multiplexer; add-on data module; asynchronous disconnected mode; mensaje de acuse de desbloqueo de grupo de circuitos (s); modo desconectado asíncrono para el mantenimiento (s) [MGU]

ADMACS Apple document management and control system

ADMD administration management domain, domaine de gestion administratif (f), domaine de gestion d'adminstration (f)

ADMIS Analysis - Design - Management of Information systems

Admitech Association pour le développement du mécénat basé sur l'innovation et la technologie (f)

ADML application description markup language (Corel *tn*)

ADMS automatic digital message switch

ADMSC automatic digital message switching center

ADMT attach/detach monitor terminal

ADMT/V attach/detach monitor in the mobile terminal/visited network

ADMV attach/detach monitor visited network

ADN abbreviated dialling number; Actualité du Net (f) (Net news); address complete - no-charge; signal de numéro complet sans taxation (f), señal de dirección completa sin tasación (s); analyseur différentiel numérique (f)

ADO ActiveX data object (Microsoft)

ADo Anschlußdose (g) (line socket)

ADOIT automatically directed outgoing inter-toll trunk

ADONIS article delivery over network information systems

ADOX ADO extensions

ADP answerer detection pattern; automated/automatic data processing, traitement automatique de l'information (f), proceso automático de datos/de la información (s); assemblage-désassemblage de paquets (f) [PAD]

ADPCM adaptive differential (delta) pulse code modulation (AT&T), modulation par impulsion codée adaptative différenciée (f), modulación por impulso codificado adaptable diferenciado (s), Pulscode-Modulation mit differenzierter Anpassung (g)

ADPE automatic data processing equipment

ADPFH average of daily peak full hour

ADPH average daily peak hour

ADPSC Automatic Data Processing Steering Committee (UK)

ADPSSO automatic data processing system security officer

ADQ code almost differential quasi-ternary code

ADR address, adresse (f); Astra Digital Radio, Radio Digital de Astra (s); ampliación de dirección de red (s) [NAE]; semi-loop loss, half-loop loss (f) [SLL/HLL]

ADR/DMX ADR digital music express

ADRET Association pour le développement de recherches sur les télécommunications et le service public (f)

ADRIM Association pour le développement et la recherche en informatique médicale (f)

ADROIT Adverse Drug Reactions On-line Information Tracking (UK)

ADS application development solutions

(AT&T); automatic data set; automatic dependent surveillance

ADSC address status changed; Adobe document structuring conventions *tn*

ADSI active directory services interface; administrative design service information, analogue display services interface

ADSIA Allied Data Systems Interoperability Agency (NATO)

ADSL asymmetric/asynchronous digital subscriber line

ADSO application development system – online

ADSR attack - decay - sustain - release

ADSTAR automated document storage and retrieval

ADSU ATM data service unit

ADT abstract data type; access data terminal; American District Telegraph; application data types; applied diagnostic techniques; asynchronous data transmission; attach/detach terminal; give tokens acknowledgement (s)

ADT/V attach/detach in the mobile terminal/visited network

ADTF ACR decrease time factor

ADTS automated digital terminal system (AT&T)

ADTT advanced digital television technologies, tecnología avanzada de televisión (s)

ADU accumulation and distribution unit; assistance directe aux utilisateurs (f); attenuation distortion unit; automatic dialling unit

ADV Abteilung für Datenverarbeitung (g) (data processing department; *also* AfDV); access data visited; attach/detach visited network; Automatisierte und Automatische Datenverarbeitung (g) [ADP]

ADVM adaptive delta modulation voice modem

ADX address-complete signal – coinbox, señal de dirección completa – teléfono de previo pago (s); Astra (satellite) 1D eXtender

ADZ advise, rendez compte (f)

AE above or equal; acoustic emission; atomic emission; activity end; application entity/(executive), entité d'application (f); Ascii Express *tn*; associated equipment; acceso de entrada (s) [IA]; anfitrión externo (s) [EH]; appel entrant (f); atenuación para la estabilidad (s)

AEA activity end acknowledgement; American Electronics Association

AEB Alliance for Electronic Business (UK); analogue expansion bus (Dialogic *tn*); mensaje de acuse de establecimiento de bucle (s) [LPA]

AEC acoustic echo canceller/ control; advanced equipment controller

A$_{ECHO}$ echo loss, affaiblissement d'écho (f), atenuación del eco (s)

AECI Association of Electronic Contractors (Ireland)

AECS advanced equipment control system; (US) Aeronautical Emergency Communications System; automated equipment control system

AED Algol extended for design, automated/(automatic) engineering design; atribución de enlaces de datos de señalización (s)

AEDT accumulated exchange downtime

AEDV allowed environment data visited network

AEE Asociación Electrónica Española (s)

AEEA Association européenne pour l'enseignement de l'astronomie (f)

AEEC Airlines Electronics Engineering Committee, Comité de Ingeniería Electrónica de las Líneas Aéreas (s)

AEF acciones de entidad funcional (s) [FEA], additional elementary functions; address extension facility/field; Ausschuß für Einheiten und Formelgrößen (g) (Committee for units and sizes)

AEG Academic Equipment Grant (Sun)

AEGIS Airborne early-warning/ground integration segment (NATO)

AEI after-etch inspection; automated equipment interface

AEK Anschalteeinheit KarTel (g) (line group and call processor)

AEL acuse de recibo de entrega en línea (s) [ODA]

AELE affaiblissement d'effet local pour la personne qui écoute (f) [LSTR]

AELM affaiblissement d'effet local par la méthode de masquage (f) [STMR]

AELP atenuación por efecto local de la potencia (s)

AEM analytical electron microscopy; applications Explorer Mission; assemblage d'entités de maintenance (f) [MEA]

AEN articulation rating; articulation reference equivalent; affaiblissement équivalent pour la netteté (f)

AEP AppleTalk Echo Protocol *tn*; application environment profile; señal de acuse de recibo de enlace de reserva preparado (s) [SRA]

AEPE atenuación del eco para la persona que escucha (s) [LE]

AEPM atenuación de equilibrado en posición medida (s) [TBRL]

AER señal de acuse de recibo de paso de emergencia a enlace de reserva (s) [ECA]

AERA Automated en route air traffic control, ATC en route automatisé (f)

AERM alignment error rate monitor/monitoring

AEROSAT Aeronautical Satellite, satellite aéronautique (f)

AES aircraft earth station; application environment specification; Audio Engineering Society Inc (US); auger electron/emission spectroscopy; acuse de establecimiento (s) [SA]; activación de enlaces de señalización (s) [LSLA]

AES/EBU Audio Engineering Society/European Broadcasting Union

AESC automatic electronic switching centre

Aesdica Asociación Española de Distribuidores de Cable (s)

AESOP An evolutionary system for on-line programming; Artificial Earth Satellite Observation Programme, Programa de observación de la tierra mediante satélites artificiales (s)

AESS affaiblissement d'équilibrage du signal de sortie (f) [OSB]

AET atenuación de equilibrado del terminal (s) [TBRL]

AEW airborne early-warning

AF address field; assigned frequency; audiofrequency; auxiliary carry flag; adresse fonctionale (f) (functional address)

AFA accelerated file access; adaptive frequency allocation; acuse (de recibo) de fin de actividad (s) [AEA]; Alte Funktionsaufteilung (g) (old distribution of functions); Association des Fournisseurs d'Accès (f)

AFACTS automated facilities test system

AFADS automatic force adjustment data system

AFAIK as far as I know (autant que je sache)

AFC address-complete subscriber-free signal – charge; signal de numéro complet, ligne d'abonné libre avec taxation (f); señal de dirección completa, abonado libre con tasación (s); Ansi Factory *tn*; application foundation classes; automatic font change; automatic frequency control

AFCCE Association of Federal Communications Consulting Engineers, Asociación de Ingenieros Asesores en Comunicaciones Federales (s)

AFCEA (US) Armed Forces Communications and Electronics Association

AFCEE Association Française pour le Commerce et les Echanges Electroniques (f)

AFCET Association Française pour la Cybernétique Economique et Technique (f)

AFCS (US) Air Force Communications Service; automatic flight control system

AFCT adresse fonctionnelle de circuit terminal (f)

AFD application framework definition; automatic file distribution; demand assignment (f)

AfDV Abteilung für Datenverarbeitung (g) (*cf* ADV)

AFDW active framework for data warehousing (Microsoft)

AFFS Amiga fast file system

AFI Association de la Fête de l'Internet (f); authority and format identifier (ISO)

AFII Association for Font Information and Interchange

AFIPS American Federation of Information Processing Societies

AFIRM automated fingerprint image reporting and match (UK)

AFIS automated fingerprint identification system

AFJ April Fool's Joke

AFK away from keyboard

AFL abstract family of languages

AFM Adobe Font Metrics *tn*; atomic force microscopy; audiofrequency modulation

AFN address-complete subscriber-free signal – no charge; signal de numéro complet ligne d'abonné libre sans taxation (f); señal de dirección completa, abonado libre sin tasación (s); all figure (telephone) number/ numbering

Afnor Association Française de Normalisation (f)

AFP Agence France Presse; AppleTalk Filing Protocol *tn*

AFPI Association Française des Professionnels de l'Internet (f)

AFPT automatic frequency planning tool

AFPU African Postal Union

AFR automatic fingerprint recognition; awaiting forward release

AFRS Armed Forces Radio Service (US)

AFS Andrew filing system (after *Andrew* Carnegie); arranque en frío solamente (s) [CSO]

AFT analogue facility terminal

AFTN/CDIN aeronautical fixed telecommunications network/common data interchange network

AFTP anonymous FTP

AFUU Association Française des Utilisateurs d'Unix (f)

AFVSt Auslands-und-fernvermittlungsstelle (g) (international and national switching centre)

AFX address-complete subscriber-free signal – coinbox; signal de numéro complet ligne d'abonné libre – publiphone (f); señal de dirección completa, abonado libre teléfono de previo pago (s)

AFY2KWG (US) Air Force Y2K Working Group

Ag silver (used as a conductor)

AG again/try again; Antigua URL suffix; appellation globale (f) [GT]

AG/EEE above ground – electronic equipment enclosures

AGAMP automatic gain adjusting amplifier

AGC AudioGraphic conferencing *tn*; automatic gain control

AGCH access grant channel

AGE aerospace ground equipment

A-GEMTF Advanced GEM Task Force

AGF additional global functions

AGIS Apex Global Information Services

AGL Atelier de Genie Logiciel (f) (a French IPSE *qv*)

AGM analogue group module

AGM theory Alchourron, Gardenfors, Makinson (AI)

AGP accelerated/(advanced) graphics port

AGPO angle-gate pull-off

AGRAS Anti-glare/anti-reflection/anti-static CRTs (NEC *tn*)

AGRU automatische Ansage geänderter Rufnummer (g) (automatic subscriber number change message)

AGT AudioGraphic terminal

AGU address-generation unit

AGV autonomous/automated guided vehicle (mobile robot)

AGW Abis Gateway modem

Ah ampere hour

AH access handler; authentication header

AHA advanced heuristic analysis

AHCIET Asociación Hispanoamericana de Centros de Investigación y Empresas de Telecomunicaciones (s)

AHDL analogue hardware design language

AHFG ATM-attached host functional group

AHLF additional high-layer functions

ahm ampere-hour meter

AHM access handler mobile terminal

AHPL a hardware programming language

AHT access handler in mobile terminal; access handler in terminating network

AHV access handler in visited network

AI action indicator; activity interrupt; analogue input; articulation index; artificial intelligence; alfabeto internacional [IA]

AI #5 alphabet international No.5 (IA5)

AIA activity interrupt acknowledgement; Aerospace Industries Association; applications integrated/integration architecture (DEC)

AIB automatic intercept bureau

AIC automatic intercept centre; AIX windows interface composer (IBM); awaiting incoming continuity; aviso del importe de la comunicación [AOC]

AID access identifier; algebraic interpretative dialogue; area identification/identifier

Aida functional version of Dictionary APL

AIDS an infected disk syndrome; Apple infected disk syndrome

AIDX derogatory reference to AIX (IBM version of Unix)

AIEA Agence Internationale de l'Energie Atomique [IAEA] (f)

AIEE American Institute of Electrical Engineers (merged with IEEE)

AIES artificial intelligence expert system

AIFF audio interchange file format (Apple Inc), format de fichiers audionumériques (f)

AIG address indicator/indicating group; árbol de información de la guía (s) [DIT]

AII active input interface; affaiblissement d'impédance longitudinale (f) [LIL], atenuación de impedancia longitudinal (s) [LIL]

AIIM Association for Information and Image Management

AILE affaiblissement d'impédance longitudinale à l'entrée (f) [ILIL], atenuación de impedancia longitudinal de entrada (s) [ILIL]

AILF Association des Informaticiens de Langue Française (f)

AIM Advanced Informatics in Medicine (CEC R&D Programme, DG XIII); amplitude intensity modulation; AOL Instant Messenger *tn*; Association for Information Management; asynchronous interface module; ATM inverse multiplexer; awaiting incoming message

AIMACO air material command compiler

AIMUX ATM inverse multiplexing

AIN advanced intelligent network (Bell Atlantic *tn*)

AIOD automatic ID of outward dialling/out-dialled calls

AIP advanced information processing; advanced infrastructure planning

AIR additive increase rate; Agencia de inteligencia de red (s) (Intelligent network agency); Asociación Internacional de Radiodifusión (s)

AIRCOMNET (US) Air Force Command and Administrative Network

AIRF additive increase rate factor

AIS alarm indication signal; automated identification system (NCIS UK); automated information system; automatic intercept system; affaiblissement d'insertion de l'équivalent sonore (f) [LIL]

AISB Association of Imaging Service Bureaux

AISM Association internationale de signalisation maritime (f)

AISP Association of Information Systems Professionals

AIT advanced intelligent tape (Sony *tn*)

AITG automatic interface test generation

AITLM aplicación de interfuncionamiento telemático (s) [TIA]

AITRC Applied Information Technologies Research Center

AITS acknowledged information transfer service (ITU-T)

AIU alarm interface unit

AIUI as I understand it, comme je l'ai compris (f)

AIUR air interface user rate

AIV arquitectura de interfuncionamiento videotex (s) [VIA]

AIX Advanced Interactive eXecutive (IBM version of Unix *tn*)

AJ analogue junction; anti-jamming

AJM analogue junction module

Ajss Jitteramplitude Spitze-Spitze (g) (jitter amplitude peak-peak)

AK data acknowledgement

AK TPDU data acknowledgement TPDU

AKC Ascending Kleene Chain

AKCL Austin Kyoto Common Lisp

AKL ANDORRA kernel language

AKM apogee kick motor

AKRO acknowledgement receipt of

AKZ Amtskennziffer (g) (PSTN local access code)

Al aluminium (hardens when used with Si)

AL additional listing; analogue link; analogue loopback; application line; artificial life; assembly language, langage assembleur (f), lenguaje ensamblador (s), linguagem de 'assembly' (p); Amtsleitung (g) (local loop); anfitrión local (s) [LH]; Anschlußleitung (g) (subscriber line); local acknowledgement time, tiempo de acuse de recibo local (s)

ALA alarm log analysis

Aladin A Language for Attributed Definition

ALARP as low as reasonably possible

AlAs aluminium arsenide

A-Law CEPT standard companding algorithm

ALBO automatic line build-out

ALC airline line control; arithmetic and logic circuits; assembly language compiler; automatic load/level control

ALD automatic logic diagram

Aldat a database language

ALDES algorithm description (language)

ALDI associated long-distance interstate message

ALDiSP applicative language for digital signal processing

ALE address latch enable; application logic element; atomic layer epitaxy; automatic link establishment

ALEC a language with (an) extensible compiler

ALEPH a language encouraging program hierarchy

Alex a language with exception handling

ALEXIS Alex input specification

ALF absorption limiting frequency; algebraic logic functional (language); auditory location finder – for the deaf (Napier University)

Alfl a lazy function language

Alg Auslandsleitung gehend (g) (international line outgoing)

ALGOL generic algorithm/algorithmic language family

ALI Acer Labs Inc (Taiwan); automatic location identification

ALIAS algorithmic assembly language

ALIBI adaptive location of inter-networked bases of information

ALICE Alaska integrated communications system

ALINK active link (HTML)

AliS alternate lighting of surfaces (Fujitsu *tn*)

ALIT automatic line insulation test/testing

ALIVE artificial life interactive video environment

ALIWEB Archie-like indexing in the web

Alk Auslandsleitung kommend (g) (international line incoming)

Allegro code name for 1999 Apple Mac OS

ALLF additional low-layer functions

ALL-IN-1 DEC system

AllVr Allverstärker (g) (general purpose repeater)

ALM assembly language for multics

ALN adaptive logic network; asynchronous learning network; dirección completa sin tasación – abonado libre [AFN]

ALNA antenna-mounted LNA

alnico aluminium-nickel-cobalt

ALOHA random access control technique used with maritime satellite communications systems

ALOHANet early version of Ethernet, University of Hawaii

ALP abstract local primitive; address complete signal coinbox – subscriber free; alarm list presentation; a list processing (language)

ALPAK ALTRAN used with a subroutine package

ALPC adaptive linear predictive coding

ALPHA DEC processor chip *tn*

Alpha AXP 64-bit RISC scalable processor family (DEC *tn*)

alpha test test of software preceding beta release

ALR Advanced Logic Research Inc; alarm log retrieval

ALRU automatic line record update

ALS advanced light source; advanced

low-power Schottky; automated library system; signal de numéro complet ligne d'abonné libre – sans taxation [AFN]

ALSTTL advanced low-power Schottky transistor-transistor logic

Alt alternating

ALT address complete signal charge – subscriber free; alternate (key), (radar) altimeter

Alta Vista Internet search engine (Digital *tn*)

ALTEL Association of Long Distance Telephone Companies (US)

ALTS analogue line termination subsystem; automatic line test set

ALU arithmetic and logic unit; Association of Lisp Users

ALVEY UK programme for R&D in computing and technology (*now* IED)

Am americium

AM amplitude modulation; application module; auxiliary marker; almacenador de mensajes (s) [MS]

AM SMIP dispositivo de almacenamiento de mensajes del sistema de mensajería interpersonal (s) [IPMS MS]

AM/FM automatic mapping/facility management

AM/PM/VSB amplitude modulation/phase modulation/vestigial s-band

AMA American Management Association; automatic message accounting

AMACS automatic message accounting collection system

AMAEC análisis de los modos de avería, sus efectos y su criticidad (s) [FMECA]

A-mail air mail: data transfer via TV channel provider

Amanda functional programming language derived from Miranda

AMANDA automated messaging and directory assistance

AMARC automatic message accounting recording center

AMARS automatic message accounting recording system

AMASE automatic message accounting standard entry

AMAZE automatic map and zap equation (entry)

AMB auto-manual bridge control; atenuación de mitad de bucle (s) [HLL]

AMBIT algebraic manipulation by identity translation (language)

AMBIT/G AMBIT for graphs

AMBIT/L AMBIT for lists

AMBIT/S AMBIT for strings

AMC administrative management complex; airborne molecular contamination; auto-manual center

AMD active matrix display; Advanced Micro Devices (US corp); air movement device (IBM-speak for a fan); alarme de maintenance différée (f) [DMA], alarma de mantenimiento diferido (s) [DMA]

AMDE analyse des modes de panne et de leurs effets (f) [FMEA]

AMDEC analyse des modes de panne de leurs effets et de leur criticité (f) [FMECA]

AMDF acceso múltiple por división de frecuencia (s) [FDMA]

AMDP acceso múltiple con detección de portadora (s) [CSMA]

AMDT accumulated major disturbance; acceso múltiple por división en el tiempo (s) [TDMA]

AMDT/IDP acceso múltiple por división en el tiempo/interpolación digital de la palabra (s) [TDMS/DSI]

AME amplitude modulation equivalent

AMEL active matrix electroluminescent

Ameol A Most Excellent Off-Line Reader *tn*

Ameritech American Information Technologies (Corp)

AMEX American Stock Exchange; American Express

AMF adaptador multifuncional [MTA]; Asian Multimedia Forum

AMH automated material handling

AMHA a mon humble avis (f) (in my humble opinion–IMHO)

AMHS automated material handling systems

AMI alarme de maintenance immédiate (f) [PMA], alarma de mantenimiento inmediato (s) [PMA]; alternate mark inversion signal; American Megatrends Inc

AMI BIOS American Megatrends Inc BIOS

Aminet Amiga network *tn*

AMIS Atari message and information system *tn*

AmiTCP Amiga Internet browser

AML actual measured loss; a manufacturing language; automatic model linking

AML/E AML entry (level)

AMLCD active matrix LCD (IBM *tn*)

AMM analogue multimeter; associated maintenance module

AMMA Advanced Memory Management Architecture (Everex Systems *tn*)

AMN abstract machine notation

amoeba derogatory name for Commodore Amiga computer

AMON ATM monitoring

amorçage boot-up

AMP active monitor present (Token Ring); administrative module processor; algebraic manipulation package

amper ampersand

AMPPL associative memory parallel processing language

AMPS advanced mobile phone service/system (TIA 553); acuse de recibo del mensaje de prueba de enlace de señalización (s) [SLTA]

AMPSK amplitude-modulated phase-shift keying

AMPSSO automated message processing system security officer

AMR audio and modem riser (module)

AMRT accés multiple par répartition dans le temps (f) [TDMA]

AMRT/CNC TDMA/digital speech interpolation (f)

AMS Advanced Micro Systems; Andrew message system *tn* (*cf* AFS)

AMSAT American Satellite Corp (radio)

AMSC American Mobile Satellite Corporation

am-si amorphous silicon (in LCD panels)

AMSISDN additional MS-ISDN

AMSL above mean sea-level

AMSTRAD Alan Michael Sugar Trading (computer manufacturer)

AMSU auto-manual switching unit

AMT advanced manufacturing technology

AMTRAN automatic mathematical translation

AMTS automated maritime telecommunications system

AMU alarm monitor unit; atomic mass unit; automatic maintenance unit

AMVFT amplitude-modulated voice-frequency telegraph

AMWI active message waiting indicator

A-N analogique-numérique (f) [A/D]

AN apertura numérica (s) [NA]

ANA assigned night answer; automatic number announcer/announcement

ANALIT analysis of automatic line insulation tests

ANAPROP anomalous propagation (of radio/TV signals)

ANBFM adaptive narrowband FM modem

ANC all-number calling, answer signal charge; signal de réponse avec taxation; señal de respuesta con tasación

ANCE Agence Nationale pour la Création d'Entreprise (f)

ANCOVA analysis of co-variance

AND access network domain; address complete signal – no charge; automatic network dialling

ANDF architecture-neutral distribution format/frame

ANDORRA parallel logic programming

16

language (with OR-parallelism AND-parallelism)
ANE Anarchy 'n' Explosives
ANHR active node hub router
ANI automatic number identification; acceso numérico integrado (s) [IDA]
ANIF automatic number identification failure
ANL Argonne National Laboratory; automatic noise limiter
ANLP applied natural language processing
A_nlp non-linear processing loss
ANLT ALVEY natural language tool (v IED)
AnlUe Anlaßübertragung (g) (line unit for triggering)
ANM answer message (call control message)
ANMA Apple Network Managers' Association tn
ANMCC Alternate National Military Command Center (US)
ANN answer (signal) – no charge; signal de réponse sans taxation (f); señal de respuesta sin tasación (s); artificial neural network
Anna annotated Ada
Annoybot an annoying robot
Annoyware shareware with persistent 'nag screen' reminder
ANOI A New Order of Intelligence
ANOVA analysis of variance
ANP AAL2 negotiation procedure; antenna near product/project; Anpassungssatz (g) [matching interface unit]
ANPC antenna near part/product complementary equipment cabinet
ANPE Anpassungseinrichtung (g) [switching equipment]
ANPS automatic (car) number plate scanner (UK police)
AnpUe Anpassungsübertragung (g) [line adaptor unit]
ANR automatic network routing; awaiting number received

ANS Advanced Network Services Inc (US); announcement subsystem; signalling analysis
ANSA Adaptive Network Security Alliance; advanced network systems architecture
ANSF American National Science Foundation (once overseer of Internet)
ANSI American National Standards Institute
ANSI/SPARC ANSI/Standards Planning and Requirements Committee
ANSI/SYS ANSI/system device driver, screen and keyboard
ANSI/X3 ANSI/Committee on Computers and Information Processing
ANSIC ANSI's C
AnSS Ansagesatz (g) [line voice message unit]
AnsVr Ansageverstärker (g) [voice message amplifier]
ANT A-number transfer
ANTIOPE Acquisition Numérique et Télévisualisation d'Images Organisées en Page d'écriture (f) (French Teletext)
ANTLR another tool for language recognition
ANU answer signal-unqualified
A-number telephone number call from intelligent network
AO analogue output; application object
AO/DI always on/dynamic ISDN
AOA Anarchists of America
AOC advice of charge; aeronautical/aviation/aircraft operational capability; awaiting outgoing continuity
AOCC advice of charge (supplementary service)
AOCE Apple Open Collaboration Environment tn
AoCI advice of charge information – supplementary service
AODI always-on dynamic ISDN
AOE application operating environment (AT&T)
AOI active output interface (used in UNI PMD specs for copper/fibre)

AOJ add one and do not jump

AOL America OnLine (ISP)

AOM automatic operation and maintenance

AONALS type A off-network access lines

AOQL average outgoing quality level

AOS add one and do not skip; add or subtract; algebraic operating system

AOSS auxiliary operator service system

AOT Art of Technology

AOTC Australian and Overseas Telecommunications Corp

AOTT automatic outgoing trunk test

AOV air-operated valve

AOWC all-optical wavelength conversion laser

AP access point; adhesion promoter; adjunct processor; anomalous propagation; antenna part; application processor; array processor; analizador de prueba (s) [TA]; arquitectura de presentación (s) [PA]

AP(S) application program (structure)

AP/Dos advanced pick – disk operating system

APA active position addressing, adressage de position active (f), direccionamiento de posición activa (s); adaptive packet assembly; advanced performance algorithm; all-points-addressable; application portability architecture; arithmetic processing accelerator

Apache type of Web server *tn*

APAD high loss operation

APAL array processor assembly language

APAR authorized program analysis report (IBM)

APAREL a parse request language

APAREN address parity enable (IBM)

APaRT automated packet recognition/translation

APB active position backward; advanced PCI bridge (Sun *tn*); A-number presentation barring

APC adaptive predictive coding/codec; advanced process control; AMARC protocol converter; American Power Conversion (UPS); Association for Progressive Communications; asynchronous programming interface; average power control

APCD add-on pollution control devices

APCE aparato de pruebas de compensadores de eco instalado en la estación (s) [ISET]

APCEE aparato de prueba de compensadores de eco en estación (s) [ISET]

APCI application protocol control information

APCO applications configurator

APCUG Association of PC User Groups

APCVD atmospheric pressure chemical vapour deposition

APD active-position-down; amplitude probability distribution; avalanche photo-diode

APDA Apple Programmers' and Developers' Association

APDL algorithmic processor description language

APDQH average of daily peak quarterly-defined hour

APDU application protocol data unit

APEC advanced process equipment control

APF access points facility; Vermittlung oder Multiplexer (g); active position forward; all PINs fail

APG available power gain

APH active position home

API advanced programming interface; American Pirate Industries; application programming interface; atmospheric pressure ionization

APIC advanced programmable interrupt controller (Intel *tn*)

APICS American Production and Inventory Control Society

APIMS atmospheric pressure ionization mass spectroscopy

APL a programming language (for

mathematics); Abschlußpunkt des allgemeinen Netzes (g) [terminal point for the general network]; average picture level

APLGOL APL variant with ALGOL-similar control structure

APLWEB Web-to-APL translator

APM acoustic plate mode; atmospheric passivation module; advanced power management (IBM)

APMR señal de acuse de recibo de paso manual a enlace de reserva (s) [COA]

APN access point name

A$_{PNL}$ atenuación por tratamiento no lineal (s) [A$_{nlp}$]

APNSS PBX protocol

app an application/executable

APP application date; application portability profile

APPC advanced program-to-program communications (IBM *tn*)

APPEL a P3P preference exchange language

APPI advanced peer-to-peer interworking

APPIX automated part and product information exchange

applet small embedded (OLE/Java) application

AppleTalk proprietary LAN connecting Macs

APPLOG application log

APPN Advanced Peer-To-Peer Networking (IBM *tn*)

apps applications

APR active position return; automatic picture replacement (Scitex *tn*); aparato de prueba (s); automatische Prüfung (g) (automatic testing); señal de acuse de recibo de paso a enlace de reserva (s) [COA]

APRE Automatische Prüfeinrichtung (g) (automatic test system)

APRIL Association pour la Promotion et la Recherche en Informatique Libre (f)

APRP adaptive pattern recognition processing

AprPl automatischer Prüfplatz (g) (automatic test position)

APS advanced photo system (IBM); advanced printing system (IBM); application process subsystem; automatic protection switch/switching; auxiliary power supply; Associação Portuguesa de Software (p) (*v* ASSOFT)

APSCC Asia-Pacific Satellite Communications Council

APSE Ada programming support environment

APSI application platform service interface

APSS automatic protection switching system

APT address pass through; advanced parallel technology; advanced photoscale technology (Brother *tn*); Advanced Pirate Technology; advanced probe technology; American Phone Terrorists; application pilot transfer; Asia-Pacific Telecommunity; audio-processing technology; automatic picture transmission; automatic programming tools; automatically programmed tools; arquitectura de protocolo telemático (s) [TPA]; señal de acuse de recibo de prohibición de transferencia (s) [TPA]

APTU Africa Postal and Telecommunication Union

APU active position up; analytic processing unit; application processing unit; arithmetic processor unit

AQ adaptador a interfaz (s) [QA]

AQI ACCESS query interface (Microsoft)

AQL acceptable quality level

AQlg Auslandsquerleitung gehend (g) (secondary international route outgoing)

Aqlk Auslandsquerleitung kommend (g) (secondary international route incoming)

AQS automatic [share] quote system

Ar argon

A$_R$ remote acknowledge time

AR accusé de réception (f) [ACK]; activity resume; affaiblissement de transmission réel (f) [AL]; acuse de recibo de datos (s) [ACK]; acuse (de recibo) de resincronización (s) [RA]; adaptador de red (s) [NA]; aspect ratio (width x height); authentication register; awaiting reply

ARA AppleTalk remote access; aptitud de recepción alterada (s) [RAJ]

ARABSAT Arab Satellite Communications Organization

ARAEN reference apparatus for the determination of transmission performance ratings, appareil de référence pour la détermination des affaiblissements équivalents pour la netteté (f)

ARALIA affectation des ressources sur les arcs logiques intra-ZAA (f)

ARAMS automated reliability availability and maintainability standard

ArAnS Anruf-Ansagesatz (g) [telephone announcement]

ARAS anti-reflective/anti-static (monitor screen)

ARB adaptive rate-based congestion control; all routes busy; Application Review Board; mensaje de acuse de bloqueo (s) [BLA]; señal de acuse (de recibo) de bloqueo (s) [BLA]

ARBG mensaje de acuse de bloqueo de grupo de circuitos (s) [CGBA]

ARBM appareil de référence pour la production de bruit modulé (f) [MNRU]

ARC advanced RISC computing (specification); Ames Research Center; anti-reflective coating (on monitor screen – cf ARAS); attached resources computing (ARCnet); accusé de réception de connexion (f)

ARCA advanced RISC computing architecture

ARCB mensaje de acuse de recibo de reiniciación de banda (s) [RBA]

ARCBR señal de acuse de recibo de reiniciación de banda (s) [RBA]

ARC-E ARC extracting utility (cf arc)

ARCnet attached resource computer network (Datapoint Corporation tn)

ARCTC ARC test chip (for ARC processor core)

ARCV Association of Really Cruel Viruses

ARD accusé de réception de données (f) [AK]; Arbeitsgemeinschaft der Rundfunkanstalten Deutschlands (g); mensaje de acuse de desbloqueo (s) [UBA]

ARDE aspect-ratio-dependent etching

ARDG mensaje de acuse de desbloqueo de grupo de circuitos (s) [CGUA]

ARE all routes explorer; accusé de réception d'établissement (f)

AREAN reference system for determining articulation reference equivalents

AREG apparatus repair strategy evaluation guidelines

ARENTO Association of Southeast Asian Nations [ASEAN]

AREV Advanced Revelation (Revelation Software tn)

AREXX Amiga REXX tn

ARF alternative routing from

ARFCN absolute radio frequency channel number

arg argument

ARG ATM Requirements Group; mensaje de acuse de reinicialización de grupo de circuitos (s) [GRA]

ARIB Association of Radio Industries Bureau (Japan)

ARIES ATM Research and Industrial Enterprise Study

ARIMA auto-regressive integrated moving average (cf ARMA)

ARIN American Registry of Internet Numbers

ARIS Audichron Recorded Information System tn

ARK analogue remote concentrator

ARL acceptable reliability level; adjusted ring length; ASSET re-use library

ARLL advanced run-length limited

ARM RISC Advanced RISC Machines Ltd

arm armature

ARM active re-configuring message; advanced RISC machine (Acorn Risc machine); annotated reference manual; asynchronous response mode

ARMA Association of Records Managers and Administrators; auto-regressive moving average (*cf* ARIMA)

ARMM automated retroactive minimal moderation (Usenet robot)

ARO after receipt of order

ARP address resolution protocol (within TCT/IP); AppleTalk address resolution protocol *tn*; arithmetic processor; automatic receiver program; automatic resolution protocol

ARP PPDU abnormal release provider PPDU

ARPA Advanced Research Projects Agency (previously DARPA)

ARPANET Advanced Research Projects Agency Network (US DoD)

ARPAP aluminium-resin-polythene-aluminium-polythene

ARPL adjust requested privilege level

ARQ automatic request (for repetition/retransmission); error correction by detection and repetition, correction d'erreurs par détection et répétition (f), corrección de errores por detección y repetición (s)

arr arrestor/arrester

ARR automatic re-routing

ARRE alarm receiving and reporting equipment

ARRG mensaje de acuse de reinicialización de grupo de circuitos (s) [GRA]

ARRL American Radio Relay League

ARRM aparato de referencia para ruido modulado (s) [MNRU]

ARS action request system; angle-resolved scattering; alternative (automatic) route selection; señal de

acuse de recibo de retorno al enlace de servicio (s) [CBA]

ARSB Automated Repair Service Bureau

ARSR air route surveillance radar

ART additional reference carrier transmission; a real-time (functional) language; alarm reporting telephone; alternative routing to; automatic reporting telephone; Agence de Régulation des Télécommunications (f); Autorité de régulation des télécommunications (f)

ARTA Apple real-time architecture *tn*

ARTCAS air route traffic control automatic/automated system

ARTCC Air Route Traffic Control Centre

ARTIC annual rural telematics workshop; a real-time interface co-processor (IBM)

ARTS asynchronous remote take-over server

ARTT asynchronous remote take-over terminal

ARU audio response unit

ARU PPDU abnormal release user PPDU

ARX accusé de réception de données expres (f) [EA]

As arsenic

AS access server; activity start; address strobe; advanced Schottky; affaiblissement pour la stabilité (f) [SL]; alignment signal; Anarchy Society; articulation score; autonomous system; alarme de service (f) [SA], alarma de servicio (s) [SA]; acceso de salida [OA]; Anrufsucher (g) [line-finder]

AS/400 Application System/400 series (mini-computers) (IBM *tn*)

AS/RS automated storage and retrieval system

AS/U advanced server for UNIX

AS3AP ANSI SQL standard scalable and portable

ASA adaptive simulated annealing; agent de système d'annuaire (f) [DSA]

ASAP advanced stepper application

program; automatic switching and processing

ASAT anti-satellite capability

ASB atenuación en semibucle (s) [HLL; *v* SLL]

ASBU Arab States Broadcasting Union

ASC abnormal station condition; Accredited Standards Committee; application specific computer; AUTODIN switching center; automatic digital network switching centre; automatic sensitivity control; Auto Size and Centring (Sony *tn*); average system cost/content

ASCC automatic sequence-controlled calculator (Harvard)

ASCI advanced speech call/control item

ASCII American standard code for information interchange (ISO 7)

ASCIIZ ASCII string ended by NULL (*ie* zero byte character)

ASCVL allowed speech coder version list

ASD Advanced Semiconductor Development; adverse state detector; anti-cosmic defence, automatic/automated skip driver (Microsoft Windows 98)

ASDL abstract-type and scheme definition language

ASDS ACCUNET Spectrum of Digital Services

ASE Advanced Software Environment (Nixdorf *tn*); application service element, élément de service d'application (f)

ASEL side-tone loudness rating (f) [STLR]

ASET adaptive sub-band excited transform (coding) (GTE Corp); automated security enhancement tool

ASETA Association of the Andean Subregional Agreement's State Telecommunications Enterprises, Asociación de Empresas de Telecomunicaciones del Acuerdo Subregional Andino (s)

ASF aspect source flag; NetShow audio format *tn*

ASG agente de sistema de guía (s) [DSA]

ASHRAE American Society of Heating Refrigeration and Air-conditioning

ASIC application-specific integrated circuit

ASIS Ada semantic interface specification

ASIU as I understand it

ASL adaptive speed levelling; algebraic specification language; Anschlußleitung (g) [switching centre/exchange line]

ASLIB Association of Special Libraries and Information Bureaux (*prev* Association for Information Management)

ASLM Apple shared library manager *tn*

ASLT advanced solid logic technology

ASM algorithmic state machine; application system modification; assembler; automatic systems management

ASME American Society of Mechanical Engineers

ASMET African Satellite Meteorological Education and Training

ASMI aerodrome/airfield surface movement indicator, radar de contrôle des mouvements de circulation au sol (f), indicador de movimientos en superficie de aeródromo (s)

ASN autonomous system number; advance shipping notification

ASN 1 abstract syntax notation one (ISO/ITU-T standard)

ASO active sideband optimum (video recording circuitry); automatic shut-off

ASOC administrative service oversight center; Air Support Operating Centre

ASOMEDIOS Asociación Nacional de Medios de Comunicación (s) (Colombia)

ASOS Automated Surface Observing System (US weather service)

ASP abstract service primitive; access service provider (of office applications online); active server page (on MS

Internet Information server); adjunct service point; advanced strip and passivation; advanced strip processor; aggregated switch procurement; antenna specialist panel; Apple Talk Session Protocol *tn*; applet security profile; assignment source point; Association of Shareware Professionals; authorized service provider; automatic speech processing; processamento automático fala (p); agent de sécurité privée (f)

ASPEN Automatic System for Performance Evaluation of the Internet

ASPI advanced SCSI programming interface

ASPIC Application Service Provider Industry Consortium (US); author's symbolic pre-press interfacing codes

ASPJ airborne self-protection jammer

ASPROM Association pour la promotion de la micro-informatique (f)

ASPS advanced signal processing system

ASR access service request; address space register; airport surveillance radar, radar de surveillance d'aéroport (f), radar de vigilancia de aeropuerto (s); answer/seizure ratio; automatic send/receive, envío y recepción automáticos (s); automatic selective reply, réponse sélective automatique (f); automatic server re-start; automatic speech recognition, reconnaissance automatique de la parole (f), reconhecimento automático da fala (p), automatische Spracherkennung (g)

ASRS aviation safety reporting system

ASS assembler, assembleur (f), ensamblador (s)

ASSET Amsterdam Stock Exchange Trading System; Asset Source for Software Engineering Technology (US DoD)

ASSI advanced SCSI programming interface

ASSIST additional support and information search tools; analytical

services statistical information system (UK DSS)

ASSODEL Asociación Nazionale Distribuzione Electrónica (i)

ASSOFT Associação Portuguesa de Software (p) (*v* APS)

ASSP application specific standard product

ASSS atenuación de simetría de las señales de salida (s) [OSA]

ASSUME Association of Statistics Specialists Using Microsoft Excel

AST abstract syntax tree; active service technology; announcement service terminal; anti-sidetone; application specific tools, herramientas específicas para aplicaciones (s); Arbeitsfeldsteuerung (g) (section control unit); sistemas perícias vocacionados (p); AST Research Inc; asynchronous system trap; automatic scan tracking

AST/SYS access storage and transfer system, système d'accés de mémorization et de transfert (f), sistema de acceso, almacenamiento y transferencia (s)

ASTARTE Accès sécurisé télématique au réseau de téléaction (f); Avion Station Relais de Transmissions Exceptionnelles (f)

ASTC airport surface traffic control, contrôle du trafic aéroportuaire de surface (f)

AST-DR announcement service terminal – digital recording

ASTeX Applied Science and Technology (US corp)

ASTM American Society for Testing and Materials

ASTP Asociación de Servidores de Telecomunicaciones de Pichincha (s) (Ecuador)

ASTRA application of space techniques to aviation

ASTRAIA-ELDORA analyse stéréoscopique par radar à impulsion aéroporté-Electra Doppler Radar (f)

Astree annuaire commercial électronique (f)

ASTREE automatisation du suivi en temps réel (f) (SNCF), automatic real-time (train) tracking

ASTRES advanced storage and retrieval system

ASTRO advanced semi-custom technology for RF system integration on silicon

ASTTL advanced Schottky transistor-transistor logic

ASTU Help desk and user technical support unit, service accueil et support [technique] aux utilisateurs (f), Servicio de Asistencia y Apoyo Técnicos a los Usuarios (s), Serviço 'Acolhimento e Apoio aos Utilizadores' (p), Dienststelle Benutzerbetreuung und Unterstützung (g), Dienststelle Betreuung und Unterstützung (g)

ASU absent subscriber service

ASV air-to-surface vessel radar, radar aéroporté de surveillance maritime (f), Radar aire-superficie para vigilancia marítima (s), Luft-Überwasser-Radar (g)

ASVD analogue simultaneous voice and data

ASW anti-submarine warfare, guerre anti-sous-marine (f), lutte anti-sous-marine (f), lucha anti-submarina (s), guerra anti-submarina (s); Kriegführung gegen U-Boote (g)

ASWS automated surface weather system, système automatisé d'observation des conditions météorologiques au sol (f)

AT access tandem; Advanced Technology (bus connecting motherboard to peripherals); aerial tape armour; 'ATtention' modem control language (Hayes *tn*); adaptateur de terminal (f) [TA], adaptador de terminal (s) [TA]; affaiblissement d'adaptation transversale (f) [RL]; Área de Telecomunicações (p)

AT&T American Telephone and Telegraph (*aka* Ma Bell)

ATA advanced technology attachment (ANSI standard); AT attachment; automatic trouble analysis; Agence télégraphique albanaise (f) (Albanian Telegraph Agency); Asociación Telerradiodifusoras Argentinas (s); señal de acuse de recibo de autorización de transferencia (s) [TAA]

ATA-2 *aka* Fast ATA, Fast ATA-2

ATACS Army Tactical Communication System

ATALA Association pour le traitement automatique des langues (f)

ATAPI advanced technology attachment packet interface

ATARS automatic traffic advisory and resolution system, service consultatif et résolutif automatique de la circulation (f)

ATAS advanced technology alert system, système de prévision technologique avancée (f); automatic telephone answering system; sistema de alerta temprana en materia de tecnología (s)

ATB all trunks busy; automated tickets and boarding pass

ATBA automatic test break and access

ATC additional trunk capacity; address translation chip; address translation controller (for ATM; Fujitsu *tn*); Advanced Technology Centre (University of Warwick, UK); air traffic control; Association of Commercial Television in Europe, Association des télévisions commerciales européennes (f), Asociación de televisión comercial europea (s), Associação de Televisão Comercial na Europa (p); ATM transfer capability; auto-tuned filter combiner; señal de acuse de recibo de transferencia de tráfico (s) [LTA]

ATCA Allied Tactical Communications Agency (Nato)

ATCAA air traffic control assigned air space, espace assigné par l'ATC pour permettre une ségrégation du trafic (f)

ATCAP ATC automation panel, groupe

d'experts sur l'automatization du contrôle de la circulation aérienne (f)

ATCAS ATC automatic/automated system, système ATC automatisé (f), sistema de automación del control de tránsito aéreo (s)

ATCC ATC centre, centre de contrôle de la circulation aérienne (f), centre de contrôle du trafic aérien (f), centro de control de la circulación aérea (s)

ATCL affaiblissement de transfert de conversion longitudinale (f), atenuación de transferencia de conversión longitudinal (s)

ATCRBS air-traffic control radar beacon system; infrastructure radar secondaire de surveillance (f); sistema de respondedores para el control de tránsito aéreo (s)

ATCRU ATC radar unit, station radar de contrôle de la circulation aérienne (f)

ATCSCC ATC systems command centre

ATCT affaiblissement de transfert de conversion transversale (f), atenuación de transferencia de conversión transversal (s)

ATD asynchronous time division; attention dial (Hayes-defined modem command); authorized terminal distributor

ATDM asynchronous time-division multiplexing, Multiplexagem Temporal Assíncrona (p)

ATDP attention dial pulse (Hayes-defined modem command)

ATDT automatic test dial tone

ATE ATM terminating equipment (SONET); automatic test equipment, equipo de prueba automática (s)

ATECO automatic telegram transmission with computers

ATF automatic tracking facility (*ie* satellite tracker – 'spy' – in cab), dispositif de poursuite automatique (f)

ATFT amorphous TFT

ATG Advanced Technology Group; automatic test generation

ATH abbreviated trouble history; attention hang-up

ATI Activists Times Incorporated; accord sur les technologies de l'information (f); alfabeto telegráfico internacional (s) [ITA]

ATI x alfabeto telegráfico internacional N^o x (s) [ITA x]

ATIC asignación en el tiempo con interpolación de muestras (s) (time assignment with sample interpolation)

ATIS Alliance for Telecommunications Industry Solutions; automatic transmitter identification system

ATKINS ATK-instituutti (Finland) (Institut de l'informatique)

ATL active template library (MS Visual C++); automated tape library; affaiblissement de transfert longitudinal (f) [LTL], atenuación de transferencia longitudinal (s) [LTL]

ATLAS abbreviated test language for all systems

ATLAS 400 Note service de messagerie X.400 de France Telecom

ATLM agente telemático (s) [TLMA]

atm atmosphere

ATM Adobe type manager; agent de transfert de messages (f) [MTA], agente de transferencia de mensajes (s) [MTS]; adaptadores del terminal multiprotocolo (s) [MTA]; alternative test method; asynchronous transfer mode (broadband ISDN); automated teller machine (IEEE specification); pluralité des codes de signaux de sélection (f)

ATM cell basic data packet handled by ATM (1) network

ATME CCITT auto-transmission measuring and signalling test equipment, aparato automático de medidas de transmisión (s)

ATMF ATM Forum

ATMS automatic transmission measuring system

ATN aeronautical telecommunications network; attention! (IEEE-488);

architecture technique nationale (f) (France Télécom)

A_{TNL} affaiblissement par traitement non linéaire (f) [A_{nlp}]

ATO Advanced Technologies Operation (Canon *tn*)

atom irreducible unit of data (used in PROLOG); 16-bit integer

ATOM first commercial microcomputer (Acorn, 1980)

ATP acceptance test procedure; advanced technology program

ATPG automatic test pattern generation

ATPS AppleTalk printing services *tn*

ATR answering time recorder; attenuated total reflectance; automatic terminal recognition

ATR-FTIR attenuated total reflectance Fourier transform infrared spectroscopy

ATRS automated trouble-reporting system

ATS abstract test suite; administrative terminal system; air traffic services; alarm termination subsystem; Apple terminal system *tn*, application technology satellite; automatic trunk synchronizer; half-loop loss (f); atribución de terminales de señalización (s) [LSTA]

ATSC Australian Telecommunication Standardization Committee

ATSE application time-sharing software engineering

ATSI Asian Telecommunications Systems Institution

ATSU Association of Time-Sharing Users

att attachment

ATT attended public telephone; attenuated; automatic toll ticketing; Microsoft Exchange encrypted file attachment; affaiblissement de transfert transversal (f) [TTL], atenuación de transferencia transversal (s) [TTL]

ATTC automatic transmission test and control

ATTIS AT&T Information System

ATTO avalanche transit-time oscillator

ATUG Australian Telecommunications Users' Group

ATUR automatic telephone using radio

ATURS automatic traffic usage recording system

ATV Aktiv Television (Turkey)

ATVA American Television Alliance

ATX automatic telex exchange

ATZ attention restore configuration profile non-volatile RAM; modem automatic test

Au gold

AU access unit; administrative unit; application unit; arithmetic and logic unit (*v* ALU); audio format

AU MIP agente de usuario de mensajería (s) [IPM UA]

AU PTR administrative unit pointer

AU SMIP agente de usuario del sistema de mensajería interpersonal (s) [IPMSUA]

AUA agent d'usager d'annuaire (f) [DUA]

AUC authentication centre

AUDIT automated data input terminal

AUDIX audio information exchange

AUG agente de usuario de guía (s) [DUA]

AUI attachment unit interface (IEEE 802.3); adaptable user interface; AUI connector (D-connector connect Ethernet to network); Association des utilisateurs Internet (f)

AUP acceptable use/user policy (on transit networks)

AURORA automatic roaming radio

AUT(H) authentication

AUTODIN automatic digital network/exchange

AUTOEXEC automatic execution of switch-on/restart batch file

AutoID automatic identification (automatic entry of data)

AUTOSEVOCOM Automatic Secure Voice Communication Network

AUTOVON automatic voice network

aux auxiliary device (usually COM1)

AV audio-visual; authenticity verification; adaptación de velocidad (s) [RA]

AVA attribute value assertion; audio-visual authoring (IBM)

AVC audio visual connection (IBM); automatic volume control

AVCS Advanced Vidicon Camera System *tn*

AVD alternate voice data/alternating voice and data

AVEM acuse de verificación de encaminamiento por la PTM (s) [MRVA]

AVHRR advanced very high-resolution radiometer

AVI audio/video interleave (Microsoft *tn*), audio video mélangés (f)

AVL tree height-balanced tree/binary search tree

AVN Ameritech Virtual Network

AVP advanced vertical processor

AVS application/advanced visualization system

AVSSCS Audio-Visual Service Specific Convergence Sublayer (ATM Forum)

AVSt Auslandsvermittlungsstelle (g) [international switching centre]

AVT audio-visual terminal

AWACS airborne warning and control system

AWB Aglets Workbench *tn*; automatic white balance

AWE Advanced WavEffect *tn*; Advanced Windowing Extensions (Microsoft *tn*); asymptotic waveform evaluation

AWG American wire gage

AWGN additive white Gaussian noise

AWHQ Alternate War HQ (Nato)

AWK Aho Weinburger Kernighan (g) [data-handling language UNIX dialect]

AWM Automated Workflow Manager

AWS Advanced Workstations and Systems (IBM *tn*)

AWT Abstract Windows/windowing Toolkit (Java)

AWW approximate Wilkinson Wallström (method)

AX architecture extended; automatic transmission

Axe Ericsson digital switches *tn*

AXP almost exactly prism

AXS AXE software management *tn*

AXSH ActiveX scripting hosts (Microsoft Windows)

AZ azimuth

AZ/EL azimuth/elevation

AZERTY European keyboard layout; alternative to QWERTY

Azl Abzweigleitung (g) (line to second phone)); Auslandszentralvermitttlungsleitung (g) (line of main international switching centre)

AZVSt Auslandszentralvermittlungsstelle (g) [main international switching centre]

B

B bel; byte (8 bits); (second) 'floppy' disk drive; language derived from BCPL, superseded by C

B channel bearer channel

B HLI broadband high layer information

B ICI broadband inter-carrier interface

B ISSI broadband inter-switching system interface

B LLI broadband low layer information

B number number used in intelligent network

B&K Brüel & Kjaer (telephonometry tester)

B/L backlog

B/R bridge/router ('brouter')

B/W bothway, in beiden Richtungen (g)

B2X binary to hexadecimal (REXX)

B3ZS bipolar with 3-zero substitution

B6ZS bipolar with 6-zero substitution

B8ZS bipolar 8-zero substitution (Bell Labs line coding)

B911 basic 911

BA Basisanschluß (g), baseband interface; broadcast control channel (BCCH) allocation; banda ancha (s) [WB]

Babbage name of an MOHLL

BABT British Approval Board for Telecommunications

BAC bus arbiter/controller

BACAIC Boeing Airplane Company Algebraic Interpreter Coding *tn*

BACnet building automation and control network

BACP bandwidth allocation control protocol

BACS bankers' automated clearance system

BACT best available control technology

BAD broken as (*ie* badly) designed

BAIC barring all incoming calls (supplementary service)

BAJN bits de afluentes disponibles para justificación negativa (s) [BTNJ]

BAJP bits de afluentes disponibles para justificación positiva (s) [BTPJ]

BAK binary adaptation kit (Microsoft *tn*)

BAL Basic assembly language; boîtier d'alimentation locale (f), local PS module

BALGOL ALGOL on Burroughs 220

BALI Backweb authoring language interface

balise HTML tag (f)

BALM block and list manipulation

BALUN balance to unbalance transformer (*aka* bazooka)

BAM Basic access method; Boyan action module; BTS alarm management

BAMAF Bellcore automatic message accounting format *tn*

BAN boundary access node; bandera (s) [F]

BANCS Bell Administrative Network Communications System *tn*; Bell Application Network Control System *tn*

Band-X bandwidth exchange

Bang ! in old web addresses

BANM Bell Atlantic Nynex Mobil *tn*

BAOC barring all outgoing-calls supplementary service

BAP bandwidth allocation protocol; Brain Aid Prolog; broadband application part/protocol

BAR boîte à rythmes (f), rhythm composer/machine

BARB board arbiter on (Sun) SuperSPARC module

BARBICAN battleground automatic radar-bearing; intercept; classification and analysis

BARRNET Bay Area Research Network

BARTS Bell Atlantic regional time-sharing

BAS basic activity subset; BASIC/QBasic file; bit-rate allocation signal

BASC Bad Austrian Swapping Crew; basic service code

BASE Bank America Systems Engineering; Boston Area Semiconductor Education (Council)

basecom base communications

BASH Bourne again shell (Unix)

BASIC Beginner's All-purpose Symbolic Instruction Code

BA size buffer allocation size

BASM built-in assembler

BAT batch-file (extension); block address translation

BATRU Bringing ATM to residential users (France Telecom)

baud baud rate (of change of state; after JME Baudot)

Baudot Baudot code

BAW bulk acoustic wave

BB backbone (of telephone circuit); Bad Brains; baseband; bulletin board

BBA Belgian Broadband Association

BBC broadband bearer capability

BBD bucket-brigade device

BBL (I'll) be back later

BBN backbone node; Bayesian Belief Network

BBS bulletin board software/system

BBU battery back-up (unit)

BBX baseband switch

Bc committed burst rate/size

BC bearer capability; bias contrast

BCAD business computer-aided manufacturing

BCAF bearer control agent function

BCBF branch on chip box full

BCC base transceiver station colour code; basic connection component; 'blind' (carbon) copy (e-mail); blocked calls checked; block calls cleared; block check character (BISYNC); block completed counter; Broadcasting Complaints Commission (UK); broadcast control channel

BCCD bulk-channel charge-coupled device

BCCH broadcast control channel

BCD back-channel carrier detect; binary-coded decimal (code), binaire codé décimal (f); blocked calls delayed

BCD/B binary coded decimal/binary

BCD/Q binary coded decimal/quaternary

BCE business support system (BSS) central equipment

BCF base (station) control function; bearer control function; budget cross-flow model (analytical tool adapted in RACE I)

BCH Bose-Chaudhuri Hocquenghem (code)

b-channel bearer channel

BCI Battery Council International; Bell Canada International; bit-count integrity

BCIE bearer capability information element

BCL batch command language

BCLB broadband connectionless bearer service (v ATM)

BCM base transceiver station (BTS) configuration management

BCN beacon

BCNF Boyce-Codd normal form

BCNU be seein' you

BCO battery cut-off

BCOB broadband connection-oriented bearer service

BCON board connection

BCP basic call process; batch communications program; best current practice (described in RFCs); Bézier control point; bridging control protocol (RFC 1639); bulk copy program; business continuity plan

BCPL systems programming language

BCPN business customer premises network

BCR Business Communications Review; byte count register

BCRCI Brigade Centrale de la Répression de la Criminalité Informatique (f)

BCS bandwidth connectivity services; bar code sorter; basic combined subset; batch change supplement; binary-coded signalling; binary compatibility standard;

block control signal; British Computer Society; Bull Cabling System *tn*; business communication system

BCSM basic call state machine/model

BCST broadcast

BCTS backward/back-channel clear-to-send

BCU buffer control unit

BCUSM basic call unrelated state model

BDA BIOS data area

BDAM basic direct access method

BDC back-up domain controller

BDCS broadband digital cross-connect system

BDE Borland database engine *tn*

BDEV behaviour-level deviation

BDF backward differentiation formulae (methods)

BDI Borland database interface *tn*

BDIR bus direction

B-display bearing display (on radar screen)

BDK (Java) Bean Development Kit (Sun Microsystems *tn*)

BDL block diagram compiler

BDM battery distribution module

BDO boîte de distribution optique (f), optical splitter

BDP business document pipeline

BDPA Black Data Processing Associates

BDPSK binary differential PSK

BDR bell doesn't ring; bus device request

BDS base de données de service (f), service data point; Black Dragon Society; (échange) bidirectionnel simultané (f), two-way simultaneous, modo bidireccional simultáneo [TRWS]; Brownian Dynamics Simulation

BDT billing data transmitter

BDTF bulk data transfer facility

BDTS bulk data transfer subsystem

BDV breakdown voltage

BDW buried distribution wire

B_E excess burst size

BE band elimination; basculement (f), component; below or equal; Benutzerforum (g), ISDN user forum;

Blockeinleitung (g), start of block; excess burst rate

bean re-usable visual/non-visual component (Java/Sun)

BeBox microcomputer by Be Inc *tn*

BEC basic error correction; bearer service code; bus extension card

BECN backward explicit congestion notification

BEDO burst extended data out (DRAM)

Beerware form of shareware paid for in kind (*eg* beer)

BEF band elimination filter

BEG back-end generator; bearer service group

BEL bell character (*equiv* to ASCII code 7)

BellCoRe Bell Communications Research Inc

BENU bull's eye non-uniformity

BEOL back-end of line

BeOS Be operating system *tn*

BER basic encoding rules, reglas básicas de codificación (s); bit error rate/ratio, tasa de errores en los bits (s)

BERKOM Berliner Komunikations (g) (German B-ISDN trial)

BERPM basic exchange rate planning model

BERT bit error rate/ratio test/tester

BERTS basic exchange radio telecommunications service

BESOI bonded and etchback silicon on insulator

BET between

Beta object-oriented language featuring concurrency; Sony VCR format *tn*

beta test test-release of software prior to final release (*cf* alpha test)

Betacam sub-broadcast standard video recorder system (Sony *tn*)

BETRS Basic Exchange Telecommunications Radio Service (US)

Betsie BBC educational text to speech Internet enhancer

BEUA Business Equipment Users' Association (UK)

BEUI BIOS extended user interface
BEX broadband exchange
BEZS bandwidth-efficient zero suppression
BF bad flag; branching filter; basse fréquence (f), low frequency
BFBS British Forces Broadcasting Services
BFF binary file format (IBM); bit fault frequency
BFGS Broyden-Fletcher-Goldfarb-Shannon optimization algorithm
BFI bad frame indication; brute force and ignorance
BFICC British Facsimile Industry Consultative Committee
BFL buffered field-effect-transistor logic
BFO beat frequency oscillator
BFR bridged frequency ringing
BFS Bayern Fernsehen (Bavarian TV)
BFSK binary frequency shift keying
BFT binary file transfer/transmission, transferencia de fichero binario (s)
BFTP background file transfer protocol; batch file transfer protocol
BFU battery/blower fan unit
BG border gateway
BG/EEE below ground – electronic equipment enclosures
BGA ball grid array; message d'accusé de réception de blocage de groupe de circuits (f) [CGBA]
BGCOLOR background colour
BGE branch if greater or equal
BGP border gateway protocol (RFC 1267/1268)
BGR Baugruppe (g) (assembly/module)
BGT branch if greater than
BGV background video
BGW billing gateway
BgZ Beginzeichen (g), proceed-to-send signal
BHC busy hour call
BHCA busy hour call attempt (of processor capacity)
BHI branch if higher

BHIA British Healthcare Internet Association
BHIS branch if higher or same
BHL busy hour load
BHLI broadband high-layer information
BHS Black Hand Society
BHT Brinell hardness test
BHW bit test and set hardware (register)
Bi canal B indicado (s), B-channel indicated
BI all barring of incoming calls (supplementary service); burn-in; backplane interconnect; backward indicator; beginning inventory; binary input; business information (system); Breidbart Index; ISDN interworking unit port; ISDN port
BIA bit indicateur vers l'avant (f), forward indicator bit
BIA-A BIA attendu (f), backward indicator bit expected
BIA-E BIA émis (f), backward indicator bit to be transmitted
BIA-R BIA reçu (f), backward indicator bit received
BIB-T backward indicator bit to be transmitted
BIB-X backward indicator bit to be expected
BIB backward indicator bit; British Interactive Broadcasting
BIBSYS Bibliotek System (Norway)
BIC barred incoming calls; bearer identification code, code d'identification du porteur/de la porteuse (f)
B-ICI SAAL B-ICI signalling ATM adaptation layer
B-ICI broadband intercarrier interface/B-ISDN inter-exchange carrier interface
BiCMOS bipolar CMOS, CMOS-Schaltkreise und hochintegrierter bipolaren (g)
BIC-roam barring of incoming calls when roaming outside home PLMN
BID bit indicador directo/(hacia adelante) (s) [FIB], forward indicator bit; block interaction diagram

BIDE bit indicador directo (s), forward indicator bit expected

BiDi bi-directional

BIDR bit indicador recibido (s), forward indicator bit received

BIDS Bath Information and Data Service (UK); broadband integrated distributed star (UK)

BIDT bit indicador transmitido (s), forward indicator bit transmitted

BIF benchmark interchange format; binary image format

BIFET bi-polar field-effect transistor

Big Blue nickname for IBM

big iron term for large expensive, number-crunching computers

BIG base de información de gestión (s) [MIB], management information base; base de información de guías/base de información de la guía (s), directory information base

BIGFON Breitbandiges Intergriertes Glasfaser Fernmelde Orts Netz (g), Deutsche Bundespost wideband integrated communication system

BIH Bureau International de l'Heure (f), International time bureau

BII bit indicador inverso (s) [BIB], backward indicator bit

BIIC bus interconnect interface chip

BIIE BII directo (s) [BIBX], backward indicator bit to be expected

BIIR BII recibido (s) [BIBR], backward indicator bit received

BIIT BII transmitido (s) [BIBT], backward indicator bit transmitted

BIJ bit de control de justificación (s), justification control bit

BIK trade fair for telecommunications and computers (Leipzig, 1996)

BIL band interleaved by line; basic impulse insulation level

Bi-level binary scan where pixels are either black or white

BIM battery interconnection module; beginning of information marker; binary intensity mask

BIMA British Interactive Multimedia Association

BiMOS bipolar metal-oxide semiconductor

BIN Business Information Network

BINAC binary automatic computer

BIND Berkeley Internet Name Domain

Bindery Novell database *tn*

binhex binary/hexadecimal

BIOS basic input/output system, sistema básico de entrada/salida (s)

BIP billing information processor; bit interleaved parity (SONET)

BIPM Bureau international des poids et mesures (f), international bureau of weights and measures

BIPV bit interleaved parity violation

B-IR Bell–Independent Relations

BIR bit indicateur vers l'arrière (f), backward indicator bit

BIR-A BIR attendu (f) [BIBX], backward indicator bit expected

BIR-E BIR à émettre (f) [BIBT], backward indicator bit to be transmitted

BIR-R BIR reçu (f) [BIBR], backward indicator bit received

bis indicates a revised ITU/CCITT standard

BIS bande intermédiaire satellite (f) intermediate satellite band; border intermediate system (ATM Forum); brought into service; business information/initiation system/services; business information system

BISCOM business information systems-communications systems

BISCUS business information systems-customer service

B-ISDN broadband integrated services digital network (600 Mbit/s)

B-ISDN PRM B-ISDN protocol reference model

B-ISPBX private branch exchange for B-ISDN

B-ISSI broadband inter-switching system interface

BIST built-in self-test

BISTSS business information system – trunks and special services

B-ISUP broadband ISDN user part

BISYNC binary synchronous communications protocol

bit binary digit (smallest binary unit)

BIT bulk ion temperature; built-in test

bitblt bit-block transfer

BITBLT bit boundary block transfer

Bit-D bit de confirmación de entrega (s), delivery confirmation bit

BITE backward interworking telephone events; built-in test equipment

Bit-M bit 'más datos' (s) [M-bit], more-data bit

bitmap array of bits mapped to each other (to form an image)

BITNET Because It's Time Network (US academic network)

BITNIC BITNET Network Information Center

BITS building integrated timing supply

BIU bus interface unit

BIV borde inferior de ventana (s), lower window edge

BIX Byte Information Exchange

Bj canal B en uso (s), B-channel in use

BJ Bubble Jet (Canon printer *tn*)

BJT bipolar junction transistor

B-Kanal Nutzkanal (g), information channel 64kb/s

BKBK break

B-KEY key for B-number ciphering

BKI break-in

bkr (circuit) breaker

BL bit line; Blue Lightning (microprocessor chip *tn*); búsqueda de línea (s), line hunting [LH]

BLA blocking-acknowledgement signal/message, message/signal d'accusé de reception de blocage (f)

black Ethernet *aka* thin Ethernet

BLAISE British Library Automated Information Service

BLAST block asynchronous transmission (Communications Research Group); Boolean logic and state transfer

BLD beam lead device

B-LE broadband – local exchange (B-ISDN)

BLE branch if less or equal; señal de bloqueo y desbloqueo-emisión (s), software-generated group blocking message

Bleam to transmit and send data

BLER block error/block error rate

BLEU Blind Landing Experimental Unit (RAF)

BLEVE boiling liquid expanding vapour explosion

BLF busy line field

BLG blocage de groupe (de circuits) (f), circuit-group blocking (message)

BLH hardware failure-oriented group blocking message (f)

BLIND blind from panel

BLISS Basic language for implementation of system software

blit to move a block of bits or bitmap

blitting transferring a bitmap

BLLF basic low-layer functions

B-LLI broadband – low layer information

BLM maintenance-oriented group blocking message (f) [MGB]

BLMC buried logic macro-cell

BLO blocking; blocking message, message de blocage (f), mensaje de bloqueo (s); blocking signal, signal de blocage (f), señal de bloqueo (s)

Blob binary large object

BLOD far-end blocking

BLOM manual blocking

BLOS branch if lower or same; system blocking

Blosim block-diagram simulator

BLQS band limited quasi-synchronous

BLR señal de bloqueo y desbloqueo-recepción (s), blocking and unblocking signal reception

BLS Business Listing Service; señal de bloqueo y desbloqueo-emisión (s), blocking and unblocking signal sending;

software-generated group blocking message (f)

BLT branch if less than

BLU basic link unit

blue bomb technique causing network PC to crash (*aka* Win Nuke)

BLV busy line verified

Bm bearer mobile; Boyer-Moore algorithm; building module; full-rate traffic channel

BM bande magnétique (f), magnetic tape

BMA maintenance-oriented group-blocking acknowledgement message (f) [MBA]

BMAP broadband modem access protocol

BMAS bit de mayor significación (s), most significant bit; business management application area in TMOS

BMASF basic module algebra specification language

BMC basic monthly change; bubble memory controller

BMDP Bio-Medical Package

BmE Betrieb mit Entdämpfung (g), operation with repeaters

BMES bit de menos significado (s), least significant bit

BMEWS ballistic missile early warning system

BMF Bird-Meertens Formalism

BMI branch if minus

BMIC BusMaster interface controller (Intel *tn*)

BMP basic mapping support; basic multilingual plane; bitmap

BMPT Bundesministerium für Post und Telekommunikation (g), Federal Ministry for Post and Telecommunications

BMS basic mapping support; battery-saving mobile station; bitmapping support

BMTI block mode terminal interface

BMU basic measurement unit; bus access matching and updating unit

BMV bloque de mensaje de verificación por redundancia cíclica (s), cyclic redundancy check message block

BMVA British Machine Vision Association

BMX business multiplexer

BN bit/bridge number

BNC 'thin-net' coaxial RG58 cable; bayonet Neill-Concelman (connector used in LANs); B-number code; bayonet nut connector

BND bandera de nuevos datos (s), new data flag

BNE branch if not equal

B-NET Berkeley network

BNF Backus normal form, Backus-Naur form

BNL Brookhaven National Research

BNN boundary network node

BNR Bell National Research

BNS bill number screening

BNT bayonet nut coupler

B-NT network termination for B-ISDN

BO all barring of outgoing calls (supplementary service); B-number origin

BOA basic office administration; message d'accusé de réception de bouclage (f) [LPA], loopback acknowledgement message

BOAM Bell owned and maintained

BOC barred outgoing calls (originating); basic operator console; Bell Operating Company (US); build-out capacitor

BOCS Berard Object and Class Specifier *tn*

BOD bandwidth on demand

BoE Betrieb ohne Entdämpfung (g), operation without repeaters

BOE buffered oxide etchant

BOF Birds of a Feather (IETF groups)

BOFH bastard operator from hell

Bogo-sort a bad, awkward algorithm for sorting

BOHICA bend over here it comes again

Bohr-bug a fixable bug appearing regularly for no apparent reason

BOIC barring of outgoing international calls (supplementary service)

BOIC-exHC barring of outgoing international calls – except those home

BOL build-out lattice

BOM beginning of message

B-ONALS type B off-network access lines

BOND bandwidth on demand (ISDN)

BONDING Bandwidth-on-Demand Interoperability Working Group

BONT broadband optical network termination

BOOTP boot/bootstrap protocol

BOP bit-orientated protocol

BOR bottom of range

BORAM block-oriented RAM

BORPQU Borland Pro Quattro *tn*

BORQU Borland Quattro *tn*

BORSHT battery overfeed ringing supervision hybrid test

BOS business and office systems

BOS11 Basic OS PDP11

BOSE Business and Office Systems Engineering

BOSS billing and order support system; binary object storage system; Book of Semiconductor Equipment and Materials International (SEMI) Standards; Bridgport Operating System Software

bot near-AI software robot data-seeking Web crawler

BOT beginning of table; beginning of tape marker

BOTE Brother of the Eye

BOTTS busy tone trunks

BOX buried oxide

bozo bit attribute bit in Apple Mac OS

BP bandpass; base pointer; binding post; break point

BPAD BISYNC packet assembler/disassembler

BPB BIOS parameter block

BPC basic physical channel

BPDU bridge protocol data unit

BPE business process engineering

BPF Berkeley packet filter

BPH break permitted here

bpi bits per inch

BPL Bedienungsplatz (g), operator console; branch if plus

BPNRZ bi-polar non-return-to-zero

BPOC Bell point of contact

bpp bits per pixel

BPP bridge port pair

BPR business process re-engineering

BPRZ bi-polar return to zero

bps bits per second

BPSK binary-phase shift keying

BPSS basic packet switching service

bpt bits per track

BPV bipolar violation

BPZ Blockprüfzeichen (g), block test signal

BQS Berkeley Quality Software

br bridge

BR bad register

BRA basic rate access

Brasilsat Brazilian communications satellite

Braun tube early name for CRT (after inventor KF Braun)

BRB base rate boundary; be right back; blocage de circuit par la réception du signal de blocage (f), circuito bloqueado por la recepción de la señal de bloqueo (s), circuit blocked by reception of blocking signal

BRC background revision control

BRD boîte de raccordement et de division (f), junction/splitter box

BRE bridge relay element

breezeway part of a composite video signal

BREMA British Radio and Electronics Equipment Manufacturers' Association

BRF bell rings faintly; benchmark report format

brg bridging

BRGC binary-reflected gray code

BRH branch and hang

BRI basic rate interface, basic rate ISDN (BT)

Brite Euram Basic Research in Industrial

Technologies for Europe – European Research in Advanced Materials

BRL balance return loss

BRM basic remote module

brouter network unit (both bridge and router)

brouteur browser

browser program to surf/browse the Internet

BRP BladeRunner Productions

BRS Big Red Switch; block received signal; B-mode receiving station; business recovery service

BRT British Rail Telecommunications

BRTN Belgische Radio en Televisie Nacht

BRTS back-channel request to send

BRU borne de raccordement usager (f), user connection terminal

BRUIN Brown University Interactive Language

BRX Branche Réseaux (f) (France Telecom)

BRZN Bibliotheksrechenzentrum Niedersachsen (g) (Germany)

BS backspace, retroceso (de un espacio) (s); backing store; backward signalling; banded signalling; base station (for RBS equipment); basic service; bearer services; bit start/stop

BSA basic serving arrangement; Broadcaster's Service Area (US cable TV); Business Software Alliance (UK); software-generated group blocking-acknowledgement message (f)

BSAM basic sequential access method

BSB British Satellite Broadcasting

BSBH busy season busy hour

BSC base station controller; binary symmetric channel; binary synchronous communications, communication binaire synchrone (f); binary synchronous control (network protocol IBM)

BSCL Bell Systems Common Language

BSCM bisynchronous communications module

BSD Berkeley Software Distribution

BSD UNIX Berkeley Software Distribution (version of UNIX)

BSE back-scattered electron detection; basic service element; Broadcast Satellite for Experiment (Japan)

BSF Bell shock force; bit scan forward

BSG basic service group

BSI bit sequence independence; British Standards Institution

BSIC base station (transceiver) identity code

BSIC-NCELL BSIC of adjacent cell

BSkyB British Sky Broadcasting (BSB and Sky merger)

BSM backward set-up message; base station management

BSN backward sequence number

BSNR backward sequence number received

BSNT backward sequence number of next SU to be transmitted

BSOC Bell Systems Operating Company

BSP Bell Systems Practice; binary space-partitioning; broadband signalling end-point; bulk synchronous parallelism

BSP tree binary space-partitioning tree

BSR bit scan reverse

BSRAM burst static RAM

BSRF Bell system reference frequency (standard: caesium beam clock); Bell System Repair Specification

BSS base station system; basic synchronized subset; block started by symbol; broadband switching system; business support system

BSSAP base station system (BSS) application part

BSSGP BSS general packet radio service (GPRS) protocol

BSSMAP BSS management application part

BSSOMAP BSS operation and maintenance application part

BST binary synchronous transmission (*v* bisync)

BSTJ Bell Systems Technical Journal

B-STP broadband signalling transfer point

BSU basic sounding unit; bearer switchover unit

BSV borde superior de ventana (s), upper window edge

BSW BTS software management

BT bothway trunk; British Telecom; buried tape armour; bridged taps; burst tolerance; busy tone

BTA basic trading area; broadband terminal adaptor (for B-ISDN)

BTAB bumped tape automated bonding

Btag beginning tag

BTAM basic telecommunications access method

BTB branch target buffer; bus terminal board

BTC Bulgarian Telecommunications Company; bit test and complement

BTE bloqueo temporal de enlaces [TTB]

B-TE broadband terminal equipment (for B-ISDN)

BTI boîtier de transition intérieur (internal transition module)

BTL Bell Telephone Laboratories

BTM benchmark timing methodology

BTMC British Telecom Mobile Communications

BTN billing telephone number

BTNA BT North America

BTNJ bits de afluentes disponibles para justificación negativa (s), bits from tributaries available for negative justification

BTOA binary to ASCII

BTOS operating system (Burroughs Corp)

BTP batch transfer program

BTPJ bits de afluentes disponibles para justificación positiva (s), bits from tributaries available for positive justification

BTR bit test and reset

btree binary tree

B-tree balanced multi-way search tree

BTRL British Telecom Research Laboratories

Btron version of Tron

BTS base transceiver station; base transceiver subsystem; bit test and set; block test signal; British Telecommunications Systems Ltd

BTT bad track table

BTU basic transmission unit

BTW by the way

BTX Bildschirmtext (German public interactive videotex facility)

Btx bildstext (g), videotex

BU branch unit

BUAF big ugly ASCII font

BUAG big ugly ASCII graphic

BUBL Bulletin Board for Libraries

BUC background update control

BUG Borland User Group

BUNCH Burroughs, Univac (Sperry, NCR, Control Data, Honeywell)

B-UNI broadband user network interface

BUS broadcast and unknown server

BUSAK busy acknowledgement

BUSRQ bus request

Butineur browser (f) (*cf* brouteur)

BUV backscatter ultra-violet spectrometer

BV breakdown voltage; busy verification

BVA British Video Association

BVH base video handler; Broadcast Video Helical scan *tn*

BVU Broadcast Video U-matic *tn*

BVW backward volume wave

Bw7R Bauweise 7R (g), equipment practice 7R

BWB Bundesamt für Wehrtechnik und Beschaffung (g), Federal office for defence technology and procurement

BWM block-write mode; broadcast warning message

BWQ buzzword quotient

BWR bandwidth ratio

BWT broadcast warning teletypewriter exchange service

BWTS bandwidth test set

BX.25 Bellcore X.25

by busy
B-Y blue minus luminance (Y)
B-Y signal component of colour TV
 chrominance signal
byacc Berkeley Yacc (*v* Zoo, Zeus)
byte set number of bits (usually 8)
BZT Bundesamt für Zulassungen in der
 Telekommunication (g), (German)
 Federal Office for Telecommunications
 Approval/Certification

C

C capacitance; conditional; consumer; container, conteneur (f), contenedor (s); coulomb; hard drive letter/notation; hexadecimal equivalent to decimal 12; programming language of UNIX OS

C++ programming language derived from/superset of C

C band microwave communication frequency (3.9-6.2 GHz)

C DV D1 canal de control en el punto de referencia V D1 (s) [C_{v1}]

C&C command and control (*v* C^2)

C&W Cable & Wireless Ltd

C/A carrier to adjacent ratio

C/I carrier-to-interference ratio; co-channel interference ratio

C/kT carrier-to-receiver noise density

C/m^3 coulomb per cubic metre

C/MFI conversion memory and fault indication

C/N carrier-to-noise ratio

C/R carrier-to-reflection ratio; command/response (field) bit, bit de commande/réponse (f)

c/s client/server

C/SCSC cost-schedule control system criteria (US gov)

C/T carrier-to-noise temperature ratio

C1-3 capa 1-3 (s) [L1-3]

C2D character to decimal (REXX)

C2X character to hexadecimal (REXX)

C^3 command, control and communications (*aka* C3)

C^{3}CM C^3 countermeasures

C5 CCITT Nº 5 signalling

C7 CCITT (ITU-T) Nº 7 common-channel signalling

CA call accepted; canal de aplicación (s), application channel; cell allocation; cellular automaton/automata; centre d'authentification (f), authentication centre; certification authority, autorité de certification (f); charging analysis; CIM architecture; collision avoidance; comienzo de actividad (s), activity start; computer animation; condición de avería (s), fault condition; connect acknowledge; contexto de aplicación (s), application context; contractual agreement; customer access

CAA Chaotic Ansi Artists; CIM applications architecture; Civil Aviation Authority (UK); commutateur à autonomie d'acheminement (f), group area exchange

CAAS computer-aided approach sequencing (aircraft landing)

CAB competitive analysis benchmarking

CABEX computer-based message switching system

CABS carrier access billing system

CAC Comité d'action commerciale (f), commercial action committee (CEPT); connection admission control; mensaje de información de control automático de la congestión (s); control automático de congestión (s), automatic congestion control (indication message); channel access control; circuit access code

CACD computer-aided chemical development

CACF control de actualización de fondo (s), background update control

cache RAM buffer for data next needed by CPU

CACP central arbitration control point

CAD call acceptance delay; computer access device; computer-aided design; computer-aided despatch; computer-aided drafting; Computergestütztes Entwerfen (g), computer-aided design; convertidor analógico-digital (s), analogue-digital converter

CADCAM computer-aided design/computer-aided manufacturing

CADD computer-aided design and drafting; computer-aided detector design

CADDIA Cooperation in Automation of Data and Documentation for Imports/exports and the management of financial control of the Agricultural Market (CEC R&D Programme)

caddy sealed CD-ROM container with metal shutter

CADE client/server application development environment

CADMAT computer-aided design manufacture and testing

CADS computer-aided departure sequencing

CADT control application development tool

CADV combined alternate data/voice

CAE client application enabler; common application environment; computer-aided engineering; control de la actividad de los enlaces de señalización (s), signalling link activity control

CAED capa de enlace de datos (s), data link layer

CAET Comité de Aprobación de Equipos de Telecom (f), Approvals committee for terminal equipment

CAF capa física (s), physical layer; constant applicative form; controlled ambient facility; customer application file

CAFM computer-aided facilities management

CAFS content-addressable file system

CAG column address generator; commande automatique de gain (f), automatic gain control; common air interface; Competitive Analysis Group; control automático de ganancia (s), automatic gain control

CAI charge advice information; commande de l'alignement initial (f), control de alineación inicial (s), initial alignment control; computer-assisted instruction

CAIDA Cooperative Association for Internet Data Analysis

CAIS collocated automatic intercept system; common Ada interface set; common Ada programming support

environment (APSE) interface specification

CaiSE Conference on Advanced Information System Engineering

CAJ metering pulse rate

CAJOLE Chris and John's Own Language

CAL client access licence (Microsoft *tn*); common applications language; computer-assisted learning; course author language

Calcomp California Computer Products Inc

CALL computer-aided language learning

call/cc call-with-current-continuation (a Lisp control)

CALRS centralized automatic loop reporting system

CALS computer-aided (acquisitions and) logistics support (US DoD)

CALVADOS French telecommunications network

CAM call accepted message; cellular automatic machine; common access method; computer-aided/assisted manufacture/manufacturing; computer-aided music; Computing Accounting Machine; content-addressable memory

CAMA centralized automatic message accounting

CAMAC computer automated measurement and control

CAMAL Cambridge Algebra system

CAMA-ONI CAMA-operator number identification

CAMEL call management language; customized applications for mobile (network) enhanced logic

CAMIL computer-assisted/managed instructional language

CAMP chemical agents manufacturing plant; Corporate Association of Microcomputer Professionals (US)

CAMR Conférence Administrative Mondiale de Radiocommunications (f) [WARC]

CAMTT-88 Conferencia Administrativa

Mundial Telegráfica y Telefónica de 1988 (s)

CAN campus area network; cancel (character); central automatique numérique (f); Computer Association of Nepal; controller area network (Bosch/Intel); convertisseur analogique/digitale (f), analogue-digital converter; coordonnées d'appareil normées (f), normalized device coordinates

CANAL command analysis

canaries extraneous, high-frequency sounds from recording channel

CANC canal de control (f), control channel

CANCA canal de control asociado (s), associated control channel

CANCC canal de control común (s), common control channel

CANCDED canal de control dedicado (s), dedicated control channel

CANCDIF canal de control de difusión (s), broadcasting control channel

CANDI capacity and network dimensioning

CANPU canal de paquetes de usuario (s), user packet channel

CANS code answer

CANT canal de tráfico (s), traffic channel

CANTAT Canada-Great Britain Transatlantic cable

CANTO Caribbean Association of National Telecommunication Organizations

CANTRAN cancel transmission

CANTV Venezuelan National Telecommunications Company

CAO circuit allocation order; completed as ordered; conception assistée par ordinateur (f), computer-aided design

CAOM administration, operation and maintenance centre (s)

CAP cable access process; calendar access protocol; CAMEL application part; carrierless amplitude (and) phase modulation; centralita automática (s),

automatic branch exchange; circuit access point; Columbia AppleTalk package *tn*; communication application platform; communications alternative provider; competitive access provider; customer administration panel

CAPI calendar application programming interface; common ISDN application programming interface; computer-aided programmed interviewing; cryptographic application programming interface

CAPL channel allocation priority level

CAPM computer-aided production management; CPU access port monitor

CAPP computer-aided process planning

CAPS call attempts per second; computer-assisted problem-solving

CAPSI centralita automática privada de servicios integrados (s), integrated services private branch exchange

CAPTAIN character and pattern telephone access information network (Japan)

CAR capa de red (s), network level; carácter (s), character; computer-aided research; content-addressable register; contents of address register; current address register

CARB central arbiter on (Sun) SuperSPARC module

CARDS Central Archive for Re-usable Defense Software (US DoD)

CARL Colorado Alliance of Research Libraries

CARLOS communications architecture for layered open systems (Esprit)

CARO Computer Anti-virus Researchers' Organization

CAROT centralized automatic reporting on trunks

CARP call accounting reconciliation process

CARRI computerized assessment of relative risk impacts

CARS cable (television) relay service (station); código de área/red de señalización (s), signalling area/network code; community antenna relay service; continuous alarm reporting service

CART Comité chargé de l'application des recommandations techniques (f), Comité de aplicación de reglamentaciones técnicas

CAS canales de agrupación de señalización (s), signalling group channel; centralized attendant services; centro de administración de servicios (s), service switching point; channel-associated signalling; circuit-associated signalling; column address strobe/select; communication applications specification (Intel); commutateur d'accés aux service (f) service switch point [SSP]; computerized auto-dial system; customer administrative system

CASA Centre for Advanced Spatial Analysis (University College London)

CASE common application service element; computer-aided software/systems engineering; Computer and Systems Engineering Plc

Casecall software allowing police to process BT call data (UK)

CASL crosstalk application scripting language (DCA)

CASSIS classified and search support information system

CAST Center for Advanced Study in Telecommunications (Ohio University); computer-aided software testing

CAT common abstract tree (language); community antenna television (cable TV); computer-aided teaching/testing; computer-aided tomography; computer-aided translation; computerized axial tomography; cumulative abbreviated trouble file

CAT-3/5 category 3/5 (cabling standard)

CATE computer-aided test engineering

catenet a concatenated network: interconnecting layered sub-networks

CATG computer-aided test generation

CATI computer-aided telephone interviewing

CATIA computer-assisted 3D interactive application (IBM *tn*)

CATLAS centralized automatic trouble-locating and analysis system

CATN computer-aided technologies network

CATNET credit-authorization terminal network (IBM)

CATNIP common architecture next generation Internet protocol (RFC 1707)

CATS computer-assisted tele-cooperation service

CATT centralized automatic toll ticketing

CATV community antenna television (broadband cable TV)

cat-wg Common Authentication Technology – Working Group (of IETF)

CAU capa de agente de usuario (s), user agent layer; Computer Anarchists Underground; crypto-ancillary unit

CAUSE Coalition Against Unsolicited E-mail; College and University System Exchange (US)

CAV constant angular velocity

CAVE computer automatic virtual environment

CAW call waiting; common aerial working

CAWC cryogenic aerosol wafer cleaning

CAX community automatic exchange

Cb columbian (old name of niobium – used in superconductors); count of bytes; C beautifier

CB cell broadcast; circuit-breaker; Citizens' Band; clear back; clear-back signal; common battery; component broker; connecting block

CB1-3 clear-back signal Nº1-Nº3

CBA call barring analysis; change-back acknowledgement signal; contador de bloques de los que se ha acusado recibo (s), block acknowledged counter

C-BAT cost-benefit analysis toolkit (RACE 1 analytic tool)

CBBS computerized BBS

CBC Canadian Broadcasting Corp; can't be called; cipher block chaining;

contador de bloques completos (s), block completed counter

CBCH cell broadcast channel/(centre)

CBCR channel byte count register

CBD call box distribution; change-back declaration signal; configuration block diagram

CB-DRX cell broadcast discontinuous reception (short message service)

CBDS connectionless broadband data service

CBDT can't break dial tone

CBEMA Computer and Business Equipment Management Association

CBGA ceramic ball grid array

CBH can't be heard

CBIR content-based information retrieval

CBK clear-back signal

CBL computer-based learning

CBM Commodore Business Machines

CBMI cell broadcast message identifier

CBMS computer-based message system

CBN call-by-name

CBNV code bit number variation

CBO caesium beam oscillator (BSRF); continuous bit-stream oriented (service)

CBPX computer PBX

CBPXI computer private branch exchange interface

CBR case-based reasoning; constant bit rate

CBS chemical bottle storage area; Columbia Broadcasting System; Common Basic Specification (UK NHS computer specification); common battery signalling; crossbar switching

CBT computer-based training; Compagnie bulgare des télécommunications

CBTA Canadian Business Telecommunications Alliance

CBV call-by-value; codificación a baja velocidad/codificación de baja velocidad (s), low-rate encoding

cbw Crypt Breaker's Workbench

CBW convert byte to word

CBX computer-based exchange; computerized branch exchange

CC call connected, comunicación establecida (s); call control; calling channel; Centro de control de tránsito aéreo (s), air traffic control; character code, código de carácter (s); charging case; chip carrier; clearing cause; cluster controller; code controller; collect call; common carrier; common control; (protocole de) confirmation d'appel (f), call confirmation; congestion control, control de congestión (s); conmutación de circuitos (s), circuit switched; connection confirm, confirmación de conexión (s); continuity check, contrôle de continuité (f); control channel; country code; courtesy copy; Cryptic Criminals

CC++ Compositional C++

CC TPDU connection confirm TPDU

CCA Cable Communications Association (UK); call control agent; carrier-controlled approach; character content architecture; code-controlled absent-subscriber service; compatibilidad de capa alta (s), high-layer compatibility

CCAF call control agent function

CCB cable connection board; code-controlled call barring; coin-collecting box; command control block; common carrier bureau; compatibilidad de capa baja (s), low-layer compatibility; componentes de conexión básicos (s), basic connection component

CCBS completion of call(s) to busy subscriber – subscriber service

CCC blind courtesy copy (f); ceramic chip carrier; Chaos Computer Club (German hackers' groups); clear channel capability; código de comienzo de campo (s), field start code; community computer centre; computer control centre

CCCH common control channel

CCCR control de congestión de conjunto de rutas de señalización (s), signalling route set congestion control

CCD centre de commutation de données (f), data switching exchange; centro/(central) de conmutación de datos (s), data switching exchange; centro/(central) de conmutación digital (s), digital switching exchange; charge-coupled device; conference call device; conjunto de contextos definido (s), defined context set; signal de demande de contrôle de continuité (f), continuity-check request signal

CCDMS centre de commutation de données mobile par satellite (f), mobile satellite data switching exchange; central de conmutación de datos del servicio móvil por satélite (s), mobile satellite data switching exchange

CCDN Corporate Consolidated Data Network (IBM)

CCE centre de commutation et d'essais (f), switching and testing centre; control de conjuntos de enlaces (s), network group control; contrôle de continuité sur circuit entrant (f), continuity check incoming; cross-connect equipment

CCEC control de cambio de estado coordinado (s), coordinated state change control

CCEI centre de commutation et d'essais international (f), international switching and testing centre

CCES common control echo suppressor

CCF call control function; campo de control facsímil (s), facsimile control field; communications control field; conditional call forwarding; configuration control function; confirmación de control de flujo (s), flow control confirmation; continuity-failure signal

CCFM call control fault management

CCFP central control fixed part

CCG common channel group

CCGB código de comienzo de grupo de bloques (s), group of block start code

CCH Comité de coordination pour l'harmonisation CEPT (f), Co-ordinating committee for the harmonization of CEPT; connections per circuit per hour; continuity-check indicator; control channel

CCI centre de commutation international (f), international switching centre; centro/(central) de conmutación internacional (s), international switching centre; common client interface; compatibilidad de capa inferior (s), low-layer compatibility; computer-computer interface; continuity-check incoming; Cyber Crime International

CCIA Computer and Communication Industry Association

CCIF International telephone consultative committee (pre-CCITT)

CCIR Comité Consultatif International des Radiocommunications (f), Zwischenstaatliche beratender Ausschuß für den Funkdienst (g), International Radio Consultative Committee

CCIR 601 CCIR standard for defining digital video

CCIRN Coordinating Committee for Intercontinental Research Networks

CCIS common channel inter-office/inter-switch signalling

CCITT Comité consultatif international de télégraphique et téléphonique (f), International Telephone and Telegraph Consultative Committee

CCITT MML CCITT man-machine language

CCL calling-party clear signal; código de comienzo de línea (s), line start code; computer control language; connection control language; continuous communications link; Coral Common LISP; customer configuration and logistics

CCLO common command language

CCM centre de commutation pour les services mobiles (f), centro de conmutación de servicio móvil (s), mobile service switching centre; circuit supervision message; communications control module; current call meter

CCM-A centre de commutation pour les services mobiles (f), centro de conmutación de los servicios móviles (s), mobile service switching centre A; CCM que controla la llamada en el traspaso (s), MSC with call control at handover; CCM al que se hace un segundo traspaso, MSC to which a subsequent handover is done; CCM al que se realiza un traspaso (s), MSC to which a handover is done

CCMC centro de conmutación de servicios móviles de cabecera (s) [MSC]; Commonwealth Cable Management Committee

CCMR CCM de rattachement (f), home mobile service switching centre

CCMS centre de commutation du service mobile par satellite (f), centro de conmutación móvil por satélite (s), mobile satellite switching centre; centro de conmutación del servicio marítimo por satélite (s), maritime satellite switching centre

CCN contact change notice; signal de contrôle de continuité négatif (f), continuity failure signal

CCNC common channel (signalling) network control; computer communications network center

CC-NDT can't call – no dial tone

CCO centre de commande d'opérations (f), operations command centre; Cisco Connection Online *tn*; centro de control operacional/de operaciones (s), operations control centre; confirmation de connexion (f), connection confirm; continuity-check outgoing

CCP call confirmation protocol, protocolo de control de la llamada (s); call control processing; capability/configuration parameter; centros de conmutación y de pruebas (s), switching and testing centres; Certificate in Computer Programming; concurrent

constraint programming; console command processor; contract configuration process; cross connection point; customer configuration program; signal de contrôle de continuité positif (f), continuity signal

CCPE control channel protocol entity

CCR centre de commande de rétablissement du service (f), restoration control point; commitment concurrency and recovery (ISO); comparison category rating (ITU test); concurrency control and recovery; condition code register; continuity-check request signal; control channel redundancy; current cell rate; customer-controlled reconfiguration (AT&T)

CCRS Centrex customer re-arrangement system *tn*; control channel redundancy switch

CCS calculus of communicating systems; calling card service; centros de control de satélites (s), satellite control centre; centum (hundreds) call seconds; common channel signalling; common command set; common communications services; compatibilidad de capa superior (s), high-layer compatibility; Computer Conservation Society; contiguous colour sequence; contrôle de continuité sur circuit sortant (f), continuity-check outgoing; customer care system; customer connectivity services

CCS^2 command control and subordinate systems

CCSA common control switching arrangement (AT&T)

CCSD cellular circuit-switched data; command communication service designator

CCSI centre de coordination du service international (f), centro de coordinación del servicio internacional (s), international service coordination centre

CCSL compatible current-sinking logic

CCSM centre de commutation du service maritime par satellite (f), maritime satellite switching centre; centro de coordinación de salvamento marítimo (s), maritime rescue coordination centre

CCSMS centro de conmutación del servicio marítimo por satélite (s), maritime satellite switching centre

CCSN common channel signalling network

CCSP contextually-communicating sequential processes

CCSS common channel signalling subsystem; common channel signalling system

CCT central control terminal; circuit, Schaltung (g); código de comienzo de trama (s), field start code; complete calls to; coupler cut through; cumulative cycle time; telephone circuit

CCTA Central Computer and Telecommunications Agency (UK, now GCIS); Comisión Coordinadora de Tecnología Adecuada (s) (Peru)

CCTAC Computer Communications Trouble Analysis Center

CCTC caja común TC (s), TC common box

CCTG configuration control task group

ccTLD country code – top level domain

CCTP cahiers de charges techniques particuliers (f), detail technical statement of requirements

CCTS Comité de Coordination pour les Télécommunications par Satellite (f), Coordinating committee for satellite telecommunications

CCTV China Central Television; closed circuit TV

CCU central control unit; channel codec unit; Computer Crime Unit (Scotland Yard, UK); crypto-control unit

CCUAP computerized cable upkeep administration programme

CCUTF Corrupt Computer Underground Task Force

CCV calling card validation

CCW cable cut-off wavelength; channel command words (IBM); counter clockwise

CCYY four-figure date format (*ie* CC for century; YY for years)

cd candela (measure of light unit); carrier detect

Cd cadmium (used in NiCad batteries); call deflection; campo de dirección (s), address field; capability data; circuit description; cohesive detection; compact disk; confirmación de desconexión (s), disconnect confirm; control device; critical dimension

CD CHRY card channel ready (IBM)

CD ROM/XA CD-ROM Extended Architecture (Philips, Sony, Microsoft)

CD/OL critical dimension overlay

CD+G compact disk and graphics

CD32 CD-ROM drive and 32-bit processor for games (Amiga Commodore *tn*)

CDA call data accumulator; capability data acknowledgement; clean dry air; command and data acquisition; Communications Decency Act (US); compound document architecture (DEC); conexión por desplazamiento de aislamiento (s), insulation-displacement connection; crash dump analyser

CDAS continuous data availability system (Ark Research Corp *tn*)

CDB common database

CD-bridge extension to CD-ROM XA standard

CDC call directing code; characteristic distortion compensation; Clark Development Company; command document continue; Computing Devices Company (Canada); confirmation de déconnexion (f), release complete message; construction design criteria; Control Data Corporation (US); contrôle dynamique de charge (f), control dinámico de carga (s), dynamic load control; Cult of the Dead Cow

CDCCP control data communications control procedure

CDCL command document capability list

CDD cell design data; command document discard; common data dictionary

CD-DA CD-ROM format standard; digital audio compact disk

CDDI copper-distributed data interface

CDE C development environment; chemical downstream etch (of silicon chip); command document end; common desktop environment; cooperative development environment

CDF channel definition format; combined distribution frame; communications data field; cumulative distribution function; cut-off decrease factor

CDFR conexión digital ficticia de referencia (s), hypothetical reference digital link

CDFS CD file system (Microsoft *tn*)

CD-G CD graphics

CDI called line identity, identité de la ligne du demandé (f), identidad de la línea llamada (s); change direction indicator; circle digit identification; collector-diffusion isolation; compact disk – interactive; contador de BID irrazonables (s), counter of unreasonable FIBs

CDIA certified document imaging architect

CDIF CASE data interchange format

CDL code (de) début de ligne (f), line start code; common/computer design language; compiler description language; control definition language

CDLI called line identity

CDLRD confirming design layout report date

CDM CD mechanism; common device model for sensor/actuator bus (SAB); companded delta modulation; content data model

CDMA code-division multiple access

CDML Claris dynamic mark-up language *tn*

CDMMF commercial data masking facility (IBM)

CDO community dial office; controlled decomposition/oxidation

CDOS concurrent DOS

C-DOT Centre for the Development of Telematics, India

CDP Certificate in Data Processing; communications data processor

CDPB command document page boundary

CDPD cellular digital packet data

CDPR customer dial pulse receiver

CD-PROM rewritable optical disk (readable by CD-ROM system)

CDPS compound document protocol specification

CDR call data records; call detail record/recording; CD – recordable; charging data record/recording; chemical distribution room; codage à débit réduit (f), low-rate encoding; command document resynchronize; connection detail record; contents of decrement register

C-DRAM cached DRAM

CD-ROM XA CD-ROM extended architecture

CD-ROM CD – read-only memory

CD-RTOS CD real-time OS

CD-RW CD – re-writable

CDS calidad de servicio (s), quality of service; Citrix Device Services; command document start; concrete data structure; current directory structure

CDSA common data security architecture

CDT Cambridge Digital Technology (UK); clases de tráfico (s), class of traffic; code (de) début de trame (f), field start code; confirmación dar testigos (s), give tokens confirm; control data terminal; credit (field), crédito (s)

CD-THOSR rewritable optical disk format *tn*

CDTV Commodore Dynamic Total Vision (CD-ROM *tn*)

CDU combining and distribution unit

CDUGD Computer Down Under Ground Digest

CDUI command document user information

CDUR chargeable duration

CD-v CD video

CDV cell delay variation; comma-delimited value; coordenadas de dispositivo virtual (s), virtual device co-ordinates

CDVT cell delay variation tolerance

CD-WO CD – write only

CD-WORM CD – write once read many (times)

CE cache enable; campo de estado (s), status field; capillary electrophoresis; centre d'enregistrement (f), billing centre; chip enable; código de enclavamiento (s), interlock code; Compact Edition (Microsoft Windows *tn*); compatible; computing element, élément de calcul (f); elemento de proceso (s); conditions d'écoute (f), listening conditions; connection endpoint; connection element; Consumer Electronics: mark of conformity for European electronic components (formerly CECC); método de la capacidad equivalente (s), equivalent capacity; signal de communication établie (f), call connected signal

CEA alarm transmission panel (f); electronic auto-switch

CEB computer estimating bureau; counter of unreasonable FIBs, compteur d'élements binaires vers l'avant (BIA) irrationnels (f)

CEC cell evaluation chip; Centre Européen de la Communication (f); Commission of the European Communities; congestión del equipo de conmutación (s), switching equipment congestion; console electronics controller; señal de congestión en el equipo de conmutación (s), switching equipment congestion (signal)

CEC DG Commission of the European Communities – Director General

CECC CENELEC Electronic Components Commission

CED called station identity, identification du poste demandé (f); campo de extensión de dirección (s), address extension field

CEDAR computer-enhanced digital-audio restoration

CEE cellular engineering equipment; complete exchange failure; control del estado del enlace (s), link state control; control execution environment

CEG continuous edge graphics

CEI Commission Electrotechnique Internationale (f), Comisión Electrotécnica Internacional (s), International Electrotechnical Commission; connection endpoint identifier; comparably efficient interconnection; conducted electromagnetic interference

CEI-PACT Central European Initiative on Parallel Computation

CEIR central equipment identity register

CEIRD confirming engineering information report date

CEK code exchange keying

Celeron Intel processor *tn*

CELEX EC legislation database

CELIP cellular language for image processing

CELLAS cellular assemblies

CELLCO cellular companies

CELLSIM (biological) cell simulation

CELP card-edge low-profile (socket); code-excited linear prediction/(predictor) (voice compression); computationally extended logic programming

CELTIC Concentrateur exploitant les temps d'inoccupation/d'inactivité des circuits (*cf* TASI) (f), circuit idle-time concentrator

CEM cement conduit; centre d'exploitation et de maintenance (f), operations and maintenance centre;

compatibilité électromagnétique (f), electromagnetic compatibility; conjunto de entidades de mantenimiento (s), maintenance entity assembly; continuous emissions monitoring

CEMA Canadian Electrical Manufacturers Association

CEMF counter-electromotive force

CEMGR centre d'exploitation, maintenance et gestion du réseau (f), operations administration and maintenance centre

CEN Comité Européen de Normalisation (f), European Committee for Standardization; compteur de NSR irrationnels (f), counter of unreasonable backward sequence numbers; control del encaminamiento de la señalización (s), signal routing control

CENA cellular network analyser

CEND end of charge point

CENELEC Comité Européen de Normalisation Electrotechnique (f), European Committee for Electrotechnical Standardization

Centrex centralized PBX services for business customers

Centronics international standard interface between PC and printer

CEO comprehensive electronic office (Data General)

CEOP conditional end of page

CEP circular error probable

CEPC control de encaminamiento de la PCCS (s), SCCP routing verification test

CEPER combined engineering plant exchange record

CEPIS Council of European Professional Informatics Societies

CEPS colour electronic pre-press system

CEPT Conference of European Post and Telecommunications, Conférence Européenne des Postes et des Télécommunications (f)

CER canonical encoding rules; cell error ratio

CER-DIP ceramic dual in-line package

CERF cifras específicas de red facultativas (s), optional network specific digits

CERFNET California Educational Research Network

CERMET ceramic metallized

CERN Conseil Européen pour la Recherche Nucléaire (f)

CERNET China Education and Research Network

CERT Committee on Energy, Science and Technology (US); Computer Emergency Response Team (US government)

CERT/CC CERT Co-ordination Center

CES circuit emulation service; código de enlace de señalización (s), signalling link control; coast earth station; computer election systems; connection end-point suffix

CESA Canadian Engineering Standards Association; coast earth station assignment

CESDL coast earth station low-speed data

CESG Communications Electronics Security Group (GCHQ UK)

CESI coast earth station interstation

CES-IS CES interoperability specification

CESNET Czech Educational and Scientific Network

CESP Common ESP

CESR correction d'erreurs dans le sens de la réception (f), forward error correction

CESSL cellular space simulation language

CEST coast earth station telex

CESTA Centre d'études des systèmes et technologie (f)

CETIS Centre Européen de traitement de l'information scientifique (f), European Scientific Data Processing Centre

CEV check environment visited network

CF call forwarding; can't find; carrier frequency; componentes funcionales (s), functional components; conexión física (s), physical connection; conversion facility; copy furnished; count forward; formatted content architecture levels; message de contrôle de flux de trafic de signalisation (s), signalling traffic flow control messages

CFA carrier frequency alarm; component failure analysis

CFAA Computer Fraud and Abuse Act (US 1986)

CFAO conception et fabrication assistées par ordinateur (f), computer-aided design-manufacturing

CFAR constant false-alarm rate

CFB call forwarding busy; call forwarding in mobile subscriber; cipher feed-back; configurable function block

CFC código de fin de conglomerado (s), end of cluster code; coin and fee check; conditional forward call

CFCA Communications Fraud Control Association

CFD coarse frequency discrimination; codage à faible débit (f), low-rate encoding; compact floppy disk; computational fluid dynamics; Computer Freelance Directory (UK)

CFE contractor-furnished equipment

CFF current fault file; critical flicker frequency

CFGBACK configuration back-up file (Microsoft Windows *tn*)

CFH cyclic frequency hopping

CFI CAD Framework Initiative; Canal France International

CFL call-failure signal; capacidades funcionales locales (s), local functional capabilities; Computing for Labour (Party) (UK)

CFM CCBS facility message; code fragment manager (Apple); compander and frequency modulation; confirmación de mensaje (s), message confirmation; contamination-free manufacturing; mensaje de confusión (s), confusion message

CFNR call-forwarding, no reply

CFNRc call forwarding mobile subscriber not reachable (supplementary service)

CFNRy call forwarding mobile subscriber no reply (supplementary service)

C-format composite format (broadcast-standard tape on reels)

CFP communicating functional processes; Computers, Freedom and Privacy conference, 1994; constraint functional programming; control fixed point; formatted processable content architecture levels, niveles de arquitectura de contenido formatado procesable (s)

CFR call failure rate; circuit fictif de référence (f), circuito ficticio de referencia (s), hypothetical reference circuit; Code of Federal Regulations (US); communications fictives de référence (f), conexión ficticia de referencia (s), hypothetical reference connection; confirmation to receive

CFRS communication fictive de référence pour la signalisation (f), conexión ficticia de referencia para la señalización/conexión hipotética de referencia para la señalización (s), hypothetical signalling reference connection

CFS caching file system; call-forwarding on mobile subscriber busy; calls for service signal (Nº 1 EAX); common file system; common functional specifications; communications frame structure

CFT cross-file transfer

CFTG circuit facility trunk group

CFTS control del flujo de tráfico de señalización (s), signalling traffic flow control

CFU call forwarding unconditional

CFUR calidad de funcionamiento de la red (s), network performance

CFW call forwarding

CG common ground; control gate; control global (s), global control

CGA carrier group alarm; colour graphics adaptor (original IBM specification)

CGB circuit-group blocking

CGBA circuit-group blocking acknowledgement message

CGC circuit-group-congestion signal, signal d'encombrement du faisceau des circuits (f), señal de congestión en el haz de circuitos (s); circuit group control

CGE common graphics environment

CGEM centre de gestion, d'exploitation et de maintenance (f), operations, administration and maintenance centre

CGF charging gateway function

CGG centros de gestión de GCU (s), CUG management centre

CGGL code-generator generator language

CGI cell global identity; common gateway interface; Computer Generation Incorporated; computer-generated imagery; computer graphics interface (ISO/IEC 9636 standard)

CGM computer graphics metafile (ISO 8632 standard)

CGN connector group network

CGNC connector group network controller

CGPM general conference on weights and measures

CGRP circuit group

CGRR circuit-group reset receipt

CGRS circuit-group reset sending

CGSA cellular geographical statistical service area; Computer Graphics Suppliers' Association (UK)

CGSET circuit-group set

CGU circuit-group unblocking message

CGUA circuit-group unblocking acknowledgement message

CH can't hear

CHA command handling application

CHAN channel

CHAP challenge handshake authentication protocol (RFC 1334); channel application profile

char character

CHARM parallel programming language based on C

CHART Computers in the History of Art (UK organization)

CHASM Cheap Assembler (shareware MS-DOS assembler)

CHAT conversational hypertext access technology

chatter bounce: unwanted opening/closing of relay contacts

CHC (señal de) congestión del haz de circuitos (s), circuit group congestion signal

CHCK channel check

CHCP change code page (system command DOS, O/S2)

CHDB compatible high-density bipolar code

CHDIR change directory

CHDL computer hardware description language

CHFN change finger (Unix)

CHG charge; charging message

CHGRP change group

Chicago early development code-name for Microsoft Windows 95

CHILL CCITT high-level language

CHIM computer and human integrated manufacturing

CHINA Communist Hackers in North America

chip common name for integrated circuit

CHIP come home, I'm pregnant; constraint handling in Prolog

chipset group of microprocessor and support chips

Chiptest early version of what became IBM's Deep Blue

CHKDSK check disk (MS-DOS utility)

CHM change-over and changeback messages; channel module

chmod change mode (Unix command)

CHN change

CHO choke packet

CHOP channel operation

chorus sound enhancement by a doubling effect

CHORUS OS developed at INRIA

CHP charging point

CHR character; channel reliability; communication history report

chroma chrominance

chromakey colour separation overlay

CHRP common hardware reference platform

CHS charging subsystem; cylinders-heads-sectors (of a hard drive)

CHT call hold and trace; call-holding time

CHV card-holder verification

CI call identity; command identifier; component interface; computer interconnect; concatenation indication; confidence interval; congestion indication/indicator; CUG index

CIA channel interface adaptor; current instruction address

CIAC code d'identification de l'autorité chargée de la comptabilité (f), código de identificación de la autoridad encargada de la contabilidad (s), accounting authority identification code; Computer Incident Advisory Capability (US Department of Energy)

CIAO Critical Infrastructure Assurance Office (US)

CIAS circuit inventory and analysis system

CIB centralized intercept bureau

CIC carrier identification code; Carrier Information Center; clean-room interface chamber; centro internacional de conmutación (s), international switching centre; circuit code; circuit identification/identity code, code d'identification de circuit (f), código de identificación de circuito (s); CSNet Information Center; trunk terminal circuit identity/identification code N° 7; content indicator code

CICC Centre for International

Cooperation for Computerization – Asian consortium

CICD centre international de coordination pour les transmissions de données (f), centro internacional de conmutación de datos (s), international data coordinating centre

CICERO Control Information system Concepts based on Encapsulated Real-time Objects

CICN mensaje de código de identificación de circuito no equipado (s), unequipped circuit identification code message

CICNE código de identificación de circuito no equipado (s), unequipped circuit identification code message

CICP centros internacionales de conmutación y de pruebas (s), international switching and testing centre

CICS customer information control system (IBM)

CICS/VS customer information control system/virtual storage (IBM)

CICT code d'identification de centre de transit (f), código de identificación de centro de tránsito (s), transit centre identification code

CID cabinet identifier; call instance data; charge-injection device, configuration installation distribution; craft interface device

CIDR classless inter-domain routing (RFC 1520)

CIE Commission Internationale de l'Éclairage (f) (colour specification); computer-integrated engineering

CIELAB colour model

CIELUV colour model

CIF campo de información facsímil (s), facsimile information field; Caltech intermediate form; captive installation function; carriage-insurance-freight; cells in frames (CIF Alliance protocol); centro de información de fallos (s), fault reporting centre; combined interface

unit; common intermediate format
(CCITT H261 on videoconferencing)

CIFAX ciphered facsimile

CIFS common Internet file system

CIG calling subscriber identification

CIGALE transmission network-packet-switched system (f)

CIGRE Conférence internationale des grands réseaux électriques à haute tension (Paris) (f), Conferencia Internacional de las Grandes Redes Eléctricas de Alta Tensión (s), International Conference on Large High-voltage Electric Systems

CIGREF Club Informatique des Grandes Entreprises Françaises (f)

CIH Chen Ing-Han (creator of CIH virus, April 1999)

CII call identity index; Compagnie Internationale pour l'Informatique (f); contador de NSI irrazonables (s), counter of unreasonable BSN

CIIL control interface intermediate language

CIL circuito intercentral de llegada (entrante) (s), incoming trunk circuit; common intermediate language; Component Integration Laboratories; Computers in Libraries (UK); communication identification line, renglón de identificación de la comunicación

cim Computers in Manufacturing (conference 1996, UK)

CIM common information model; CompuServe information manager; computer input (on) microfilm; computer-integrated manufacturing; señal de conexión imposible (s), connection-not-possible signal

CIMC centro internacional de mantenimiento de la conmutación (s), international switching maintenance centre

CIME customer installation maintenance entities

CIM-OSA computer-integrated

manufacturing – open systems architecture (ESPRIT program)

CIMT centro internacional de mantenimiento de la transmisión (s), international transmission maintenance centre; computer-integrated manufacturing and technology

CIN señal de conexión infructuosa (s), connection-not-successful signal

CINÉ message de code d'identification de circuit non équipé (f), unequipped circuit identification code message

CIO confirming informal order

CIOCS communication input/output control system

CIP calendar interoperability protocol; call in progress; carrier identification parameter; command interface port; common indexing protocol; communications improvement programmes; congestion indication primitive (f); control de interrupción del procesador (s), processor outage control

CIPH cipher/ciphering

ciphony ciphered telephony

CIP-L computer-aided intuition-guided programming language

CIPT computing intelligent processing technology, technique informatique de traitement intelligent (f), técnica informática de tratamiento inteligente (s), intelligente Datenverarbeitungstechnologie (g)

CIQM Centre for Information Quality Management (UK)

CIR calling-line identity request signal; channel interference ratio; circuit reliability; centro inteligente de la red (s), intelligent network centre; code d'identification de réseau (f), código de identificación de red (s), network identification code; committed information rate; computer-integrated research; current instruction register

CIRC cross-interleaved Reed-Salomon code

CIRD code d'identification de réseau

pour données (f), código de identificación de red de datos (s), data network identification code

CIRI código de identificación de red RDSI (s), ISDN network identification code

CIRL código de identificación de red liberante (s), clearing network identification code

CIRM Comité international radio-maritime (f), International Marine Radio Association

CIRP code d'identification de réseau privé (f), private data network identification code

CIRPD código de identificación de red privada de datos (s), private data network identification code

CIRS cross-interleaved Reed-Salomon

CIRT código de identificación de red de tránsito (s), transit network identification code; código de identificación de red télex (s), telex network identification code

CIRTT code d'identification de réseau de télex temporaire (s), telex temporary network identification code

CIRU Computer Industry Research Unit

CIS campo de información de señalización (s), signalling information field; CASE integrated/integration services; channel and isolation supervision; circuito interurbano de salida (s), outgoing trunk circuit; code d'identification de support (f), bearer identification code; commercial instruction set; compressed image sequence; CompuServe Information Service; contact image scanner/sensor; cooperative information system; customer intercept services

CISA Certified Information Systems Auditor

CISC complex instruction set computer, système de calcul à instructions complexes (f), microprocesseur à jeu d'instructions complexes (f), ordenador de juego de instrucciones complejo (s)

CISI Compagnie Internationale de Services en Informatique (f)

CISO código de identificación de soporte (s), bearer identification code

CISPR Comité international spécial des perturbations radioélectriques (f), International Special Committee on Radio Interference

CISS computer-imaging sperm selection

CIT centro internacional de televisión (s), International TV Centre; circular information type; código de intervalo de tiempo (s), time-slot code; Compagnie Industrielle des Telecommunications (f); computer-integrated telephony; craft interface terminal

CITADEL Citoyens et Internautes tous Associés pour la Défense des Libertés (f)

CITEL Comisión Interamericana de Telecomunicaciones (s), Inter-American telecommunications conference

CIU channel interface unit; communication interface unit

CIUS Conseil international des Unions scientifiques (f)

CIVR computer and interactive voice response

CIWS concentrator isolation working subsystem

CIX Commercial Internet Exchange (organization of ISPs); Computerlink Information Exchange

CJK Chinese - Japanese - Korean

CJKV Chinese - Japanese - Korean - Vietnamese

CJLI command job language interpreter

CK check bits; check sum

CKD centre for key distribution (*ie* encryption keys); count key data

CKO checking operator

CKS Celestial Knights

CKSN ciphering key sequence number

ckt circuit

CKT-ID circuit identification

CL message d'ordre de connexion de

liaison sémaphore de données (f),
signalling-data-link-connection-order
message; canal lógico (s), logical
channel; central local (f), local exchange;
centre line; clear confirmation,
confirmation de libération (f),
confirmación de liberación (s);
connectionless (service); control channel
of line system

CL1 congestion level 1

CLA custom logic array

CLAMS Customers' Lobby Against
Monopolies

CLAN cordless local area network

CLAO compleción de llamadas a
abonado ocupado (s), completion of calls
to busy subscriber

CLAR channel local address register

CLASP computer language for
aeronautics and programming

CLASS Central Livestock Auction
Satellite Sales; centralized/custom local
area selective signalling; custom local
area signalling service

CLAUDE case study on layered
protocols and architecture definition

CLC cancel lot cycle; clear carry flag;
clear confirmation; control link

CLCC ceramic-leaded chip carrier

CLCD clear confirmation delay

CLCI common-language circuit identity

CLD clear direction flag;
communications logistic depot

CLDN calling line directory number

CLE customer located equipment

clear delete data from memory

CLEC competitive local exchange carrier
(US)

CLEI common language equipment
identification/identity

Clémentine (French) neighbourhood
mobility service

CLEO clear language for expressing
orders (ICL); common language
equipment order

CLF clear-forward signal

CLFI common-language facility identity

CLHEP (C++) class library for high-
energy physics

CLI call-level interface; calling-line-
identity (message), identité de la ligne du
demandeur (f), identidad de la línea que
llama (s); clear indication; clear interrupt
flag; command length indicator;
command line interface/interpreter;
(signal de) connexion de la liaison
impossible (f), connection-not-successful
signal; Comisión de Libertades e
Informática (s), Commission for
Liberties and Informatics

CLIC closed-loop intensity control

CliCC Common Lisp-to-C compiler

CLID calling line identification

CLIP caller line identity/identification
presentation (ISDN); cellular logic image
processor

CLIPS C language integrated production
system

CLIR calling line identification
restriction (supplementary service)

CLISP conversational LISP

CLIST command list

CLIT current location information
(mobile) terminal

CLL cellular local loop; control de
llamada (s), call control

CLLI common language location
identification/identity

CLLM consolidated link layer
management

CLM clock module

CLN signal de connexion de la liaison
non réalisée (f), connection-not-
successful signal

CLNP connectionless network protocol

CLNS connectionless network
service/server

CLO signal d'ordre de connexion de
liaison sémaphore de données (f),
signalling-data-link-connection-order
signal; circuit layout order

CLOAX corrugated-laminated co-axial
cable

clone hardware/software imitating that produced by another maker

CLOS common LISP object system

CLP cell loss priority; connectionless protocol; constraint logic programming

CLR cell loss ratio/rate; central logic rack; circuit loudness rating; clearance; clear request; combined line and recording; Consortium for Lexical Research; signal de connexion de la liaison réalisée (f), connection-successful signal

CLRC circuit layout record card

CLRD clear request delay

CLS clear screen (DOS command); connectionless service (s)

CLSM call leg state model

CLT call limit timer; Compagnie Luxembourgeoise de Télédiffusion (f)

CLTP connectionless transport protocol

CLTS clear task switch (flag)

CLU calling-line identity unavailable signal

CLUI command line user-interface

CLUK Cyber-Liberties UK

CLUSIF Club de la Sécurité des Systèmes d'Information Français (f)

CLUT colour look-up table

CLV códigos de longitud variable (s), variable length codes; constant linear velocity

CLVR crystal log video receiver (microwave receiver)

CM cassette module; categoría de mensaje (s), message category; central memory; centre maritime (f), centro marítimo (s), maritime centre; centro de mantenimiento (s), maintenance centre; command module; conditional mandatory parameter; configuration/connection management; continuity message; control memory/module; control monitor

CMA Computer Misuse Act, 1990 (UK); concert multi-thread architecture; connection management architecture

CMAA centre de maintenance d'accès

d'abonné (f), centro de mantenimiento de accesos de abonado (s), subscriber access maintenance centre

CMAS cellular management application area in TMOS

CMB cyclic redundancy check message block

CMBD joint committee on circuit noise and availability

CMC cable maintenance centre; call modification complete/completed message; Canadian Marconi Company; capacity modular control; cassette module controller; cellular mobile carrier; coherent multi-channel (in optical transmission); combine multicast control; communication; complement carry (flag); computer-mediated communications; CUG management centre

CMC7 seven-segment standard font used in LCD/character recognition

CMD circuit mode data; command

CMDF combined main distribution frame

CMDR command reject (X.25)

CMDS centralized message data system

CME circuit multiplication equipment; connection management entity

cmf cymo-motive force

CMF señalización por código multifrecuencia (s), multi-frequency code

CMG Computer Management Group (US); control-moment gyroscope

CMI cable microcell integrator; code/coded mark inversion, codage par inversion (f), inversión de marcas codificadas (s); centre de maintenance internationale (f), centro de mantenimiento internacional (s), international maintenance centre; coding method identifier; Committee on protection of telecommunications lines underground; composante moyenne de l'image (f), average picture level; computer-managed instruction; control mode idle

CMI/HIC cable microcell integrator/headend interface converter

CMIP common management information/interface protocol

CMIS common management information services/system/specification

CMISE common management information/interface service element

CML champ de commande multi-liaison (f), multi-link control field; current mode logic; conceptual modelling language

CMM capability and maturity model; channel mode modify; computer main memory; concentration main module

CMMS computerized maintenance management software

CMMU cache/memory management unit (Motorola)

CMODE calling mode

CMOS complementary metal oxide semiconductor

CMOT CMIP/CMIS over TCP/IP

CMP camp-on; chemical mechanical polishing; communications plenum cable

CMPS compare word string; connection-mode packet-switching

CMR call modification request message; cancel move request; CBX management register; common-mode rejection ratio; cell misinsertion rate; centralized mail remittance; Communications Moon Relay; communications riser cable

CMRJ call modification reject message

CMRR common-mode rejection ratio

CMS cellular mobile (telephone) system; circuit maintenance system; circuit multiplication system; code management system; colour/monochrome state; communications management system; compiler monitor system; conversation monitoring system; coordinate measuring system

CMSG C message weighting

CMT cellular mobile telephone; character mode translator; configuración multiterminal (s), multi-terminal configuration; control de la memoria

tampón (s), buffer (memory) control; Country Music Television

CMT-LI centres de maintenance de la transmission pour la ligne international (f), centros de mantenimiento de la transmisión para la línea internacional (s), transmission maintenance point (international line)

CMTT Joint CCIR/CCITT Study Group on Transmission of Sound Broadcasting and Television Systems over Long Distances

CMU communications management unit

CMVC configuration management version control (IBM)

CMW compartmented mode workstation

CMX concentration module extension; Corel exchange format file

CMY cyan, magenta, yellow

CMYK cyan, magenta, yellow, black (black = K, since B = blue in RGB model)

CN check not OK; combined delivery/non-delivery note/(notification), notificación combinada de entrega/no entrega (s); coin trunk; connect

C-n container-n

CNA cellular network administration; Certified NetWare Administrator (Novell); communications network application; correction note for application systems

CNAI cellular network administration interface

CNAM cellular network activity manager

CNAME canonical name

CNAPS co-processing node architecture for parallel systems (Adaptive Solutions Inc *tn*)

CNBT capacidades no básicas de terminal (s), non-basic terminal capacities

CNC cellular network configuration; computer numerical control; condensation nucleus counter

CNCC customer network control center

CNCD centre national de commutation

de données (f), centro nacional de conmutación de datos (s), national data switching exchange

CNCL Commission nationale de la communication et des libertés (f) (French equivalent of US FCC)

CNCM cellular network configuration management

CND call number display

CNE centre network environment; Certified NetWare Engineer (Novell); código de identificación de red liberante (s), clearing network identification code; computer network exploitation

CNED Centre National d'Etudes et d'Enseignement à Distance (f)

CNET Centre National d'Etudes des Telecommunications (f)

CNET Clnet the Computer Network

CNETP consolidated new equipment training plan

CNF conjunctive normal form

CNG calling (tone)

CNI Certified Novell Instructor; changed number interception; Coalition for Networked Information (US); common network interface; conventional network interface; correction note – issue

CNIC clearing network identification code

CNIDR clearing-house for networked information and data retrieval

CNIL Commission Nationale de l'Informatique et des Libertés (f)

CNL constant net loss

CNM circuit network management message group; communications network management

CNMA Communications Network for Manufacturing Applications (UK)

CNN Cable News Network; cellular neural network; composite network node; conduit numérique non fourni (f), digital path not provided (signal)

CNO cellular network operations; Custom Network Option (AT&T *tn*)

CNOTCH C message weighting with notch filter

CNP cellular network performance; concentration numérique de la parole (f), digital speech interpolation; connection-not-possible-signal

CNPM cellular network performance management

CNPR cellular network performance reporting

CNR carrier-to-thermal-noise ratio; combat-net radar/radio; complex node representation (ATM Forum); (Italian) National Research Council

CNRI Corporation for National Research Initiatives (US)

CNRS Centre National de Recherche Scientifique (f)

CNS central nervous system; Certified Novell Salesperson; complementary network service; connection-not-successful-signal; connection-oriented network service

CNSP semi-permanent digital connection

CNSS core nodal switching subsystem

CNT centro nacional de televisión (s), national TV centre

CNVT convert

CNX certified network expert; conexión (s), connection

CO central office; charging origin; coinbox line; crystal oscillator

COA Certificate of Authenticity; change-over acknowledgement signal

COAM centro de operaciones, administración y mantenimiento (s), operations, administration and maintenance centre; customer-owned and maintained (equipment)

COAST cache on a stick

COB chip-on-board

COBE Cosmic Background Explorer (NASA satellite), Satellite d'exploration du fond cosmique (f), Explorador del Fondo Cósmico (s)

COBOL common business-oriented language

cobweb web page not updated for a long time

COC centre of competence; code de canal sémaphore (f), signalling link code; compiler object code; cost of consumables; signalling channel code (Nº 7)

COCA command category

COCAA Centre opérationnel combiné d'appui aérien (f) (combined air support operations centre)

CoCom Coordinating Committee on Multilateral Export Controls (US)

COCOMO constructive cost model

COD click of death (associated with 'super-floppy/Zip' drives); connection-oriented data

CODA Carnegie Mellon advanced distributed file system

CODAN carrier-operated device anti-noise

CODASYL Conference on Data Systems Languages (COBOL; US DoD)

CODATA Confederation of Design and Technology Associations

CODE client-server open development environment; crackers of digital equipment

codec coder-decoder

CODIPHASE computing-coherent digital phased-array system

CODIT code division testbed (RACE II project 2020)

CODLS connection-mode data link service

COE contrôle de l'encombrement (f), congestion control

COED computer-operated electronic display; computer-optimized experimental design

COEES central office equipment engineering system

COF cause of failure; confusion signal, signal de confusion (f), señal de confusión (s); coordination of flash services; coordination of functions;

cursor off, curseur arrêté (f), cursor inactivo (s)

COFDM coded orthogonal frequency-division multiplexing

COFF common object file format

COG centralized ordering group

COGO coordinate geometry

COGS cost of goods sold

COI canal de operaciones insertadas (s), embedded operations channel

COIN coin phone operational and information network system

CO-IPX connection-oriented IPX

COL señal de colgar (s), clear-back signal

COLD computer-output laser disk

COLI connect line identity

Colossus early electronic digital computer

COLP connected line identification presentation (supplementary service)

COLR circuit order layout record; connected line identification (presentation) restriction

COLS computing communications for online systems

Colt City of London Telecommunications (UK)

com port serial communications port (in DOS system)

COM alarm switching type; centro de operación y mantenimiento (s), operations and maintenance centre, centro de operación administración y mantenimiento (s), operations administration and maintenance centre; complete; component object model (Microsoft; *cf* CORBA); computer output on microfilm/-fiche; continuation of message

COM1 first serial communication port (*cf* com port)

COM2 communication multiplied by two (TASI system)

COMAL common algorithmic language (enhanced BASIC in Danish schools)

COMASIII computerized maintenance and administration support III

COMB transmitter combine

Comcat component category (Microsoft Windows system file)

COMDEX Communications Development Exposition; Computer Dealers' Exposition

COMDLG common dialog (Microsoft Windows system file)

COMED combined map and electronics display

COMET computer-operated management evaluation technique; Cornell Macintosh terminal emulator

COMETT Community Action Programme in Education and Training for Technology

COMFOR commercial wire centre forecast program

COMINT communications intelligence

COMJAM communications jamming

COMMON area area of storage accessible from sub-routines in Fortran

Common LISP version of LISP integrating FranzLisp and MACLisp

comopt moving pictures combined with optical sound track

COMP computers (USENET category)

COMPAC Commonwealth Trans-Pacific Telephone Cable

compander compressor/expander device

Compaq Compaq Computer Corporation

compole commutating pole

COMPUSEC computer security

COMSAT Communications Satellite Corporation

COMSEC communications security

COMSL communication system simulation language

COMSTAR common system for technical analysis and testing

COMTELCA Comisión Técnica Regional de Telecomunicaciones (s)

CON concentrator; connect; connect message, message de connexion (f); console; content; continuity signal, señal de continuidad (s); cursor on, curseur en marche (f), cursor activo (s); MS-DOS name for screen and keyboard

CON N connection board

CONATEL Consejo Nacional de Telecomunicaciones (s) (Spain)

CONCISE Cosine Network's Central Information Service for Europe

condela connection definition language

CONECS connectorized exchange cable splicing

CONEX connectivity exchange

CONF conference call add-on; conference calling; 'confirmation' mode

CONFIG configuration

CONFIG.SYS system configuration (text) file

CONG congestion

CONLAN consensus language (hardware description language)

CONN/conn connect/connected/connection

CONNACK connect acknowledgement

Co-NP complementary non-deterministic polynomial

CONS connection-oriented network service (*cf* CLNS)

CONTEL Continental Telecom Inc

Conus continental United States (*ie* excludes Hawaii and Alaska)

COO changeover-order signal; chief operating officer; cost of ownership

cooC concurrent object-oriented C

cookie small ID tag left by a website (properly 'a state object')

Cool combined object-oriented language; concurrent object-oriented language

COOLS community on-line system

COP character-oriented protocols; code of practice; connection-oriented protocols; control port

COPAN command post alerting network

COPC control orientado a la conexión de la PCCS (s), SCCP routing verification test

COPS concept-oriented programming

system; configuration and operation support software

COPTAC Conférence des Postes et Télécommunications de l'Afrique centrale (f)

COPUOS Committee on the peaceful uses of outer space

COR confirmation of receipt

CORAL class-oriented ring associated language; common real-time applications language

CORAL 66 UK military programming language based on Algol 66

CORBA common object request broker architecture

CORC Cornell Compiler

CORDIS Community Research and Development Information Services

CORDS coordination of record and database system

CORE composite object reference; corrected overall reference equivalent; Criminals of Radical Extremes; Internet Council of Registrars

COREINAP core INAP

CoREN Corporation for Research and Enterprise Network

CORNET Corporate Network signalling protocol (Siemens *tn*)

CORODIM correlation of the recognition of degradation with intelligibility measurements

CORP Creators of Revolutionary Pictures

CORSA Cosmic Radiation Satellite (Japan)

COS change of subscribers; class of service; code-operated switch; compatible OS; Corporation for Open Systems (US); Cray OS *tn*

COSAM co-site analytical model

COSATI Committee on Scientific and Technological Information (US)

COSBA Computer Services and Bureaux Association

COSE common open software environment; common operating/open

system environment; ensemble de spécification d'interfonctionnement (f), interworking specification suite

cosh hyperbolic cosine

COSINE Cooperation for open systems interconnection networking in Europe

COSMIC common sense main interconnecting; Computer Software Management And Information Centre (NASA)

COSMOS computer system for mainframe operations

Cosnav Russian navigation satellite system

COSP central office signalling panel

COSPAR Committee on Space Research

COSPAS-SARSAT International Satellite System for Search and Rescue

COSS common object services specification

COST Cooperation in the Field of Scientific and Technical research (EU)

CoSysOp co-systems operator

cot co-tangent

COT central office terminal; class of traffic; continuity; continuity signal, signal de continuité (f), señal de continuidad (s); customer office terminal; customer-owned tooling

COTC Canadian Overseas Telecommunications Corp; class-of-traffic check

CO-TP connection-oriented transaction processing (ECMA standard)

CoTRA Computer Threat Research Association

coulomb SI unit of electrical charge

COV signal de passage sur liaison de réserve (f), change-over signal

COVIRA Computer Vision in Radiology (EU program)

COW character-oriented Windows interface (IBM OS/2)

COWSEL controlled working space language (Edinburgh University)

cp UNIX command to copy a file

CP call proceeding; call progress;

capacidad portadora (s), bearer capacity; card punch; central processor; centre primaire (f), centro primario (s), primary centre; chip place; conmutación de paquetes (s), packet-switched; control program; process capability; processable content architecture levels

CP/M computer program/maintenance; control program for microprocessors, programme de contrôle pour microprocesseur (f), programa de control para microprocesadores (s); Control Program Monitor (Digital OS; DOS precursor *tn*)

CP PPDU connect presentation PPDU

CPA PPDU connect presentation accept PPDU

CPA calling party answer; centralized/bulk power architecture; central processor A-side; codificación predictiva adaptativa (s), adaptive-predictive coding; conmutación de protección automática (s), automatic protection switching; control por programa almacenado (s), stored program control; co-polar attenuation; critical path analysis

CPACC control de prueba de acceso y congestión de conjunto de rutas de señalización (s), signalling route set congestion test control

CPAN comprehensive Perl archive network

CPAS cellular priority access services

CP-B central processor B-side

CPC calling party's category; call processing control; cellular phone company; computer program component; constant point calculation; continuar para corregir (s), continue to correct; control de prueba de conjunto de rutas de señalización (s), signalling route set test control

CPCH calling party cannot hear

CPCS check processing control system (IBM); common part convergence sub-layer (ATM)

CPCS-SDU common part convergence sub-layer – service data unit

CPD central pulse distributor; code du point de destination (f), código del punto de destino (s), destination point code; concurrent product development; cumulative probability distribution

CPE central processing element; communications participation and education program; customer provided/premises equipment

CPES control de prueba de enlaces de señalización (s), signalling link test control; control prueba estado de subsistema (s), subsystem status test

CPF campo de parámetro facilidad (s), facility parameter field; control program facility

CPFF cost plus fixed fee

CPFR calling party forced release (on DPO)

CPG call progress charge (information); call progress message (signal); central processor group

CPGA ceramic pin grid array

CPH characters per hour; cost per hour

CPI cable pair identification; call progress indicator; common part indicator; common programming interface (IBM); computer-to-PBX interface; Corrupt Programming International

CPI-C common programming interface for communications (IBM)

CPIF cost plus incentive fee

Cpk process capability index

CPL capability password level; combined programming language (precursor of BCPL and C)

C-Plane control plane

CPM called party free message; call protocol method; cost per hour; counts per minute; critical path method

CPMP carrier performance measurement plan

CPMS cable pressure monitoring system

CPN calling party number; cordonnées

de projection normées (f), normalized projection co-ordinates; customer premises network

CPO code du point d'origine (f), código del punto de origen (s), originating point code

CPODA contention priority-oriented demand assignment

CPOL communications procedure-oriented language

CPP conductive plastic potentiometer; critical path planning

CPP+NSS código de punto primario + número de subsistema (s)

CPPI Consultative Panel on Public Information (UN)

CPPS computing critical path planning and scheduling

CPR (señal de) confirmación para recibir (s), confirmation of receipt signal; continuous progress indicator

CPR PPDU connect presentation reject PPDU

cps characters per second; cycles per second

CPS call processing subsystem; commande des procédures de signalisation (f), control de procedimiento de señalización (s), signalling procedure control; continuation passing style

CPS 1 candidate protocol suite Nº 1

CPS+NSS código de punto secundario + número de subsistema (s), primary point code + subsystem number

CPSA control punto de señalización admitido (s), control point signalling allowed

CPSC control punto de señalización congestionado (s), congested signalling point code

CPSI code de point de signalisation international (f), código de punto de señalización internacional (s), international signalling point code

CPSK coherent PSK

CPSP Computer Professionals for Social

Responsibility (US); control punto de señalización prohibido (s), control point signalling congested

CPT códigos de punto de traducción (s), translation point codes; command pass-through; compatibility tests; comptage (*ie* as in charging meters); conditional probability tables; critical path technique

CPTP central privada de telefonía pública (s)

CPU central processing/(processor) unit; communications processor utility/unit

CPU-bound computation bound

CPUG closed private mobile radio (PMR) user group

CQ call to all stations; centre quaternaire (f), quaternary centre

CQAO contrôle de qualité assisté par ordinateur (f), computer-assisted quality control

CQM circuit group query message; computerized quality management

CQN closed-queuing network

CQP circuit group query

CQR circuit group query response message

CR call reference; call request; card reader; carriage return; channel/circuit reliability; clear record; connection request; connexion de réseau (f), conexión de red (s), network connection; contador de retransmisiones (s), retransmission counter; control de recepción (s), reception control; criterio de ruido (s), noise criterion

CR TPDU connection request TPDU

CR/LF carriage return/line feed

CRA cable remote antenna; Computer Research Association; confirmación de rearranque (s), restart confirmation

CRACF call/connection-related radio access control function

CRADA cooperative research and development agreement

CRAF Committee on Radio Astronomy Frequencies

CRAG Cellular Radio Advisory Group (UK)

CRAM card random access memory

Craps Campaign for the Re-instatement of AOL Pricing Structure

CRC capacidad de recepción comprometida (s), receiving ability jeopardized; communications relay centre; Communications Research Center (Canada); contrôle de redondance cyclique (f), verificación por redundancia cíclica (procedimiento de) (s), Prüfung der zyklischen Redundanz (g), cyclic redundancy check

CRCC cyclic redundancy check character

CRCH conexión rechazada (s), connection refused

CRD código de identificación de red de datos (s), data network identification code

CRE cell re-establishment; contrôle de continuité renouvelé sur circuit entrant (f), continuity re-check incoming; corrected reference equivalent(s)

CRED credit card calling, llamada con tarjeta de crédito (s)

CREDIT CRT-based text editor (Intel *tn*)

CREF computer-ready electronic files; control de revisión de fondo (s), in-depth review; connection refused

CREG concentrated range, extension and gain

CREMISI information on parallel computing at CNR

CREN Corporation for Research and Educational Networking (US)

CREST comprehensive radar effects trainer (Ferranti Computer Systems)

C-Ret colour resolution enhancement (Hewlett Packard *tn*)

CREW concurrent read exclusive write

CRF cell relay function; connection-related function

CRF(VC) virtual channel connection-related function

CRF(VP) virtual path connection-related function

CRFMP cable repair force management plan

CRG charging; charge information message

CRI cellular radio interface; centre radiophonique international (f), centro radiofónico internacional (s), international sound programme centre; colour reproduction indices; confirmation de réinitialisation (f), confirmación de reinicialización (s), reset confirmation; continuity-recheck incoming; control and radio interface

CRIM Centre de recherche informatique de Montréal (f)

CRIMP cross impact model (analytical tool in RACE I)

CRIS command retrieval information system; current research information system; customer record information system

crit critical

CRITICOM critical intelligence communications

CRL Carnegie Representation Language; certificate revocation list; coded run length; common representation language; Computing Research Laboratory (US)

crlf carriage return/line feed

CRM cell rate margin; closed user group selection and validation; cost/resource model; code de réseau mobile (f), mobile network code; configuration reporting management; missing RM-cell count

CRN centre radiophonique national (f), centro radiofónico nacional (s), national sound programme centre; checkpoint reference number; Computer Reseller News; señal de congestión en la red nacional (s), national-network congestion signal

CRO cathode-ray oscilloscope; continuity-recheck outgoing

CROM control ROM

cron UNIX clock daemon

CROS coffret de raccordement optique sécurisé (f), fail-safe optical connection cubicle

CROW concurrent read – owner write

CRP call request packet; centro de reserva de programas (s), programme booking centre; command repeat; common reference point

CRPL Central Radio Propagation Laboratory

CRPS control reanudación del punto de señalización (s) [TPRC], signalling point restart control

CRQ call request

CRS cell relay service; circuit reset; computer reservation system; contrôle de continuité renouvelé sur circuit sortant (f), continuity-recheck outgoing [CRO]

CRS4 Center for advanced studies (R&D) Sardinia

CRSAB Centralized Repair Service Answering Bureau

CRT bits de contrôle de trame (s), check bits; cathode-ray tube; channel rate and type; Companhia Riograndense de Telecomunicações (p); continuous ring tone

CRTC Canadian Radio-television and Telecommunications Commission, Conseil de la radiodiffusion et des télécommunications canadien (f)

CRV contact resistance variation

cryptosystem cryptographic system

CRYSTAL concurrent representation of your space-time algorithms

Cs caesium (used in photo-electric devices and atomic clocks)

CS call segment; calls per second; canal de servicio (s), service channel; capability set (enhanced INAP); centre secondaire (f), secondary centre; channel sequence; chip select; circuit-switched; clear screen, effaçage écran (f); borrado de la pantalla (s); clear to send; code segment; common-channel signalling; communication satellite; conexión de sesión (s), session connection; configuration station; convergence sub-layer; cycles per second

CS1/2 capability set 1/2

CS2 Computing Surface 2 – super-computer at the Lawrence Livermore Laboratory

CSA Asociación de Servicios de Cálculo (s) (Spain); calendaring and scheduling API (IBM); call segment association; called subscriber answer; Canadian Standards Association; CIM systems architecture; command session abort; Computer Services Association (UK, 1975); Covenant, Sword and Arm of the Lord

CSAPI common speller application program interface (Microsoft *tn*)

CSAR channel system address register

csc cosecant

CSC cell site controller; circuit supervision control; circuit-switching centre; common signalling channel; Computer Sciences Corporation (US); control signalling code; customer support centre

CSCA computer-supported collaborative argumentation

C-SCANS client-systems computer access networks

CSCC command session change control

CSCP criptosistema de claves públicas (s), public key crypto-system

CSCV call segment connection view

CSCW Computer Supported Co-operative Workshop (Apple & CCTA in UK)

CSD centre de signalisation des dérangements (f), fault-reporting centre; circuit-switched data; corrective service diskette (IBM)

CSDC circuit-switched digital capacity/capability; customer-switched digital capability

CSDN Chinese software distribution network; circuit switched data/digital network

CSE Comité de spécifications des

équipements (f); command session end; Communications Satellite For Experimental Purposes (Japan); control systems engineering department

CSEr command session suspend

CSES consecutive severely-errored seconds

CSF caracteristica de sensibilidad en función de la frecuencia (s), sensitivity/frequency characteristics; Computer Suppliers' Federation; concesión de subsistema fuera de servicio (s), subsystem out of service grant; critical success factor; cut sheet feeder

CSG constructive solid geometry; Consulting Services Group (Lotus)

CSH called subscriber held

cshrc C shell run commands

CSI called subscriber identification; CAMEL service indicator; centre de commutation international (f), international switching centre; CompuServe Inc; Computer Security Institute (US); control sequence introducer/inducer; signalling capability

CSID calling station ID; call subscriber ID; character set identifier

CSK-D code sending key-set – digital

CSL canal de señalización de línea (s), line signalling channel; Computer Science Laboratory (Sony); computer sensitive/structure language; control and simulation language (obsolete); current-steering logic

CSLC coherent side-lobe canceller

CSLIP compressed serial line interface protocol

CSM call set-up message; call supervision message; Commission for Synoptic Meteorology; communications services manager; computer systems manufacturing; control de submuestro (s), sub-sample control; Corrupt Society Magazine

CSMA/CD carrier-sense, multiple-access/collision detection (v Ethernet)

CSMP continuous system modelling program

CSN common services network

CSNET Computer + Science NETwork (+ BITNET became CREN)

CSNet US network linking ArpaNet with other scientists

CSO cold-start-only; colour separation overlay; Complementary Solutions Organization; customer service order

Csound system for programming complex sound and music

CSP central switching point; Centro Supercalcolo Piemonte (i) (Turin); Certified Systems Professional; chip scale packaging; commercial sub-routine package; communicating sequential processes; CompuCom speed protocol chip scale package *tn*

CSPC control sin conexión de la PCSS (s), SCCP connectionless control

CSPD circuit-switched packet data

CSPDN circuit-switched public data network

CSPED concurrent semiconductor production and equipment development

CSPP Computer Systems Policy Project (US)

CSR CCBS facility message; code sender-receiver; common services rack; continuous speech recognition

CSRE corrected send reference equivalent

CSS campo de subservicio (s), sub-service field; cascading style sheets (HTML feature); cell site switch; centralized structure store; circuit-switching system; command session start, instrucción comienzo de sesión (s); common services subsystem; Computer Search and Selection (contract agency); computer special system/subsystem; connection-successful signal; contents scrambling system (CD/DVD encryption); continuous system simulator; control signalling subsystem

CSS/II computer system simulator II

CSSA Computer Services and Software Association (UK); control de subsistema admitido (s), subsystem allowed

CSSE computer services system engine

CSSF concesión subsistema fuera de servicio (s), subsystem out-of-service grant

CSSG Customer Service Solutions Group

CSSL continuous system simulation language

CSSM client-server systems management (IBM)

CSSN circuit state sequence number, numéro d'ordre de l'état du circuit (f)

CSSP control de subsistema prohibido (s), subsystem prohibited

CSSr command suspended session reactivate

CST carrier power supply transistorized; CIM systems technology; computer-supported telephony

CSTA computer-supported telephony applications

CSTD command session typed data; Committee on Science and Technology for Development (US)

CSTN circuit-switched telephone network

CSTO Computing Systems Technology Office (US DoD)

CSTR Centre for Speech Technology Research (Edinburgh University); continuous stirred tank reactor

CSTW command session two-way simultaneous

CSU category support services updating; channel service unit; check signal unit; circuit switching unit; common services unit; customer service unit

CSU/DSU channel service unit/data service unit

CSUI command session user information

CSV comma separated value/variable; common services verbs (interface) (Microsoft Windows)

CSVC closed-user-group selection and validation response

CSVDM continuously variable slope delta modulation

CSVR closed user group selection and validation request

CT call transfer (supplementary service); canal troncal (s), trunk call; canales terrenales (s), terrestrial channels; capacidades de transacción (s), transaction capacity; Cellular Telecom (UK); channel tester/type; circuit terminal; complete translation; computer telephony; conformance test; continuity transceiver; control de transmisión (s), transmission control; control type; cordless telephone; current transformer; cycle time; (international) transit centre, connexion de transport (f), conexión de transporte (s)

CT 1/2/3 first-/second-/third-generation cordless telephone standard

CTA Cable Television Association (UK); computer time of arrival; Computer Traders Alliance (UK); control de la autorización de transferencia (s), transfer-allowed control

CTAK cipher text auto key

CTB cipher type byte

CTBM chief testboard man

CTC centralized test center; cluster tool controller; combiner tuner controller; Compañía de Teléfonos de Chile (s); continue to correct; control de transferencia controlada (s), transfer-controlled control; Cornell Theory Center (US National Science Foundation)

CTCA channel-to channel adaptor

CTCP client-to-client protocol

CTCR complaint-to-completion ratio

CTCSS continuous tone-coded squelch system (TP (1))

CTD cell transfer delay; centro terciario (s), tertiary centre; charge transfer device; channel-translating equipment; circuit de trafic entrant (f), incoming trunk circuit; continuity tone detector;

cumulative transit delay; coefficient of thermal expansion

CTF call transfer fixed

CTI capacité de transfert d'information (f), information transfer capability; centre de transit international (f), centro terminal internacional (s), international transit centre; centralized ticket investigation; computer telephony integration, couplage de la téléphonie et de l'informatique; centre télévisuel international (f), international TV centre; co-axial transceiver interface; commutateur de transit international (f), international transit exchange; Computers in Teaching Initiative (UK); cycle time improvement; co-axial transceiver interface

CTI4G 4th generation international transit exchange

CTIA Cellular Telecommunications Industry Association (US)

CTIDM communication télévisuelle internationale à destinations multiples (f), liaison télévisuelle internationale à destinations multiples (f), conexión internacional de televisión con destinos múltiples (s), international multiple destination TV connection

CTIS centre de transmission international par satellite (f), centro de transmisión internacional por satélite (s), international satellite transmission centre

CTL checkout test language; circuito telefónico local (s), local telephone circuit; compiler target language; complementary transistor logic; computational tree language; control de tratamiento de llamadas (s), call-processing control; control key lettering

CTM capa de transferencia de mensaje (s), message transfer layer; centralized technical management; channel tester module; complete treatment module; contact trunk module; continuity transceiver module; cordless telephone mobility; cordless terminal mobility

CTMC cluster tool modular communications

CTMS carrier transmission measuring/maintenance system

CTN centre national de terminaison/(terminal) (f), terminal national centre; centre télévisuel national (f), national TV centre

CTN4G 4th generation national transit exchange

CTO call transfer outside; cut-off

C-to-C cassette-to-cassette

CTOS computerized tomography OS; Convergent Technologies OS (US corp)

CTP control de prohibición de transferencia (s), transfer-prohibited control

CTPA co-axial to twisted pair adaptor

CTR cell traffic recording; central transceiver; centre tertiaire (f), tertiary centre; common technical requirement/regulation (EU); common technical regulation; comparative tracking resistance; control de transferencia restringida (s), transfer-prohibited control; response for continue to correct

CTRAP customer trouble report analysis plan

Ctrl control (key)

CTRON Communications Realtime Operating System Nucleus; version of Tron

CTS cable terminal/turning section; cell traffic statistics; channel time slot; circuit de trafic (f), outgoing trunk circuit; clear to send (RS 232C signal); Communications Technology Satellite (Canada); communications test system; configuration technical specialists; conformance testing service (EC R&D Programme, DG XIII); cordless telephony system

CTSI central terminal signalling interface

CTSS compatible time-sharing system; Cray time sharing system *tn*

CTS-SN CTS service node

CTS-WAN conformance testing system WAN

CTT cartridge-tape transport

CTTC cartridge-tape transport controller

CTTH copper to the home

CTTN cable trunk ticket number

CTTY console teletype

CTU Caribbean Telecommunication Union; cartridge tape unit

CTUS channel tester support

CTV call transfer user variable

CTX Centrex system number; charge code

CTx cordless telephone Nº x

CTXCO Centrex Central Office

CTXCU Centrex customer

CTY console teletype

cu call Unix

Cu copper

CU clases de usuario (s), user class (character); commitment unit; computer underground; control unit; 'see you'

CUA Computer Users' Association, Asociación de usuarios de ordenadores (s) (Spain); common user access (IBM)

CUB central utility building

CUBES capacity utilization bottleneck efficiency system

CUCH Curry-Church (lambda-calculus)

CUD Celerities Utilities Division; Computer Underground Digest

CUDAT common-user data

CUE computer update equipment, système de mise à jour du calculateur (f); Custom Updates and Extras (Egghead Software *tn*)

CUG closed user group; closed-circuit user group

CUG/OA closed user group with outgoing access

CUI Centre Universitaire d'Informatique (f); character-based user interface; common user interface (IBM)

CUL 'see you later'

CULA 'see you later, alligator'

CUPID completely universal processor I/O design (AST Research *tn*)

CUPL Cornell University programming language

CURTS common user radio transmission sounding system

CUSF call-unrelated service function

CUSP commonly-used system program

CUSUM cumulative sum

CUT Campaign for Unmetered Telecommunications (UK); control unit terminal; Coordinated Universal Time (referred to GMT)

CUTE Clarkston University terminal emulator

CUTS computer users' tape system; cassette users' tape specification

CV capacitance-to-voltage; circuito virtual (conmutado) (s), virtual circuit; communications virtuelles (f), virtual communications; constant vertex/ices (3D graphics); contenedor virtual (s), virtual container

CVC chemical vapour cleaning

CVCM collected volatile condensable materials

CVD chemical vapour deposition

CVF compressed volume file (Microsoft DriveSpace)

CVGA colour video graphics array

CVI centro de videoconferencia internacional (s), international video conference centre

CVIA Computer Virus Industry Association

CVM closed user-group validation check message; Community Voice Mail (US)

CVP circuit virtuel permanent (f), circuito virtual permanente (s), permanent virtual circuit

CVPN cellular virtual private network

CVS closed user group selection and validation check request

CVSD continuously-variable slope differential (pulse modulation)

CVSDM continuously-variable slope delta modulation

CVT circuit validation test; constant-voltage transformer

CVW CodeView for Windows

cw British cable standard notation; carrier wave; composite wave; continuous wave

Cw clockwise

CW call waiting (supplementary service); continuous wave (Morse signalling); Cyber Warriors

CW radar continuous-wave radar

CWC Cable and Wireless Communications; cellular wideband communication

CWI call waiting indication; (Dutch) National Research Center for Mathematics and Computer Science

CWIC compiler for writing and implementing compilers

CWIS campus-wide information service

CWP communicating word processor

CX coinbox set; coin-collecting box; composite signalling

CXN mensaje de conexión (s), connection signal

CXR carrier

CYBERNET Control Data Corporation network

CYBERNYM abbreviation used in telecommunications and related fields

Cyberthécaire bibliothécaire du Web (f), Webmaster

CYC Encyclopaedia Britannica encoding project

CycL language (used with CYC project)

CYCLADES French packet-switched network

Cyclo Cyclomatic complexity tool

CYM cycle model

CYMK cyan – yellow – magenta – black (colour set)

CYTA Cyprus Telecommunications Authority

Cz Czochralski process

CZCS coastal zone color scanner

CZRS code de zone/réseau sémaphore (f), código de zona/red de señalización (s), signalling area/network code; código de zona de señalización/identificación de red (Argentina) (s), signalling area code/network identification

D

D data (and signalling) channel; D-channel (signalling); dedicated; delay; disconnect; document, document (f), documento (s); ISDN signalling channel

D* specific detectivity

D/B die bonding

D/C DLCI/control (LAPF {1})

D/I develop inspect

D/L data link, liaisons de données (f), enlace de datos (s); liaison codée (f); down-link

D/S Drhystone per second

D/U delay unit

D1 GSM network in Germany (T-Mobil)

D-1,2,3,5 broadcast-standard digital component video in differing physical formats

D2 GSM network in Germany (Mannesmann Mobilfunk)

D2C decimal to character (REXX)

D2-Kanal Zeichengabekanal im ISDN 64 kbit/s (g), 64 kbit/s ISDN signalling channel

D2-MAC standard television signal transmission plus concurrent digital stream at 10 Mbit/s

D2T2 dye-diffusion thermal transfer

D2X decimal to hexadecimal (REXX)

da deca/deka 10 (*ie* delta – δ)

D4 T1 transmission framing and synchronization method

DA data available; datos acelerados (s) [ED], expedited data; demand assignment; demande d'appel (f), call request [CR]; descarte de actividad (s), activity discard [AD]; desk accessory (in Apple Mac system); destination address; device address; digit absorbing; digital access; digit analysis; directory assistance

D-A digital-to-analogue, digital-analógico (s) [D/A]; doesn't answer

DAA data access arrangement; difficile à atteindre (f), hard to reach [HTR]; Distributed Application Architecture (HP-Sun)

DAADS data acquisition and display subsystem, sous-système d'acquisition et d'affichage des données (f)

DAB data access board; digital audio broadcasting

DABLE database with an overview of Europe's 500 largest companies

DAC data acquisition and control; data analysis and control; digital-to-analogue converter, convertisseur numérique-analogique (f); discretionary access control; dual attachment concentrator

DACC directory assistance call completion

DACS digital access cross-connect system (AT&T); directory assistance charging system

DACTL Declarative Alvey Compiler Target Language (University of East Anglia)

DAD desktop applications director (Corel *tn*); digital audio disk; direccionamiento ampliado (s), extended addressing [EA]

DAD-E délimitations alignements détection d'erreur (transmission) (f), delimitation, alignment, error detection (transmitting)

DAD-R delimitación, alineación y detección de errores (recepción) (s), delimitation, alignment, error detection (reception)

DADS digitally-assisted display system (*ie* taxi meter, US)

DADT delimitación, alineación y detección de errores (transmisión) (s), delimitation, alignment, error detection (transmitting)

DAEDR delimitation, alignment, error detection (reception)

DAEDT delimitation, alignment, error detection (transmitting)

daemon background command-handling process in Unix

DAG data address generator; Defense

Agencies Group; direccionamiento ampliado (s), extended addressing (calling); directed acyclic graph

DAI distributed artificial intelligence

DAIS Defense Automatic Integrated Switching system (US); distributed automatic intercept system

daisychain means of connecting a number of devices to one controller

DAIV data area initializer and verifier

DAK deny all knowledge

DAL data access language(s); data access line; Drahtlose Anschlußleitung (g), ligne sans fil (f), terminal inalámbrico (s), ligação sem fio (p), wireless connection subscriber line

DALE descripción ampliada de la línea de exploración (s) [ELD]

DAM data acquisition and monitoring; digital audio music; direct access memory (*now* RAM)

DAMA demand assignment/(assigned) multiple access

DAMP demand analysis manufacturing planning

DAMPS digital advanced mobile phone system (US)

DAMQAM dynamically-adaptive multicarrier QAM

DAMS data access management system

DAMSU digital auto-manual switching unit

DAN Daten-Anpassungsteil (g), data adaptor in DTU; direct access node

DANTE Delivery of Advanced Network Technology to Europe

DAO data access object (Microsoft Visual Basic/Jet); dessin assisté par ordinateur (f), computer-aided design

DAP data access protocol (DEC's DNA – application layer *tn*); deformation of vertical aligned phases; developer assistance programme; Dial-a-Phone (UK company); directory access protocol; distance/speed analysis provisioning; distributed array processor; document application profile

DAPIE developers' application programming interface extensions

DAPO Digital Advance Production Order

DAPS dernier arrivé – premier sorti (f), last in – first out [LIFO]

DARE differential analyser replacement

DARI database application remote interface (IBM)

DARLS dynamic adaptive routing and load-sharing

DARMS digital audio reconstruction technology

DARPA Defense Advanced Research Projects Agency (*prev* ARPA) (US)

DART document archive retrieval and transformation

DARU distributed automatic response unit

DAS data acquisition system, système d'acquisition de données (f), sistema de adquisición de datos (s); Datensammlungssystem (g); data analysis software; data assistance system; data auxiliary set; device access software; digital access service; digital-analogue simulation/simulator, simulador digital-analógico (s); direct absorption spectroscopy; directory assistance system; distributed application server; distributor and scanner; dual attachment station

DASD direct access storage device (*ie* a disk drive, IBM)

DASE distributed application support environment

DASH digital audio stationary head, direct access storage handler

DASL Datapoint Advanced System Language *tn*

DASP digital assisted services program

DASS demand assignment signalling and switching; digital access signalling system, sistema de señalización de acceso digital (s), digitales Zugriffsignalisierungssystem (g)

DASSL differential algebraic system solver

DAS-WAT distributor and scanner – watchdog timer

DAT Deutsch-Atlantische Telegraphen-AG (g); disk array technology; digital audio tape, cassette audionumérique (f), cinta de audio digital (s), Digitaltonbandkassette (g); dynamic address translation, traduction dynamique d'adresses (f), traductor dinámico de direcciones (s); distorsión armónica total (s), total harmonic distortion [THD]

DATA Design and Technology Association (UK)

DATABUS Datapoint Business Language *tn*

DATACOM Data Communications Corp

Datagram packet of information (in packet-switching system)

Datakit AT&T packet switching system *tn*

DATAM document architecture transfer and manipulation

Datanet MCI Communications Corp network

DATAPAC Canadian public data packet-switched network

Dataplex multiplexing of data signals

DATEC US data technical support group

DATEL RCA global communications data transmission service using telephone circuits

Datex échange automatique des données entre centres ATC (f), data exchange between ATC centres

DATEX-P packet switching network (Germany)

DATS digital-analogue test system

DATU direct access testing unit

DAU direct access unit

DAV data above voice; datos de alta velocidad (s), high-speed data (exchange) [HSD]

DAVC delayed automatic volume control

DAVIC Digital Audio-Visual Council

DAVID digital audio/video interactive decoder

DAW digital audio workstations

DAZIX Daisy/Cadnetix Corp

dB decibel

DB database; data buffer; data bus connector; Deutsche Bundesbahn (g) (Federal Railways); Deutsche Bundespost (g) (German Post Office/telecoms); dirección del bloque (s), block address [BA]; document bulk transfer class; dummy burst

DB SAP digital loop carrier broadcast SAP

DB2 IBM database management system *tn*

dBa adjusted-weight noise power in dB

DBA database administrator/(administration); data book archive (file format); digital business architecture; dynamic bandwidth allocation; message d'accusé de réception (de) déblocage (f), unblocking acknowledgement message [UBA]

dBa(F1A) noise power measured by a set with F1A-line weighting

dBa(HA1) noise power measured by a set with HA1-receiver weighting

dBa0 noise power measured at zero transmission level point

DBAS database administration system

dBASE Borland database software *tn*

DBASEII Ashton-Tate database program

DBBA message d'accusé de réception (de) déblocage (f), unblocking acknowledgement message [UBA]

dBc dB relative to carrier power

DBC digital business centre

DB-CES dynamic bandwidth – circuit emulation service

DBCS delivery bar code sorter; double-byte character set

DBDL database definition language

DBF dBase format; déblocage de faisceau de circuits (f), circuit group unblocking message [CGU]; digitale Bezugsfolge (g), digital reference sequence

DBFA database function area

DBFH Deutsche Bundespost Telekom France Telecom and BT

DBG message de déblocage de groupe de circuits (f), circuit group unblocking message [CGU]

DBH hardware failure-oriented group unblocking message (f) [HUA]

dBi decibel referred to isotropic radiator

DBIN Description bibliographique internationale normalisée (f), International standard bibliographic description

DBIS Dun & Bradstreet Information Services

D-bit delivery confirmation bit; X.25 delivery-confirmation facility

DBL double connection; mensaje de desbloqueo a la dirección (s), message de déblocage (f), unblocking message [UBL]

dBm dB referred to 1 milliwatt/per milliwatt

DBM database in mobile terminal; database manager; document bulk transfer and manipulation class; maintenance-oriented group unblocking message (f) [MGU]; mobile data function

dBm(psoph) noise power in dBm measured by set with psophometric weighting

dBm0 noise power in dBm referred to or measured at 0TLP

DBML database mark-up language

DBMS database management system

dBmV dB referred to 1 millivolt across 75 ohms

DBO drop build-out capacitor; signal de déblocage (f), señal de desbloqueo (s), unblocking message [UBL]

DBP Deutsche Bundespost (g) (Federal Post Office)

dBr decibel and reference point/relative level (power difference in dB between any point and a reference point), niveau relatif (f), nivel relativo (s)

DBR deterministic bit rate; distributed Bragg reflector (cavity for lasers)

DBRI dual basic rate interface

dBrn decibels above reference noise (level)

dBrn(f$_1$-f$_2$) flat noise in dBrn

dBrnC noise power in dBrn measured by a set with C-message weighting

dBrnC0 noise power in dBrnC referred to or measured at 0TLP

DBS database server; direct broadcast satellite (on Ku-band frequencies); duplex bus selector; software-generated group unblocking message (f) [SGU]

DBTG Database Task Group (established by CODASYL)

DBU distribution unit

dBv dB relative to 1 volt peak-to-peak

DBV débit binaire variable (f), variable bit rate [VBR]

dBW decibels referred to 1 watt

dBx dB above reference coupling

dBμ V EMF dB microvolts per microvolt, electro-motive force

dBμ micro decibels

dBμV decibels per micro volt

dBμV/m dB microvolts per meter

DC TPDU disconnect confirm TPDU

DC data channel; datos de capacidad (s), data capacity; device control; device co-ordinates; direct current; directional coupler; confirmation de déconnexion (f), disconnect confirm (signal); Dublin Core (structure for categorizing electronic documents)

D$_C$ diamètre du coeur [D$_{Co}$]

DC/DC direct current/direct current

DC1-4 device control 1-4

DCA data communications architecture, architecture télématique (f), arquitectura telemática (s), Telematikarchitektur (g); Defense Communications Agency (US, *now* DISA); Digital Communications Associates; differential channel allocation; direct chip attachment; document content architecture (IBM); dynamic channel allocation

DCALGOL data communications ALGOL

DCAM direct chip attach module

DCATS double-contained acid transfer system

DCB disk co-processor board (Novell)

dcc double-cotton covered

DCC data collection computer; data country code; digital colour code; digital compact cassette; direct client-to-client (IRC protocol); Gleichspannungsregler (g), direct current controller

DCCC double-current cable code

DCCF discounted cumulative cash flow

DCCH dedicated control channel

DCD data carrier detect/(detector), détection de porteuse (f)

DCDC DC/DC converter

DCDL digital control design language

DCDR data collection and data relay

DCE data circuit-terminating equipment; data communications equipment (*ie* a modem); digital communications equipment; distributed computing environment (OSF UNIX system), spécification d'OSF (f), entorno informático distribuido (s)

DCF data communication facility (IBM); data count field; default call forwarding; trama de datos codificados para facsímil (s), facsimile coded data [FCD]

DCF block data communication function block

DCFL direct-coupled field effect transistor logic

DCG definite clause grammar; domain coordination group

DCH digital control handler

D-channel single low-bandwidth channel (part of an ISDN)

DCI data channel interface; display control interface; delimitador de codificación de imagen (s) [PCD]; presentation capabilities descriptor; designador de código internacional (s), international code designator [ICD]

DCI-AG data channel interface – analogue G.703

DCL data control language, lenguaje de control de datos (s); delayed call limit; Delphi Common LISP; Digital Command Language (DEC *tn*); display communication log; direct communications link

D$_{Clav}$ average cladding diameter, diámetro medio del revestimiento (s)

DCLP señal de dirección completa abonado libre teléfono de previo pago (s), address-complete subscriber-free signal – coinbox [AFX]

DCLST señal de dirección completa abonado libre sin tasación (s), address-complete subscriber-free signal – no charge [AFN]

DCLT señal de dirección completa abonado libre con tasación (s), address-complete subscriber-free signal – charge [AFC]

DCLU digital carrier line unit

DCLZ data compression Lempel-Ziv, compresión de datos Lempel-Ziv (s)

DCM data communications module; diagnostic control module; diámetro del campo modal/(de modo) (s), mode field diameter; digital carrier module; digital circuit multiplication; distributed computing model (Bull *tn*)

DCME digital circuit multiplication equipment

DCMG DCME gain

DCML differential current mode logic

DCMS digital circuit multiplicated system; distributed call measurement system

DCMU digital concentrator measurement unit

DCN disconnect, desconectar (s); data communications network; dirección completa sin tasación (s), address complete – no charge [ADN]; disconnect frame

DCO SAP DLC connection-oriented service access point

D_{Co} core diameter, diamètre du noyau (f), diámetro del núcleo (s)

DCO demande de connexion (f), connection request; digital central office; distributed component object (model)

DCO-CS digital central office – carrier switch

DCOM Distributed COM (Microsoft Windows NT 4 *tn*)

DCOM95 Distributed COM (Microsoft Windows 95 *tn*)

DCP data collection platform; data coordinating point; definitional constraint programming; demand for critical parts; digital communications protocol; (señal de) dirección completa; teléfono de previo pago (s), address complete signal – coinbox; distribution common point; distributed communications processor; duplex central processor

DCPBH double-channel planar buried heterostructure

DCPM differential pulse code modulation

DCPN domestic customer premises network

DCPSK differentially coherent PSK

DCR degradation category rating (ITU protocol)

DCS data collection system; Data Communication Service (packet-switching network) Belgium; data communications subsystem/service; data control system; Defense Communications System (US); defined context set; desktop colour separation (Quark *tn*); digital cellular system; digital command signal; digital communications service; digital cordless standard; digital cross-connect system; distributed computing system; distributed control system; dynamic channel selection

DCS 1800 Digital Communications System (cellular system at 1 800 MHz) *tn*

DCSS Defense Communications Satellite System; digital customized support services; discontiguous shared segments

DCST señal de dirección completa, sin tasación (s), address complete signal – no charge [ADN]

DCT data compression technology; digital carrier trunk; digital component technology (using 3/4 inch metal tape); digital cordless telephone; (señal de) dirección completa, con tasación (s), address complete signal – charge [ADC]; direct dial inwards; discrete cosine transform

DCTL direct-coupled transistor logic

DCTN Defense Commercial Telecommunications Network (US)

DCU data-cache unit; Digital Credit Union; disk/drum control unit

dcws direct-current working volts

D-D digital-digital

DD data definition; data directory; definition description (HTML); direct dialling; direct digital (control); discriminating digit; domain directory; dual/double density; referencia de destino (s), destination reference [DST-REF]

DDA defined display area; difícil de alcanzar (s), hard to reach [HTR]; digital differential analyser; disk drive adaptor; distributed data access; domain-defined attribute

DDB device descriptor block; device dependent bitmap; double-declining balance

DDBMS distributed database management system

DDC demande de déconnexion (f), released [RLSD]; direct digital control; display data channel (VESA)

DDCMP digital data communication message protocol (DEC)

DDCP direct digital colour proof(-ing)

DDCS distributed database connection services (IBM)

DDD digital diagnostic disk(-ette); direct distance dialling

DDE direct data entry; dynamic data exchange (Microsoft Windows, Macintosh, OS/2)

DDEML dynamic data exchange manager library

DDES digital data exchange standards

DDF digital distributed/distribution frame; dynamic data formatting (IBM)

DDH digital distributed hardware

DDI device driver interface; digital data interface; direct dial inward; direct dialling-in

DDIE direct digital interface equipment

DDIF digital document interchange format (DEC)

DDK device driver kit (Microsoft)

DDL data definition language, langage de définition des données (f), lenguaje de definición de datos (s) [CPM]; data/document description language; device description language; message de demandé libre (f), called party free message

DDLP database-definition language processor

DDM digital data multiplexer; distributed data management

DDMS defect data management system

DDN NIC Defense Data Network – Network Information Center

DDN Defense Department Network (US)

DDNS dynamic domain naming system

DDO direct data output; direct dialling-out

DDOV digital data over voice

DDP datagram delivery protocol (Apple Macintosh *tn*); delivery duty paid; digital data processor; distributed data processing

DDR demand refresh confirmation information; dialled digit receiver; direct domestic reception (*ie* TV)

DDR DRAM double data rate DRAM

DDRSGRAM double data rate synchronous graphics RAM

DDS Dataphone Digital Service (AT&T *tn*); digital data service/system; digital distributed software

DDS-SC digital data system (with) secondary channel (AT&T)

DDT DEC debugging tape; delayed dialling tone; don't do that!; double-diffused transistor; dynamic debugging tool; generic program which debugs other programs (as in insecticide)

DDTX downlink discontinuous transmission uplink

DDWG Digital Display Working Group

DDX digital/distributed data exchange

DDXF DISOSS document exchange facility

DDXI Japanese packet and circuit switched network

DE device end; discard eligible/eligibility; données d'exception (f), exception data; message de demande nécessaire à l'établissement émis vers l'arrière (f), backward set-up message

DEA data encryption algorithm

DEACON direct English access and control

DEAT dispositivos emisores automáticos telex (s), telex automatic emitting devices [TAED]

DEBUG MS-DOS software utility

DEC decadic; device clear; Digital Equipment Corporation (founded 1956); digital exchange controller

decal texture mapping

DECCO Defense Communications Agency Commercial Communications Office (US)

DECDTM DEC distributed transaction manager (DEC *tn*)

DECEO Defense Communication Engineering Office (US)

DE-CIX Deutschland-Commercial Internet Exchange

DECmcc DEC management control centre (DEC *tn*)

DECnet DEC products and proprietary network protocol (DEC *tn*)

DECNMS DEC network measurement system (DEC *tn*)

DECPSK differentially-encoded coherent PSK

DECT Digital European Cordless Telecommunications (air interface standard)

DECTP DEC transaction processing (DEC *tn*)

Dectra Decca Track Range (radio locating system)

DECUS Digital Equipment Computer Users Society

DED dark-emitting diode (*ie* a burnt-out LED); détecteur d'état défavorable (f) [ASD], detector de estados desfavorables (s), adverse-state detector [ASD]; distant-end disconnect

DEDS data entry and display system, système d'entrée et de présentation des données (f); discrete event dynamic simulation

DEE Datenendeinrichtung (g), data terminal equipment

de-emph de-emphasis

Deep Blue IBM's parallel processing supercomputer

def defective

DefCon US hacker conference

DEFR Dun & Bradstreet European Financial Records

defrag to defragment (a hard disk)

DEFT dynamic error-free transmission system

DEG derechos especiales de giro (s), special drawing rights [SDR]; message de demande générale (f), general request message [GRQ]

DEK data encryption key

DEL diode électroluminescente (f), diodo fotoemisor/diodo emisor de luz (s), light-emitting diode [LED]

Del delayed; delete (keyboard key)

delay delay line used for short-term signal storing

Delphi Borland programming language *tn*

DELSTR delete string (REXX)

delta PCM version of DCPM

DELTA Developing European Learning Through Technological Advance

DeM delta modulation

DEM demodulator; digital elevation model (*ie* 3D mapping)

demarc demarcation point

DEMS digital electronic message service

demux demultiplexer

DEN Daily Entertainment Network (a webzine *tn*); Digital Echo News; document enabled networking (Novell-Xerox *tn*)

DENET Danish Academic network

DEO digital end office

DEP datos de error de predicción (s), prediction error data [PED]; desensamblado/ensamblado de paquetes (s), packet assembly/dis-assembly (facility) [PAD]

DEPBA Digital Exchange Pirate Board Alliance

DEPIC dual-expanded plastic-insulated conductor

dequeue double-ended queue

DER distinguished encoding rules

DERA Defence Evaluation and Research Agency (UK)

de-rezz to disappear, fade away, dissolve (*ie* de-resolve)

DES Data Encryption Standard (IBM/US National Bureau of Standards), standard de cryptage des données (f), norma de puesta en clave de datos (s); desactivación de enlaces de señalización (s), signalling link de-activation [LSLD]; destination end station; digital echo suppressor, supresor de eco digital (s); display equipment status

Designet artificial intelligence network design

DESQview multi-tasking software for MS-DOS *tn*

DESY Deutsches Electronen Synchrotron Laboratory

DET detach; device execute trigger

DETAB decision table (COBOL)

detem detector/emitter
DETOL directly-executable test-oriented
language
DEV device
DEVCB device control bus
DEVSB device speech bus
DEX données exprès (f), exception data
[ED]
DF data field; description fonctionnelle
(f), descripción funcional (f), functional
description [FD]; device flag; direction
finding; distributed function; distribution
field/frame; Douglas fir (telephone pole)
DFA design for assembly; deterministic
finite-state automaton; distribution field
A in deterministic finite-state automaton
DFB distributed feedback laser;
distribution field B in deterministic
finite-state automaton
DFB-SIBH distributed feedback laser –
semi-insulating buried heterostructure
DFD data flow diagram
DFDDS data facility dataset services
DFE decision feedback equalizer;
desktop functional equivalent (Compaq
tn)
DfESH Design for Environment Safety
and Health
DFHSM data facility hierarchical storage
manager
DFI digital facility interface; domain-
specific part format identifier
D-flip-flop flip-flop device with one
input and two outputs
DFLP design for low power
DFM data-fax modem, design for
manufacturability (IBM)
DFMS digital facility management
system
DFO direct file output
DFP digital flat panel (monitor/screen);
Digital Free Press; distributed functional
plane
DFR design for reliability; digital fault
(transient) recorder
DFS depth-first search; Deutscher
Fernmeldesatellit (g) (German

Telecommunications Satellite
Kopernikus); directory file system
(Microsoft Windows NT); Disk Filing
System (OS of BBC microcomputer *tn*);
distributed file system
DFSA dynamic frame-length slotted
aloha
DFSG direct formed supergroup
DFSMS data facility storage management
subsystem (IBM)
DFSWI SAP DLC frame switching SAP
DFT design for test/testability; diagnostic
function test; digital facility terminal;
digital initiated file transfer; discrete
Fourier transform; distributed function
terminal; distributed initiated file transfer
DFU data file utility
DG XIII Directorate General XIII (of the
CEC) Telecommunications Information
Market and Exploitation of Research
DG datagram, datagramme (f),
datagramma (s); dead granny (response
to bogus virus alarms); differential gain;
directeur genéral (f); domaine de gestion
(f), dominio de gestión (s), management
domain [MD]
DGA accusé de réception de déblocage
de groupe de circuits (f), circuit group
unblocking acknowledgement message
[CGUA]
DGAD dominio de gestión de
administración (s), administration
management domain [ADMD]
DGC mensaje de desbloqueo de grupo de
circuitos (s), circuit group unblocking
message [CGU]
DGE mensaje de desbloqueo de grupo de
circuitos por fallo del equipo (s),
hardware failure-oriented group
unblocking message [HGU]
DGG dominio de gestión de guía (s),
directory management domain [DMD]
DGGPR dominio de gestión de guía
privado (s), private directory
management domain [PRDMD]
DgHVSt

Durchgangshauptvermittlungstelle (g), main transit switching centre

DGIS direct graphics interface specification/standard

DGL data generation language

DGM mensaje de desbloqueo de grupo (de circuitos) para mantenimiento (s), maintenance-oriented group blocking-acknowledgement message [MBA]

D$_{Gmoy}$ diamètre moyen de la gaine (f), average cladding diameter [D$_{Clav}$]

DGN Digital Freedom Network

DGPR domaine de gestion privé (f), dominio de gestión privado (s), private management domain

DGPS differential GPS

DgQl Durchgangsquerleitung (g), transit tie-line

DGSF mensaje de desbloqueo de grupo por fallo del soporte físico (s), hardware failure-oriented group unblocking message [HGU]

DGT Directorate General of Telecommunications; Direction générale des télécommunications; Dirección General de Telecomunicaciones (f) (EU)

DG-UX Data General Unix

DHA dialogue handling; message d'accusé de réception de déblocage de groupe de circuits (f), hardware-failure oriented group unblocking acknowledgement message [HUA]

DHC data highway controller

DHCF distributed host command facility

DHCP dynamic host configuration protocol (RFC 2131)

DHEMT depletion-mode high-electron mobility transistor (Fujitsu)

DHL dynamic head loading

DHS demande de mise hors service d'un sous-système (f), subsystem out of service request [SOR]; digital handshaking speed; diferencia honradamente significativa (s), honestly significant difference [HSD]

DHSD duplex high-speed data

DHT distorsion harmonique totale (f), total harmonic distortion [THD]

DHTML dynamic HTML

DI data in; de-ionized; dielectric isolation; destination index; domain identifier

DIA Defense Intelligence Agency (US); document interchange architecture (IBM)

DIA/DCA document interchange architecture/document content architecture

DIAC Directions and Implications in Advanced Computing

DIADEM Digital Information Access Demonstration Centre (Cranfield University)

DIAL digital idle asset listing; Direct Intelligent Access Listening (Lorraine Electronics *tn*)

DIANA descriptive intermediate attributed notation for Ada

DIANE Direct Information Access Network for Europe

DIAX small ISDN local exchange

DIB data input bus; device independent bitmap; diagrama de interacciones de bloque (s), block interaction diagram [BID]; directory information base, base de données de l'annuaire (f); dual independent bus

dibit set of two bits (*ie* one of four combinations)

DIBOL Digital Interactive Business Oriented Language (DEC *tn*)

DIC differential interference contrast

DICE DARPA Initiative in Concurrent Engineering

DICOM digital imaging and communications in medicine

DID data interface description; direct inward-dialling; display interface device

DIDS distributed intrusion detection test

DIF data interchange format; digital interface format; document interchange format, format d'échange de données (f), formato de conversión de datos (s)

diff changed list (*ie* a list with differences)

DIFL difusión local (s), local broadcasting

DIFM digital instantaneous frequency measurement

DIFU difusión (s), dissemination/broadcasting

Dig Disney Internet Guide *tn*

DIG design implementation guide

Digerati digital world's literati/glitterati

Digex Digital Express Group Inc (US ISP)

digicam digital camera

DigiCash digital cash (electronic payment method) *tn*

digit a DEC employee

digitizer analogue-to-digital converter

Digizine digital (*ie* electronic) magazine

DII designador de indicativo internacional (s), international code designator [ICD]

DIIG digital information infrastructure guide

DIIP direct interrupt identification port

DIKON digital Konzentrator (g)

DIL dual-in-line (device)

DILIC dual-in-line integrated circuit

DIM data in the middle; demand integration meeting; distributed integrated multimedia; document image management

DIMA data information management architecture

DIMATE depot-installed maintenance automatic test equipment

DIMDI Deutsches Institut für Meinische Dokumentation und Information (g)

DIME direct memory execute

DIMM dual in-line memory module

DIN data international exchange network; Deutsche Industrienorm (g); réseau d'échange de données (f), international data interchange network

Dingbat font with symbols, images; the images themselves (Zapf *tn*)

DINS Digital Information Network Service

DIO data input/output (line) (IEEE-488)

DIP dial-up Internet protocol; digital integrated telephone; digital path; document image processor/processing; dual in-line package/pin

DIQd disk-insulated quad

DIR directory command in MS-DOS

DIRC directional coupler

DirRuf Direktruf (g), hotline call

Dirt design in real time (UI for X Window system by R Hesketh)

DIS digital identification signal; digital information systems; dynamic impedance stabilization (CompuCom)

DISA Data Interchange Standards Association; Defense Information System Agency (*formerly* DCA, US); direct inward system access

DISC Digital International Switching Centre; disconnect (command)

DISCON Defense Integrated Secure Communications Network (Australia)

DISER digital information systems external resources

DISN Defense Information System Network (US)

DISNET Defense Integrated Secure Network (US)

DISOSS distributed office support(-ed) system (IBM)

DISPLAY digital service planning analysis

distfix distributed fixity

DISU digital international switching unit

DIT Data-processing and Telecommunications Directorate (EU), Direction de l'Informatique et des Télécommunications (f), Dirección de Informática y de Telecomunicaciones (s), Direcção de Informática e Telecomunicações (p), Direktion Informatik und Telekommunikation (g), Directoraat Automatisering en Telecommunicatie (nl); directory information tree, arbre d'information de l'annuaire/arbre de données de l'annuaire (f)

DIU data/digital interface unit

DIUD digital interface unit for digital line unit

DIV datos en la banda telefónica (s), data-in voice; DigitalVermittlungstechnik (g), digital switching techniques

DIVA data enquiry-voice answer; DIV-Ausland (g), digital international switching

DIVE direct interface video extension (OS/2 Warp)

Divf digitale Fernvermittlung (g), digital long-distance switching exchange

DIVO digitale Ortsvermittlung (g), digital local exchange

DIVON interworking demonstration via optical networks (1 year CEC Programme DG XIII)

Divx Digital Video Express (Panasonic *tn*)

DIW type-D inside wire (AT&T)

DIX Digital - Intel - Xerox

DJ digital junction

DJC distributed job control

D-Kanal Signalisierungskanal (64 kbit/s) (g), signalling channel (64 kbit/s)

DKI driver-kernel interface

DKS digital key system (Plessey)

DKZ-Nl D-Kanal-Zeichengabe Endstellenleitungen (g), D-channel signalling on lines to terminals

DL data link (layer); definition list; demande de libération (f), clear request [CLR]; distribution list; download

DL/1 data manipulation language 1 (IBM)

DLA Daten-Leitungsauswahleinheit (g), data line selector; Defense Logistic Agency (US)

DLB data line board; data link buffer

DLBI device level burn-in

DLC data link connection; data link control (protocol); digital line card; digital line control; digital loop carrier; diamond-like carbon; dynamic load

control; signalling-data-link-connection-order signal

DLCF data link control field

DLCI data link connection/control identifier

DLD data link discriminator; display list driver

DLE data link escape (carrier); direct line equipment

DLF Deutschland Funk (g); direct line filter

DLI digital line interface

DLL data link layer; desviación de llamada (s), call deflection [CD]; dial long-line equipment; dial long-line units; digital local line; dynamic link library (of files)

DLM distributed lock manager; dynamic linear model; dynamic link module; signalling-data-link-connection-order message

DLN digital library network

DLOC Dark Lords of Chaos; developed source lines of code

DLP Digital Light Processing (Texas Instruments *tn*); distributed logic programming

DLPI data link provider interface

DLR DOS LAN requester

DLRD design layout report date

DLRN destination local reference number

DLS data link buffer; data link service; data link splitter, répartiteur de liaison de données (f); digital line system; digital link service; display lot status

DLSAP data link service access point

DLSDU data link service data unit

DLSO dial line service observing

DLSw data link switching (IBM)

DLT data link layer trailer; device level test; digital linear tape; digital line termination; digital line test equipment; downloading termination procedure, procédure d'interruption du téléchargement (f)

DLTS deep-level transient spectroscopy

DLTU digital line trunk unit

DLU dependent LU requester; Digital-Anschlußeinheit (g), digital line unit

DLU–PG digital line unit – pair gain

DLUR dependent logical unit register (IBM)

DLUR/DLUS1 dependent logical unit requester/dependent logical unit server

DLUS dependent logical unit server (IBM)

DLVA detector logarithmic video amplifier

dly delay

DLZ1 digital Lempel-Ziv 1

Dm control channel (ISDN terminology applied to mobile service)

DM datos mecanografiado (s), typed data (*v* datos tipificados) [TD]; dead memory; defect management; degraded minute(s); delta modulation; digital multiplexer; disconnect(-ed) mode (X.25); dispositif de médiation (f), dispositivo(s) de mediación (s), mediation device; distributed memory; document manipulation class

DMA Defense Mapping Agency (US); deferred maintenance alarm; Direct Marketing Association (US); direct memory access; dynamic mechanical analysis; maintenance-oriented group unblocking-acknowledgement message

DMAC direct memory access controller

DMAD diagnostic machine aid digital

DMB disconnect and make busy; dynamic multipoint bridging

DMCL digital modem command access

DMD digital micro-mirror device; directory management domain, domaine de gestion de l'annuaire (f)

DME digital management education; direct machine environment (ICL *tn*); distance-measuring equipment; distributed management environment

DMEP data network modified emulator program

DMF (high) density media format (*ie* floppy disk)

DMF Distribution Media Format (Microsoft *tn*)

DMH display message helps

DMI desktop management interface; digital multiplexed interface (AT&T); durée moyenne d'une interruption (f), mean interruption delay [MID]

DMIS directory management information system

DML database management language; data manipulation language, langage de manipulation de données (f), lenguaje de manipulación de datos (s); datos en modo de línea (s), line mode data [LMD]; display message log

DMM digital multimeter

DMMS dynamic memory management system

DMO disque magnéto-optique (f)

DMOS degradation mean opinion score; differential mean opinion score; diffused metal-oxide semiconductor; double-diffused metal oxide semiconductor (transistor)

DMP distributed memory processor; dot-matrix printer

DMPC distributed memory parallel computer

DMQS display mode query and set (IBM)

DMR demultiplexing-mixing-remultiplexing (device); digital mobile radio; display move requests

DMS data management service/standard/system; data mobile station; Defense Message System (US); digital management system; diskless management service

DMSD digital multi-standard decoding

DMSS distribution marketing sales and service

DMT datos de modo de trama (s), frame mode data; design maturity test; digital multi-tone; discrete multi-tone; dynamic method table (Borland Delphi)

DMTF desktop management task force

DMU data manipulation unit; digital master unit

DMUX demultiplexer; Digital Multiplex Group (UK broadcasters)

DMV durée moyenne de vie (f), mean (service) life [ML]

DMW digital milliwatt

DMX Digital Music Express

DMY day - month - year

DMZ demilitarized zone (*ie* a protected network between the Internet and a LAN)

DN delivery (status) notification; destination network; directory number; disconnect; distribution network; distinguished name; dual numbering

DNA digital named account; digital network architecture (DEC); Digital News Associates; Digital Noise Alliance; distributed Internet architecture (Microsoft Windows *tn*); dynamic network architecture (Northern Telecom, UK)

DNC direct numerical control; dynamic network controller

DNF disjunctive normal form

DNHR dynamic non-hierarchical routing (AT&T)

DNI data network interface; digital non-interpolated, digital no interpolado (s)

DNIC data network identification code

DNIS dialled number identification service

DNOS Distributed Network Operating System (Texas Instruments *tn*)

DNPA data numbering plan area

DNR dialled number recorder, dynamic noise reduction

DNS Digital Nervous System (Microsoft *tn*); domain name system (RFC 1034/1035)

DNT do not test

DNX dynamic network cross-current

DO data out; design objective; dynamic optimization; drop-out

DOA dead-on alignment; dead on arrival (*eg* of a new PC); Distributors of Anarchy

DOAS differential optical absorption spectroscopy

DOB data option board

DOC distributed object computing; drop-out compensator; dynamic overload control

DOCS display operator control system; document organization and control system

DOCSIS data-over-cable system interface specification

DocuRT Document Recognition Technology (Xerox *tn*)

DOCUS display-oriented computer usage system

DoD 1 (US) Department of Defense 1 (early, unofficial name of Ada *qv*)

DODISS (US) Department of Defense Index of Specifications and Standards

DOE design of experiments

DOF depth of field/focus

DOI digital object identifier

DOIP dial other Internet providers (IBM)

DO-Kanal Zeichengabekanal im ISDN (g), ISDN signalling channel

DOM data on master; document object model

domain identifying part of a WWW address

DOMAIN distributed operating multi-access interactive network

DOMF distributed object management facility

DOMSAT domestic satellite system

dongle hardware device to enable software protection

donuts old term for sets of memory bits

DOORS dynamic object-oriented requirements system

DOR digital optical recording

DORA déploiement de l'optique dans le réseau d'accès (f), optical fibre deployment in the access network

DORAN Doppler ranging system *tn*

DORIS system coverage developed from MASCOT and CORE (BAe *tn*)

DORSY drop-out reduction system

DORUM draft once re-use many

DOS disk operating system (*colloquial abbrev* of MS-DOS); distributed object server (Microsoft *tn*)

Dos attack Denial of Service attack (on a network by hackers)

DOSBS data-over-speech bearer service

DOSEM DOS emulation

DoT Department of Telecommunications, India

DOT Digital Online Terminal (US home banking)

DOTS digital-optical technology system; digital office timing supply

DOV data-over-voice, datos por encima de la banda telefónica (s)

DOVAP Doppler velocity and position

DOW direct over-write

DP data processing, traitement des données (f), Datenverarbeitung (g); datos de paridad (s), parity data; decadic pulsing (signalling); demarcation point; (call-processing) detection point; device processor; dial/dialled pulse; disconnection pending; distributed processes; (módem de) dos procedimientos (s), two-procedures modem (TP modem); dot pattern; double pole

DP(S) data packet (switching)

DPA data processing activities; Data Protection Act (1984 UK); demand protocol architecture; destructive physical analysis; dial pulse access; different premises address; distributed power architecture

DPAM demand priority access method

DPAREN data parity enable (IBM)

DPB DePositByte (*ie* put something in the middle); drive parameter block

DPC destination point code; direct program control

DPCM differential/delta pulse code modulation

DPCS data personal communications service (Apple *tn*)

DPD data products division; (European) Data Protection Directive

DPDCH dedicated packet data channel

DPDT double-pole double-throw

DPE data path extender; demande pour émettre (f), request to send [RTS]

DPGS digital pair gain system

dpi dots per inch

DPI distributed protocol interface

DQDB distributed-queue dual bus

Dpl 55 Dämpfungsplan 55 (g), transmission loss plan 55

DPL descriptor privilege level

DPLM domestic public land mobile

DPM data processing manager; digital panel meter

DPMA Data Processing Management Association

DPMI MS-DOS protected mode interface (Microsoft)

DPMS display power management signalling (VESA specification); display power management support

DPN digital-path-not-provided signal

DPN-PH data packet network – packet handler

DPNSS digital private network signalling standard/system (UK)

DPO dial pulse originating

DPR Data Protection Registrar (UK)

DPRL digital property rights language (Xerox)

DPS demand protocol architecture (3Com); display process status; different premises subscriber; document processing system

DPSK differential phase-shift keying

DPSRAM dual-port static random access memory

DPST double-pole single-throw

DPT dial pulse terminating; different premises telephone number; display processing unit

DPU processeur graphique (f)

DPVSt Datenpaketvermittlungsstelle (g) (data packet switching centre)
DQ directory enquiry service
dq distorsion de quantification (f), quantizing distortion unit [qd]
DQDB distributed-queue dual bus
DQL DataEase query language *tn*; data query language
DQPSK differential quaternary/(quadrature) phase-shift keying
DR data rate/received; demand refresh; destination reference; Digital Research Inc; direct-routed; disconnect request
DR BOND dial-up router bandwidth on demand (NEC *tn*)
DR TPDU disconnect request TPDU
DRAM dynamic RAM
DRAPAC Design Rule and Process Architecture Council
DRAT SAP DLC rate adaption transfer SAP
D$_{Rav}$ average reference surface diameter, diámetro medio de la superficie de referencia (s)
DRAW direct read after write
DRCS dynamically redefinable character set
DRDW direct read during write
DRC data rate compression; data resource centre; design rule check
DRCS digital radio concentrator system (Australia); distress radio call system; dynamically-redefinable character set
DRDA distributed relational database architecture (IBM)
DR-DOS Digital Research-Disk OS *tn*
DRDPS digital research data processing system
DRDRAM DirectRambus DRAM (Rambus Corp *tn*)
DRDW direct read during write
DRE demande de renseignements d'état (f), status request; destruction removal efficiency; directional reservation equipment
DRGS direct-readout ground station

Drhystone benchmark system for measuring computer performance
DRI demande de réinitialisation (f), reset request [RSR]; Digital Research Inc
DRIFTS diffuse reflectance infrared Fourier transform spectroscopy
DRLA Daten-Richtungs- und Auswahl-einheit (g), data route and line selector
DRM demand refresh mode; Digital Radio Mondiale
D$_{Rmoy}$ diamètre moyen de la surface de référence (f), average reference surface diameter [D$_{Rav}$]
DRMU digital remote measurement unit
DRN document reference number
DRO data request output; destructive read-out; dielectric resonator oscillator
DROOL Dave's recycled object-oriented language (for games)
DRP data reception process; directional radiated power; distribution and replication protocol; distribution resource planning
DRPF decimal reference publication format
DRR demand refresh request
DRS data rate selector; Data Relay Satellite; delayed release message/(signal); (PCM) digital reference sequence; document registration system
DRT defect review tool; diagnostic rhyme test; discontinuous reception/transmission
DRV drive
DrV Druckschriftenverwaltung (g), administration for printed documents
DRX discontinuous reception (mechanism)
DS data send; data server/store; digital section; digital signal/stream; digital-carrier span; direct sequence; distributed single-layer test method; document storage; double-sided; Durchwahlsatzdienst (g), direct inward dialling circuit; (telephone) disconnected
DS-0 digital signal level 0 (64 kbit/s)

DS0-4 digital signal levels 0,1,1C,2,3,4

DS1 digital signal level 1 (1.544 Mbit/s)

DS3 PLCP DS3 physical layer convergence protocol

DS3 digital signal level 3 (44.736 {45} Mbit/s)

DSA data striping without parity; dial service architecture/assistance; digital service area; digital subtraction angiography; digital switching application; directory service agent; directory system agent; display system activity; distributed systems architecture, architecture de système distribuée (f), arquitectura de sistema distribuida (s); software-generated group unblocking-acknowledgement (f) [SUA]

DSAC digital simulated analogue computer

DSAN debug syntax analysis

DSAP destination service access point; directory-scope analysis program

DSAU digital signal access unit (DS1)

DSB Defense Science Board (US); development system bug; dial system B switchboard; double-sideband (transmission)

DSBAM double-sideband amplitude modulation

DSBEC double-sideband emitted carrier

DSBRC double-sideband reduced carrier

DSBSC double-sideband suppressed carrier

DSBTC double-sideband transmitted carrier

DSC digital selective calling; differential scanning calorimetry; direct satellite communications; données sans connexion (f), unitdata [UDT]; dynamic satellite constellation (*ie* a group of LEOS)

DSCD dual-scan colour display

DS-CDMA direct sequence CDMA

DSCS Defense Satellite Communications System (US)

DSDC direct service dial capability

DSDD double-sided, double-density (floppy disk)

DSDM drop site database manager

DSDS dataphone switched digital service

DSE data storage equipment; data-switching exchange; discriminating satellite exchange; distributed single-layer embedded test method; distributed system environment; double-switching exchange

DSEA display station emulation adaptor

DSEE domain software engineering environment

DSF dead space free; Deutsch Sport Fernsehen (g), German Sports TV

DSHD double-sided high-density (floppy disk)

DSI data subscriber interface; delivered source instruction; device support interface; digit sequence integrity; digital speech interpolation

DSig (W3) Digital Signature (Working Group)

DSIMM dynamic (RAM) single in-line memory module

DSIN digital software information

DSIR Department of Scientific and Industrial Research (UK)

DSIS distributed support information standard

DSL deep scattering layer; denotational semantics language; digital simulation language; digital subscriber line

DSLA digital software licensing architecture

DSLAM digital subscriber line access multiplexer

DSLC digital/data subscriber line carrier

DSM data structure manager; demand side management; digital standard MUMPS; digital storage medium; direct signal monitoring; distributed systems manufacturing

DSMA digital sense multiple access

DSMC direct simulation Monte Carlo; digital service monthly charge

DSM-CC digital storage media – command and control

DSN data source name; Defense Switched Network (US); Deep Space Network; digital signal – level N

DSO digital storage oscilloscope

DSOM distributed system object model

DSP device stop, arrêt dispositif (f), détención de dispositivo (s); differential signal processing; digital send processor; digital signal processing; digital signal processor; digital sound (field) processing; digital sound program; directory/display system protocol; directory synchronization protocol (Lotus); directory system protocol; domain specific part

DSPU downstream physical unit

DSQ downstream quartz

DSQD double-sided quad density (floppy disk)

DSR data set ready; data signalling rate; device status register/report; Digital Satellite Radio; dynamic service register

DSRI destination station routing indicator; digital standard relational interface

DSS data sub-set; decision support system; Digital Satellite System (Sony *tn*); digital signature standard (NIST); digital supervisory signal; direct station selection; display stocker status; domaine de sous-service (f), sub-service field [SSF]; double-sideband system

DSS1 digital subscriber signalling system Nº 1

DSSCS Defense Special Service/(Secure) Communications System (US)

DSSI digital small storage interconnect; Digital Standard Systems Interconnect (DEC)

DSSL document style semantics and specification language

DSSS direct sequence spread spectrum, séquence directe à étalement de spectre (f)

DSSSL document style semantics specification language

DST data switching exchange; device start, mise en marche dispositif (f), arranque de dispositivo (s); Dienst (g), service; dispersion-supported transmission

DSTAR digital system technical architecture research

DSTE digital/(data) subscriber terminal equipment

DSTN double super-twisted nematic; dual-scan twisted nematic (of laptop colour display screens)

DST-REF destination reference

DSU data service/store; data switching unit; digital service unit; disk subsystem unit; distributive switching unit

DSUWG Data Systems Users' Working Group

DSVD digital simultaneous voice and data

DSVT digital secure voice telephone

DSW data/device status word; direct step-on-wafer

DSX digital cross-connect frames; digital signal/(system) cross-connect; Distributed Systems Executive (IBM)

DT data; data message; data transfer, transfert de données (f); data transcript; detach timer; dial tone; di-group terminal; dynamic test

DT TPDU data TPDU

DT&E developmental test and evaluation

DT1 data form 1, send signal packets

DT1/DT2 data form 1/2

DTA data terminal adaptor; detailed traffic analysis; differential thermal analysis; direct tape access (Seagate); disk transfer area

DTALGOL decision table ALGOL

DTAM document transfer and manipulation

DTAP direct transfer application part

DTAS Decision Theory and Adaptive Systems group (Microsoft Research); digital test access system

DTAT dispositivo de transmisión automática télex (s), telex automatic emitting devices [TAED]

DTB decal texture blending

DTBP dedicated total buried plant

DTC Data Technology Corp (US); detection threshold computer; digital transmit command; digital trunk controller; di-group terminal; direct thermocouple control

DTCE data terminating circuit equipment

DTD document type definition/descriptor (in XML); double talk detector

DTDM digital terminal data module

DTE data/(digital) terminal equipment, équipement terminal de traitement de données (f), equipo terminal de datos (s); dumb terminal emulator;

Dtels Directorate of Telecommunications (UK)

DTF dial tone first; direct transfer file; distributed test facility; dynamic track following

DTFS direct tape-file system (Seagate)

DTG date-time-group; direct trunk group

DTH direct to home (satellite TV); down the hill (link)

DTI Dark Towers International; data terminal interface; data transmission interworking (unit); digital trunk interface; distortion transmission impairment

DTIF digital tabular interchange format; digital transmission interface frame

DTISC data transmission interworking selection case

DTL designated transit list; dialogue tag language (IBM); diode-transistor logic

DTLS descriptive top-level specification

DTM defect test monitor; delay time multiplier; device test module; digital terrain map

DTMF dual-tone multi-frequency (signalling)

DTMPN defect test monitor phase number

Dtn Datenübermittlungsdienst (g), data transmission service

DTN digital telephone network

DTO de-centralized toll office

DTP data transfer part; desktop publishing; direct trigger point (of network planning)

DTR data terminal ready; data transfer rate; down-time ratio

DTRM digital transceiver module

DTRX data transceiver

DTS data transmission subsystem; digital tandem switch; digital termination system; digital test sequence; digital transmission system; Diplomatic Telecommunications Service (US); droits de tirage spéciales (f), special drawing rights [SDR]

DTST dial tone speed test

DTT (bloque de) datos de transporte (s), transport data [TDT]

DTTv digital terrestrial television (UK)

DTU data terminating/transfer unit; Datenaustausch- und Übertragungssteuerung (g), data terminal unit; dato unidad (s), unitdata [UDT]; digital transmission unit; di-group terminal unit; direct to user

DTV desk-top video; digital TV

DTVC desktop video conferencing

DTVI Distributed Tutored Video Instruction (Sun Microsystems)

DTW dynamic time warping

DTWX dial teletypewriter exchange service

DTX discontinuous transmission (mechanism)

DU delivery unit; dimensioning unit; disk usage

DUA Dallas Underground Association; directory user agent

DUAT direct user access terminal

DUCE denied usage channel evaluator

DUE detection of unauthorized equipment

DUeU Datenübertragungseinrichtung (g), data communicating equipment

DUF diffusion under epitaxial film

DUHT domain update handler (in the mobile) terminal

DUI dialogue user transmission

DUIB document user information block

DUN dial-up networking (Microsoft Windows 95)

DUNCE dial-up network connection enhancement

dup duplicate

DUP data user part; dedicated user port (frame relay)

Dup-loop endless stream of duplicate/near-duplicate messages

DUS data-user station; datos de usuario de sesión (s), session user data [SUD]

DUST Datenübertragungssteuerung (g), data transmission and supply unit

DUT device under test

DUV data under voice; deep ultraviolet

DuWaUe Durchwahlübertragung (g), through-dialling line unit

DV data vector, datos de vector (s); demi-vie (f), half-life; design verification

DVA Datenverarbeitungsanlage (g), data processing unit; designed verified and assigned date; displacement velocity acceleration (of sine controller)

DVB digital video broadcasting

DVBS digital video broadcasting by satellite

DVB-S digital video broadcasting for satellite transmission

DVBST direct view bi-stable storage tube

DVC digital video cassette

DVD digital versatile disk, CD-ROM vidéo numérique (f), disque vidéo numérique (f), vidéodisque numérique (f), videodisco digital (s)

DVDC Digital Versatile Disk Committee (of the BVA)

DVE digital video effects

DVER design rule verification

DVI Digital Video Interactive *tn*

DVl Datenvermittlungsleitung (g), data exchange unit; direct voice link

DVM data-over-voice multiplexer; datos del vector de movimiento (s), motion vector data [MVD]; digital volt-meter

DVMA direct virtual memory access

DVMRP distance vector multicast routing protocol

DVN digital-video network

DVOM digital volt-ohm-meter

DVORAK European keyboard configuration (*ie* non-Qwerty)

DVS descriptive video service; design verification system; display vehicle status; distributed virtual storage

DVST direct view storage tube

DVT design verification test; digital video tape

DVTR digital video tape recorder

DVX digital voice exchange

DW device wait, dispositif en attente (f), espera de dispositivo (s); device unit; don't want; drop and block wire

D-W Durbin-Watson statistic

DWCI DEC-Windows compiler interface

DWDM dense wave/wavelength-division multiplexing

DWG domain work group

dwg drawing

DWIM do what I mean

DWLL digital wireless local loop

DWN Data Warehouse Network

DWS Durchwahlsatz (g), through-dialling unit

dx distance reception; distant; duplex

DXC data exchange control, digital cross connect

DXF drawing eXchange format

DxHAs Datex-Hauptanschluß (g), datex exchange line

DXI data exchange interface

DXI/SNI data exchange interface at the subscriber network interface

DxP Datex-Paketvermittlung (g), DATEX packet switching

DYANA dynamics analyser

DYLAN dynamic language

Dylperl dynamic linking (package for) Perl

DYNAMO dynamic models

DYP directory Yellow Pages
DYSTAL dynamic storage allocation

E

E echo channel bit (of D channel); essential, essentiel (f), esencial (s); établissement (f), establecimiento (s) [S]; exa, 10^{18} (one million million million); voltage

E channel echo channel (D channel deviant)

E&M ear and mouth (*ie* receive and transmit)

E/O electrical-to-optical

E/S entrée/sortie (f), entrada/salida (s), input/output [I/O]

E: Ersatz (g), standby

E0 en-tête 0 (f), encabezamiento 0 (s), heading code 0 [H0]

E1 encabezamiento 1 (s), heading code 1 [H1]; European standard for digital transmission (2 048 Mbps, 30-channel PCM; ITU-T Rec. G.703 & G.732)

E13B standardized font for use in MICR

E3 Electronic Entertainment Expo (Atlanta, 1998); European equivalent of DS3 (34 368 Mbit/s)

E911 enhanced 911

Ea each

EA address field extension bit, extension d'adresse (f); electronic arts; entidad de aplicación (s), application entity [AE]; equipos asociados (s), associated equipment; evolutionary algorithm; expedited data acknowledgement; extended attribute O/S2; external access; external alarm; messages d'établissement émis vers l'avant (f), forward set-up message [FSM]

EA TPDU expedited acknowledgement TPDU

EAA encaminamiento alternativo automática (s), automatic alternative routing [AAR]; équipement d'appel automatique (f), automatic calling equipment [ACE]

EAAT equal access alternative technologies

EAC EIR administration centre; external access common

EACP essai pour la vérification de l'acheminement dans le SSCS (f), SCCP routing verification test [SRVT]

EACU external alarm collection unit

EAD encaminamiento alternativo desde (s), alternative routing from [ARF]; encoded archival description (Library of Congress); extended addressing (called)

EADAS engineering and administrative data acquisition system

EADAS/NM EADAS/network management

EAF effort adjustment factor

EAG extended addressing (calling); extended affix grammar

EAGE electrical aerospace ground equipment

EAGS entidad de aplicación de gestión de sistema (s), system management application entity [SMAE]

EAH encaminamiento alternativo hacia (s), alternative routing to [ART]; external access handler

EAI external authoring interface

EAM electro-absorption modulator; explotación administración y mantenimiento (s), operation administration and maintenance; extended-answer-message indication

EAN European Academic Network; European article numbering (*aka* bar code)

EAO enseignement assistée par ordinateur (f), computer-aided learning

EAPROM electrically alterable programmable ROM

EAR external access register

EARL (English) Electronic Academic Reference Library

Early Bird first communications satellite (1964)

EARN European Academic and Research Network (part of BITNET)

EAROM electrically alterable ROM

EARS electro-acoustic rating system; electronic access to reference services

EAS/ASE European Space Agency/Agence spatiale européenne

EASD equal access service data

Easdaq European Association of Securities Dealers Automated Quotation

EASE equipment and software emulator

Easter egg jocular message inserted and hidden in a program by its developers

EASY extended access system (Astrocom Corp *tn*)

EATA enhanced Advanced Technology (AT) bus attachment

EATM entidad del agente de transferencia de mensajes (f), message transfer agent entity

EATMS electro-acoustic transmission measuring system

EATP essai pour la vérification de l'acheminement dans la SSTM (f), MTP overall transfer time [MRVT]

EAU entidad del agente de usuario (s), user agent entity [UAE]; external auxiliary unit

EAX electronic automatic telephone exchange; environmental audio extensions

EB estación (de) base (s), base station

EB&F equipment blockage and failure

EBC EISA bus controller

EBCA external branch condition address

EBCDIC extended binary-coded decimal interchange code [*also* EBDIC]

EBCI external branch condition input

EBCT electron beam computer tomography

EBD effective billing date; Empfangsbezugsdämpfung (g), receiving reference equivalent

EBDF emisión de bloqueo y desbloqueo de grupo de circuitos para el fallo del soporte físico (s), hardware failure-oriented circuit group blocking and unblocking sending [HBUS]

EBDI electronic business data interchange

EBDL emisión de bloqueo y desbloqueo de grupo de circuitos generados por el soporte lógico (s), software-generated circuit group blocking and unblocking sending [SBUS]

EBDM emisión de bloqueo y desbloqueo de grupo de circuitos para el mantenimiento (s), maintenance-oriented circuit group blocking and unblocking sending [MBUS]

E-beam electron beam

EBED error de bloque en el extremo distante (s), far end block error [FEBE]

EBES electron-beam exposure system

EBHT electron beam high-throughput lithography

EBI equivalent background input

EBIC electron beam-induced current

EBIOS extended BIOS

EBIT European broadband interconnection trial

EBL emisión de signal de blocage (f) [BLS]

EBN European Business News

EBNF extended BNF

EBONE European backbone network (European academic research)

EBR edge-bead removal; electron beam recording

EBSG elementary basic service group

EBT electronic benefits transfer

EBU European Broadcasting Union

EC Eastern Cedar (telephone pole); echo canceller/controller; electrical conductor; elemento de conexión (s), connection element; emergency correction; engineering change; equipment controller; equivalent capacity; error control/correction; exchange carriers; extra control

EC++ extended C++

Ec/No ratio of energy per modulating bit to noise spectral density

ECA emergency change-over acknowledgement signal

ECAC European Civil Aviation Conference

ECAD electronic/engineering computer-aided design

ECAE electronic computer-aided engineering

ECAP electronic circuit analysis program; electronic customers' access program

ECAT electronic card assembly and test (IBM)

ECB electrically-controlled bi-refringence; electronic code-book

EC-BCSM edge control basic call state model

ECBF équipement de connexion basse fréquence (f), equipo de conexión de baja frecuencia (s), audio-connecting equipment [ACE]

ECBS engineering of computer-based systems

ECC East Coast Crackers; electronic cash card; enter cable change; Ericsson cable cell; error checking and correction; error correction code

ECCM error de concentridad de campo modal (s), mode field concentricity error [MFCE]

ECCR equipo de control de la conmutación de restauración (s), restoration switching control equipment [RSCE]

ECCS economic hundred (centum) call seconds

ECD electronic cash disbursements; enhanced CD; error control device

ECDO electronic community dial office

ECE echo control equipment; Electronic Commerce – Europe

ECF echo control factor; enhanced connectivity facility

ECFA Electronic Communications Forwarding Act (US)

ECH echo cancellation; signal d'échec de l'appel (f), call-failure signal [CFL]

ECHO European Commission Host Organisation

ECHT European Conference on Hypertext

ECI entidad de control de imagen (s), picture control entity [PCE]; external call interface

ECIS European Committee for Inter-operable Systems

ECITC European Committee for Information Technology Certification

ECL emitter-coupled logic; error code logging; extensible control language

ECM electronic control module, módulo de regulación electrónica (s); electronic counter-measures; emergency change-over message; error-correcting memory; error-correction mode; error cuadrático medio (s), mean square error; estados de congestión múltiples (s), multiple congestion states

ECMA European Computer Manufacturers' Association (1961)

ECMA-10 ECMA data transmission standard

ECN emergency communication network; engineering change notice; Ericsson Corporate Network; explicit congestion notification

ECNE Enterprise Certified NetWare Engineer (Novell)

ECO engineering change order; electron-coupled oscillator; emergency-change-over order signal [ECO]

ECOC European Conference on Optical Communication

ECOM electronic computer-originated mail; electronic counter-countermeasures

Econet Environmental Conference Network

ECOOP European Conference on Object-Oriented Programming

ECOS extended communications OS (Harris Corp *tn*)

ECP engineering change proposal; enhanced/extended capabilities port; equipment conversion package; extended communications port; extended concurrent Prolog

ECPT electronic coinbox public telephone

ECQB electrochemical quartz crystal balance

ECR electron-cyclotron resonance; equipo de conmutación de restauración (s), restoration switching equipment [RSE]; estación de coordinación de la red (s), network coordination station [NCS]

ECRA ECR de asignación (s), network coordination station assignment

ECRC ECR común (s), network coordination station common; European Computer Industry Research Centre (Munich)

ECRH ECR de haces (s), network coordination station spot-beam [NCSS]

ECRI ECR entre estaciones (s), network coordination station interstation [NCSI]

ECRP earcap reference point; enunciado de conformidad de realización de protocolo (s), protocol implementation conformance statement [PICS]

ECS echo control subsystem; electronic cross-connect system; Enhanced Chipset (Amiga *tn*); European Communications Satellite; Experimental Communications Satellite (Japan)

ECSA Exchange Carriers Standards Association

ECSL extended computer structure language

ECSS extendable computer system simulator

ECT echo cancellation technique; explicit call transfer (supplementary service)

ECTEL European Conference of Telecommunications & Electronics Industries

ECTL electronic communal temporal lobe

ECTRA European Committee for Telecommunication Regulatory Affairs (CEPT 3)

ECTS echo-canceller testing system; European Consumer Trade Show

ECU EISA configuration utility; energy/environmental control unit

ED bit de extensión de campo de dirección (s), address field extension bit [EA]; enlace de datos (s), data link [DL]; encaminamiento directo (s), direct route; end delimiter; exception data, datos de excepción (s); expedited data

ED TPDU expedited data TPDU

EDA electronic design automation; embedded document architecture; équivalent de distorsion d'affaiblissement (f), equivalente de distorsión de atenuación (s), attenuation distortion equivalent [ADE]; exploratory data analysis

EDAC electromechanical digital adaptor circuit; error detection and correction

EDACS Enhanced Digital Access Communications System (Ericsson digital trunked radio *tn*)

EDC electronic digital computer; enhanced data correction; error detection and correction

EDD earliest due date; empaquetado/desempaquetado de datos (s), packet assembly/disassembly [PAD]; envelope delay distortion

EDDC extended distance data cable

EDE entidad de depósito y (de) entrega (s), submission and delivery entity [SDE]

EDF earliest deadline first

EDFA erbium-doped fibre amplifier

EDFG Edge Device Functional Group (ATM Forum)

EDFR enlace digital ficticio de referencia (s), hypothetical reference digital link [HRDL]

EDG external database gateway

EDGAR electronic data-gathering analysis and retrieval

EDGE enhanced data rates for GSM evolution

EDI electronic data interchange, échange

de données informatisé (f); end system identifier (ATM, ISO)

EDIBANX EDI Bank Alliance Network Exchange

EDIF electronic design interchange format

EDIFACT European Data Interchange for Administration Commerce and Transport (ISO 9735)

EDILIBE EDI for libraries and booksellers in Europe

Edison programming language for designing real-time programs

EDL edit decision list; event description language; event-driven language; experiment description language; tiempo de tránsito máximo esperado distante-local (s), expected maximum transit delay remote-to-local [ERL]

EDLC Ethernet data link control

EDLIN MS-DOS line editor

EDM electronic document management

EDMC equipo digital de multiplicación de circuitos (s), digital circuit multiplicated equipment [DCME]

EDMS electronic document management system

EDO equipment design objectives

EDP electronic data processing, traitement électronique de l'information (f), proceso de datos electrónico/proceso electrónico de datos (s), elektronische Datenverarbeitung (g); enhanced dot pitch (Hitachi monitors *tn*); event detection point; extended definition progressive (image)

EDPM electronic data processing machine

EDS electronic data switching system (German network); Electronic Data Systems (corporation owned by Ross Perot); electronic distribution system; energy-dispersive spectroscopy; exchangeable disk storage/store

EDSAC Electronic Delay Storage Automatic Calculator (1949)

EDSI enhanced small device interface; equivalent delivered source instruction

E-DSS1 European DSS1 protocol

EDSX electronic digital system cross-connect

EDT environment data terminal; énoncé de travaux (f), statement of work

eDTAU enhanced DS1 test access unit

ED-TPDU-NR ED TPDU number

EDTV Emirates Dubai Television; extended-definition TV

EDU Education Computer Systems; Einkanal-Datenüberttragungseinrichtung (g), single channel data transmission unit; equipment dependent up-time

EDUEX externe EDU (f), external EDU

EDU-OVSt EDU in der OVSt (g), EDU in local switching centre

EDVAC electronic discrete variable automatic computer (1944)

EDWG ESnet DECnet Working Group

EDX energy-dispersive X ray; event-driven executive

EDXA energy-dispersive X-ray analysis

E-E electronics-to-electronics

EE end-to-end signalling; extended edition (IBM); messages d'échec de l'établissement, émis vers l'arrière (f), unsuccessful backward-set-up acknowledgement message [UBM]

EEA Electronics Equipment Association (UK)

EEC electronic equipment; extended error correction; signal d'encombrement de l'équipement de commutation (f), switching equipment congestion [SEC]

EECT end-to-end call trace

EED error en el extremo distante [FEE]

EEDF electron energy distribution function

EEDP expanded electronic dialling plan

EEE electronic equipment enclosures

EEHO either end hop-off

EEI equipment-to equipment interface

EEL electric echo loss; Epsilon error correction; Epsilon extension language

EELS electron energy-loss spectroscopy

EEM extended memory management
EEMA European Electronic Messaging Association
EEMS enhanced expanded/(extended) memory specification/system
EEO extremely elliptical orbit
EEP error en el extremo próximo (s) [NEE]
EEPLD electrically-erasable programmable logic device
EEPROM electrically-erasable programmable ROM
EER extended entity-relationship model
EEROM electrically-erasable ROM
EES escrow encryption standard
EEST enhanced Ethernet serial transceiver
EET equipment engaged tone; Ericsson engineering tool
EETDN end-to-end transit delay notification
EETPU Electrical Electronic Telecommunications & Plumbing Union (UK)
EF éléments de fonction (f), elementos de función (s), functional element [FE]; elementary function; elemento de función utilizado entre TC y TL (s) [FE]; entidad funcional (s), functional entity; entrega física (s), physical delivery; especificación funcional (s), functional specification [FS]
Ef&I engineer furnish and install
EFA extended file attribute
EFC signal d'encombrement du faisceau des circuits (f), circuit-group-congestion signal [CGC]
EFCI explicit forward congestion indication/(-or) (ATM)
EFDS error free deci-seconds
EFF Electronic Frontier Foundation (US)
EFI échange de formulaires informatisés (f); electromechanical frequency interference; expedited flow indicator
EFL emitter-follower logic; extended FORTRAN language
EFM eight-fourteen modulation

EFO electronic flame-off
EFOCS evanescent fibre-optic chemical sensors
EFP electronic field production; Ericsson field communication platform
EFR enhanced full rate
EFRAP electronic/exchange feeder route analysis program
EFS end of frame sequence; error-free seconds
EFSM extended finite-state machine
EFT electronic funds transfer
EFT-EDI EFT – electronic data interchange
EFTIR emission Fourier transform infrared spectroscopy
EFTPOS EFT at point of sale
EFTS EFT system
EFU external fuse unit
EFV excess flow valve
EG élément d'image (f), picture element (*ie* pel) [PE]
EGA enhanced graphics adaptor (IBM)
EGC enhanced group call; entidad de gestión de capa (s), link management entity [LME]
EGP exterior/external gateway protocol (IETF)
EGREP extended global regular expression print (Unix command)
EGS équivalent global pour la sonie (f), overall loudness rating [OLR]
EGW Endgruppenwähler (g), final group selector
EH electronic highway; external host
EHEMT enhancement-mode high-electron mobility transistor (Fujitsu)
EHF extended hi-frequency; extremely-high frequency (30-300 GHz)
EHLLAPI emulator high-level language application programming interface
EHM encryption handler mobile (terminal)
EHN encryption handler network
EHO estimated hourly output
EHP electron-hole pairs

EHPT Ericsson Hewlett-Packard Telecom

EHS extremely hazardous substances

EHTS Emacs hypertext system

EI elemento de información (s), information element [IE]; ending inventory; environment identifier; equipment integration; exchange identification; exchange interface; expansion interface; external indication, äussere Indikation (g); estación fuera de las instalaciones (s), off-premises station [OPS]; external indication

Ei3 European Industry Initiative

EIA Electronics Industries Association (US)

EIAJ Electronics Industry Association of Japan

EIC equipo en las instalaciones del cliente (s), customer-premises equipment [CPE]

EICAR European Institute for Computer Anti-virus Research

EICD équipement intermédiare de commutation de données (f)

EID endpoint identifier, identificateur de point d'extrémité (f)

EIES electronic information exchange system

Eiffel object-oriented programming language (Bertrand Meyer)

EIGRP enhanced interior gateway routing protocol

EII European Information Infrastructure

EIN electronic tandem network (PBX-based voice network); European Informatics Network

EIP equipment installation procedure

EIR equipment identity register (wireless/GSM); error-indicating recording; excess information rate

EIRP effective isotropic radiated power

EIS electrochemical impedance spectroscopy; enterprise integration services; executive information system; expanded in-band signalling

EISA Electronics Industry Standards

Association; entorno de interconexión de sistemas abiertos (s), open systems interconnection environment [OSIE]; extended ISA

EISS economic impact study system

EIT encoded information type; erreur sur un intervalle de temps (f), error de/(en el) intervalo de tiempo (s), time interval error [TIE]

EITO European Information Technology Observatory

EIU extended interface unit

EIV Erdimpulsverfahren (g), earth-pulse dialling

EJF estimated junction frequency

EJS Elite Justice Society

EKF extended Kalman filter

eKi (DES) encryption Ki (encryption subscriber authentication key)

EKT Einführungskonzentrator (g), growth concentrator

EKTS electronic key telephone service/system

EL echo loss; electroluminescent (display); exchange line; experience level

EL 1 extensible language one

ELAG European Library Automation Group

ELAN emulated LAN (ATM Forum LANE)

ELAP EtherTalk Link Access Protocol (AppleTalk *tn*)

EL-AZ elevation-azimuth

ELC embedded linking and control

ELCB earth-leakage circuit-breaker

ELD electroluminescent display; extended scan line description; tiempo de tránsito máximo esperado local-distante [ELR]

ELDO European Launcher Development Organization

ELDS exchange line data service

ELF employee locator facility; executable and linking format; extremely low frequency (< 300 Hz)

Elg Endvermittlungsleitung gehend (g), terminal switching-centre line outgoing

ELI embedded Lisp interpreter

ELINT electronic intelligence

Elk Endvermittlungsleitung kommend (g), terminal switching-centre line incoming

ELK extension language kit

ELLA equipo de llamada automática (s), automatic calling equipment

ELLIS EuLisp LInda System

ELMAP exchange line multiplexing analysis program

ELN enregistreur de localisation nominal (f), home location register [HLR]

ELP English Language Programs

ELR expected maximum transit delay local-to-remote

ELS entry level system

ELSAR establecimiento de llamadas sin asignación de radiocanal (s)

ELSE rule last rule of an incomplete decision table

ELSEC electronic security

EL-SSC electronic switching system control

ELT emergency-load-transfer signal; emergency locator transmitter

ELV enregistreur de localisation pour visiteurs (f), visitor location register [VLR]

Em émission (f), emisión (s), transmit

EM electromagnetic; element management (layer); e-mail (electronic mail); end of medium; end mark; enterprise model; entité de maintenance (f), entidad de mantenimiento (s), maintenance entity [ME]; error medio (s), medium error; estación(es) móvil(es) (s), mobile stations; expanded memory; extension module; fin del medio físico (s), end of physical medium

EM+ enhanced monitor plus

EM/MU enhanced monitor/matrix unit

EMA Electronic Mail Association; Electronic Messaging Association; emergency area; enterprise management

architecture; error medio absoluto (s), mean absolute error [MAE]; extended Mercury autocode

EMAA entidad de mantenimiento de accesos de abonado (s), subscriber access maintenance entity [SAME]

EMACS editing macros; extensible portable display editor (distributed with GNU)

Em-ACT emisión de acción ACT

e-mail electronic mail (CCITT: X.400)

EMAP Ericsson MAP

EMAS Edinburgh multi-access system (University of Edinburgh)

EMB extension module bus

EMBARC electronic mail broadcast to roaming computer (Motorola)

EMC electromagnetic compatibility; enhanced memory chip; equipo de modulación de canales (s), channel-translating equipment; extended maths co-processor

EMCD equipo de multiplicación de circuitos digitales (s), digital circuit multiplication equipment [DCME]

EMCN équipement de multiplication de circuit numérique (f), digital circuit multiplication equipment [DCME]

EMCON electromagnetic emission control

EMD Edelmetall-Drehwähler (g), noble metal rotary selector

EMDA European Media Development Agency (EU fund, 1998)

EMDIR electronic mail directory (CERN)

EMDM extension module distribution magazine

EMDO EMD-Ortsvermittlungsstelle (g), EMD local PSTN switching-centre

EMEA établir (la) mode équilibrée asynchrone (f), set asynchronous balanced mode [SABM]

EMF electromagnetic field; electromotive force

EMG electro-migration; équipement de modulation de groupe (f), equipo de modulación de grupo (s), group-modulating equipment [GME]; extension module group

EMGC equipo de modulación de grupo cuaternario (s), quaternary group modulating equipment [SMTE]

EMGT equipo de modulación de grupo terciario (s), tertiary group modulating equipment [MTE]

EMI commande de l'émission (f), transmission control [TXC]; electromagnetic interference

EMIA entidad de mantenimiento de la instalación de abonado (s), entité de maintenance d'installation d'abonné (f), subscriber installation maintenance entity [SIME]

EMIP entorno de mensajería interpersonal (s), personal message environment

EMIRTEL Emirates Telecommunications Corporation

EML element management layer; expected measured loss

eMLPP enhanced Multi-Level Precedence and Pre-emption service

EMM expanded memory manager (Microsoft Windows)

EMM386 Windows 386 expanded memory manager ((Microsoft Windows swap-file)

EMMA electron microscopy and micro-analysis

EMMI electrical man-machine interface

EMML extended (MML) man-machine language

EMMS Electronic Music Management System (IBM *tn*)

EMN European Museum Network

EMO emergency off

EMP electromagnetic pulse; equipo de modulación de grupo primario (s), group translating equipment [GTE]; error medio en porcentaje (s), mean percent error [MPE]; excessive multi-posting (of e-mails); postage multiple abusif (s)

EMR enhanced metafile record, enter move request

EMRP effective monopole-radiated power; extension module regional processor

EMRPB extension module regional processor bus

EMRPM extension module regional processor module

EMRPS EMRP speech bus

EMS electronic mail/message system; element management system; equilibrium mode simulator; equipo de modulación de grupo secundario (s), secondary translation equipment [STE]; expanded memory specification

EMS memory expanded memory (Lotus, Intel, Microsoft)

EMSEC emanation security

EMT electrical metallic tubing

EMU electromagnetic unit

EMV Elektromagnetische Vertrüglichkeit (g), electromagnetic compatibilty; équipement de modulation de voie (f) [CTE]

EMWAC European Microsoft Windows NT Academic Centre (Scotland)

en ligne on line

EN egress node; Einsteller (g), controller; end node

ENA extended network-addressing

ENADS enhanced network administration system

ENPC Ericsson Network Programming Centre

ENDEC encoder/decoder

ENDS ends segment

ENDSU expedited network service data unit

ENET Ethernet

ENEX European News Exchange

ENFIA exchange network facilities for interstate access (US)

ENG electronic news gathering

ENIAC electronic numerical integrator and calculator (1944-46)

ENN establecimiento de facilidades no

normalizadas (s), non-standard facilities set-up [NSS]

ENQ enquiry, petición de respuesta (s); pregunta (s); enquiry character

ENS enterprise network services; European nervous system

ENSS exterior nodal switching subsystem

en-tête header

ENTP Ericsson Network Planning Tool

ENU essential-non-essential-update

EO end office; heading code Ø [H0]

EOA end of address

EOB end of block; end of bus

EOC embedded operations channel (SONET); end of contents; end-of-cluster code

EOCS end-office connections study

EOD end of data; end of delivery; end of dialling

EOE electronic order exchange

EOES équivalent objectif électrique pour la sonie (f), electrical objective loudness rating [EOLR]

EOF end-of-file (marker), fin de fichier (f), Dateiende (g); Enterprise Objects Framework (Next Computer *tn*)

EOFB end of facsimile block

EOI end or identity (IEEE 48); end of input

EOJ end of job

EOL end of life (of a satellite); end of line

EOLR electrical objective loudness rating

EOM end of message

EOP end of optional parameter; end of procedure

EOR end of record; end of retransmission; end of run

EORM environmental and occupational risk

EOS early operational signal; earth observation satellite/system (NASA); electrical over-stress; end of selection, fin de selección (s); équivalent objectif pour la sonie (f), overall loudness rating [OLR]; extended operating system

EOSDIS EOS data and information system (NASA)

EOSE équivalent objectif pour la sonie à l'émission (f), transmitting objective loudness rating [TOLR]

EOSR end of status request signal; équivalent objectif pour la sonie à la réception (f), receiving objective loudness rating [ROLR]

EOT end of tape; end of transfer; end of transaction; end of transmission; end of transport; end of TSDU mark

EOTT end-office toll trunking

EOUG European ORACLE Users Group

EOV end of volume, Datenträgerende (g)

EOWT engineering order wire terminal, Dienstleitungsterminal (g)

EOY end of year

EP electro-polish; elemento pictográfico (s), pictorial element; elementos de protocolo (s), protocol element; entidad de protocolo (s) [PE], protocol entity; esquema de puntos (s), dot pattern [DP; evolutionary programming; executive process; extreme pressure

EPA Elite Programming Association; Environmental Protection Agency

EPC earth potential compensation; Ericsson protocol converter *tn*

EPCS Experimental Physics Control Systems

EPD early packet discard; elemento(s) de procedimiento de documento (s); exchange parameter definition

EPER empresa privada de explotación reconocida (s), re-organized/recognized private operating agency [RPOA]

EPF evolution prospects and framework

EPG electronic programme guide (on-screen satellite TV guide)

epi epitaxially grown layer (of silicon semiconductor)

EPI elevation-position indicator

EPIC Electronic Privacy Information Center (US); explicitly parallel instruction computing (Intel IA-64 *tn*)

EPILOG extended programming in logic

EPIM enterprise product information management

EPIRB emergency position indicating radio beacon

EPL effective privilege level; electron projection lithography; experimental programming language

EPLD electronically-programmable logic device; erasable programmable logic device

EPM enhanced Editor for Presentation Manager (IBM); enterprise process management (IBM)

EPO emergency power off

EPOS electronic point-of-sale

EPP enhanced/(extended) parallel port

EPPT European Printer Performance Test (ISO 10561)

EPR earth potential rise; electronic pin registration (*not* PIN); electron-paramagnetic resonance; exploitations privées reconnues (f), re-organized/recognized private operating agency

EPRCA enhanced proportional-rate control algorithm

EPROM electrically-programmable ROM; (UV)-erasable programmable ROM

EPROM OTP EPROM one-time programmable

EPS Encapsulated PostScript

EPSCS enhanced private switched communications system (AT&T)

epsilon something negligible/vanishingly small

EPSS electronic performance support system; experimental packet-switched service

EPT elevación del potencial de tierra (s), earth potential rise; equipment performance tracking

EQ gate equivalence gate

EQ equalization (of sound)

EQG overall R.25 equivalent

EQG$_{EG}$ équivalent global pour la sonie (f), overall loudness rating [OLR]

EQG$_{IS}$ índice de sonoridad global (s), overall loudness rating [OLR]

EQTV extended quality TV

EQUAL equalize

Equel embedded Quel

Equip Education Training Quality Improvement Plan (EU)

EQUIP European Quality in Information Programmes (EU)

EQUIP C/I equipment control and integration

EQUIP RTC equipment real-time control

equiv equivalent

ER easy to reach; egress router; elemento de red (s), network element [NE]; engineering route; entity-relationship; equivalent(s) de référence (f), equivalente de referencia (s), reference equivalent [RE]; error; error en UDPT (s), error transfer protocol data unit; error register; exception report; exchange requirement; explicit rate; radio-relay system (s) [RR]

ER TPDU error transfer protocol data unit

ER/RC extended result/response code

ER25E équivalent R.25 à l'émission (f), equivalente R.25 en emisión (s), sending R.25 equivalent [SR25E]

ER25R équivalent R.25 à la réception (f), equivalente R.25 en recepción (s), receiving R.25 equivalent [RR25E]

ERA extended registry attributes

ERA diagram entity-relationship attribute diagram

ERA model ERA model

ERAM equipment reliability-availability-maintainability

ERAR error return address register

ERAS Electronic Routing and Approval System (Hughes Aircraft)

ERC equivalent release concentration; equivalentes de referencia corregidos (s), corrected reference equivalents [CRE]; European Radio Commission

ERCE équivalent de référence corrigé à l'émission (f), equivalente de referencia corregido en emisión (s), send corrected reference [SCRE]

ERCG équivalent de référence corrigé global (f), equivalente de referencia corregido global (s), corrected overall reference equivalent [CORE]

ERCIM European Research Consortium on Informatics and Mathematics

ERCR équivalent de référence corrigé à la réception (f), equivalente de referencia corregido en recepción (s), receiving corrected reference equivalent [RCRE]

ERD elastic recoil detection; entity-relationship diagram

ERE emulated radio environment; équivalent de référence à l'émission (f), equivalente de referencia en emisión (s), send reference equivalent [SRE]

EREL équivalent de référence par effet local (f), equivalente de referencia por efecto local (s), sidetone reference equivalent [STRE]

EREP environmental recording editing printing

ERF emergency restoration facility; Evangeliste Rundfunk Fernsehen (g)

ERG enhancement and review group

ERGC emisión de reinicialización de grupo de circuitos (s), circuit group reset sending [CGRS]

ERI exchange radio interface

ERIC Educational resources information centre

Ericsson LM Ericsson Telefonaktiebolaget

ERL echo return loss; Electronic Reference Library; erbium-doped laser; expected maximum transit delay remote-to-local

ERLE echo return loss enhancement

ERLL enhanced run-length limited; emulated radio lower layers

ERM enterprise reference model

ERMES European Radio Messaging System

ERN external recurrent neural network; signal d'encombrement sur le réseau

national (f), national-network-congestion signal [NNC]

ERNIE Electronic Random Number Indicating Equipment (UK)

ERO European Radiocommunications Office (CEPT 2)

EROS Ericsson operating system *tn*

erotics electronics

ERP ear reference point; effective radiated power; electronic retail payment; enterprise resource planning; equivalent radiated power; extended range pyrometer; (señal de) enlace de reserva preparado (s), standby ready signal [SBR]

ERR error; equivalent de référence à la réception (f), equivalente de referencia en recepción (s), receive reference equivalent [RRE]; error en unidad de datos de protocolo (s), protocol data unit error; response for end of retransmission

ERS electronic register-sender; emergency reporting system; emergency response service; event-reporting standard

ERT emergency response time; equivalent random theory; equivalent random traffic; estimated repair time

ERTMS European Rail and Traffic Management

ERTS error rate test set

ERTU Egyptian Radio and Television Union

ERU error return update

Es sporadic E-layer

ES earth station; echo suppressor; end-system (ISO); Energy Star (power consumption specification); engineering specification; enlace de señalización (s), signalling link [SL]; entidad de soporte (s), support entity [SE]; équivalent pour la sonie (f), loudness rating [LR]; errored second, secondes erronées (f), segundos con error (s); establecimiento (s), station; evolution strategy; expert system; Externsatz (g), inter-exchange line unit

ESA elemento de servicio de aplicación

(s), application service element [ASE]; European Space Agency

ESA/ASE European Space Agency/Agence Spatiale Européenne (f)

ESAC Electronic Surveillance Assistance Center; elementos de servicio de aplicación común (s), common application service elements [CASE]

ESAD mange de la merde et crève (f)

ESAE elemento de servicio de aplicación específica (s), specific application service element [SASE]

ESAM elemento de servicio de administración de mensajes (s), message administration service element [MASE]

ESC echo suppressor control; Egyptian Satellite Channel; EISA system component; electronic still camera; electrostatic chuck; emergency service centre; engineering service circuit; escape (key); (commande des) essais des canaux sémaphores (f), signalling link test control [SLTC]

ESC/P Epson standard code (for) printers

ESCA electron spectroscopy for chemical analysis; elemento de servicio control de asociación (s), association control service element [ACSE]; accusé de réception de message d'essai de canal sémaphore (f), signalling link test message acknowledgement [SLTA]

ESCD extended system configuration data

ESCES Experimental Space Communication Earth Station

ESCM extended services communications manager (IBM)

ESCO message d'essai de canal sémaphore (f), signalling link test message [SLTM]

ESCON Enterprise System Connectivity/(Connection) (IBM *tn*)

ESCSI embedded SCSI

ESD electronic software distribution; electronic systems development; electrostatic discharge

ESDC extended system configuration data

ESDI enhanced small-device interface

ESDM elemento de servicio depósito de mensajes (s), message delivery service element [MDSE]

ESDP Educational Services Development and Publishing

ESE équivalent pour la sonie à l'émission (f), send loudness rating [SLR]; Ersatzschalteinrichtung (g), standby change-over circuit

ESEM elemento de servicio entrega de mensajes (s), message delivery service element [MDSE]

ESEXM elemento de servicio de extracción de mensajes (s), message retrieval service element [MRSE]

ESF electrostatic field coating (on CRT); Eureka Software Factory; European Science Foundation; Extended SuperFrame (format; AT&T *tn*)

ESG Engineering Systems Group; exchange software generator

ESH Environment Safety and Health

ESI end-system identifier; enhanced serial interface (Hayes specification); European Software Institute

ESIG European SMDS Interest Group

ESIOP implémentation de la norme Corba

ES-IS end system – intermediate system

ESJ équivalent de référence pour la sonie de la jonction (f), junction loudness rating [JLR]

ESL essential service line

ESLF Eastern Seaboard Liberation Front

ESM electronic service manual; electronic support measures (interception etc); external storage module

ESML extended systems modelling language

ESMR electrically-scanning microwave radiometer

ESMR enhanced specialized mobile radio

ESN electronic security number;

electronic serial number; electronic switched network; equipment serial number

ESNC mensaje hacia atrás para información de establecimiento simple no completado (s), unsuccessful backward-set-up acknowledgement message [UBM]

ESnet Energy Science Network

ESO echo suppressor originating end; entry server offering; Ericsson Support Office

ESOC European Space Operations Centre

ESOD elemento de servicio operaciones a distancia (s), remote operation service element [ROSE]

ESONE European Standards of Nuclear Electronics

ESP econometric software package; electrostatic precaution; emulation sensing processor; encapsulating security payload; enhanced service provider; enhanced serial port; extended self-containing Prolog; extra-simple Pascal

ESPITI European Software Process Improvement Training Initiative

ESPRIT European Strategic Programme for R&D in Information technology

ESR effective series resistance; équivalent pour la sonie à la réception (f), receive loudness rating [RLR]; equivalent series resistance

ESRA extended slotted ring architecture

ESRIN European Space Research Institute

ESRM elemento de servicio de extracción de mensaje (s), elemento de servicio (de) recuperación de mensajes (s), message retrieval service element [MRSE]

ESS echo suppression subsystem; electronic switching system (Western Electric); équilibre du signal de sortie (f), output signal balance (loss) [OSB]; extended switching subsystem

ESSI European Systems and Software Initiative

ESSX electronic switching system exchange

ES-STM elemento de servicio sistema de tratamiento de mensajes (f), message handling system service element [MHS-SE]

EST echo suppressor terminating end; Ersatzschalteteil (g), standby change-over switch section

ESTEC European Space Research and Technology Centre

ESTF elemento de servicio transferencia fiable (s), reliable transfer service element [RTSE]

ESTM elemento de servicio transferencia de mensajes (s), message transfer service element [MTSE]

ESTS echo suppressor testing system

ESU electrostatic unit; empty signal unit; exchange signalling unit, unité de signalisation de central (f)

ESV experimental space vehicle (satellite)

ESVN executive secure voice network

ET Elliniki Tileorassis 1 (satellite broadcaster, Greece); engaged tone; enhanced/(enhancement) technology; equipo terminal (s), terminal equipment [TE]; estación telefotográfica (s), phototelegraph station [PS]; exchange terminal/termination

ET&Q engineering technology and quality

ET1/ET2 equipo terminal de tipo 1/tipo 2 (s), terminal equipment 1 / terminal equipment 2 [TE1/TE2]

ETA equipos de telegrafía a frecuencias armónicas (s), voice-frequency telegraph equipment [VFTE]; estación terrena de aeronave (s), aircraft earth station [AES]

ETAB Executive Technical Advisory Board

ETACS extended total access communication system (IBM)

ETAD eventos telefónicos de

interfuncionamiento hacia adelante (s), forward interworking telephone events [FITE]

e-tail electronic retailing

ETAN indication d'état ('alignement normal') (f), status indication 'N' ('normal alignment') [SIN]

ETANN electrically-trainable analogue neural network (Intel chip)

ETAP indication d'état ('hors alignement') (f), status indication 'out of alignment' [SIO]

ETAR événement téléphonique d'interfonctionnement transmis vers l'arrière (f), backward interworking telephone events [BITE]

ETAS emergency technical assistance

ETAT evento telefónico de interfuncionamiento hacia atrás [BITE]

ETAU indication d'état 'alignment urgent' [SIE]

ETAV événement téléphonique d'interfonctionnement transmis vers l'avant (f), forward interworking telephone events [FITE]

ETB end of text block; end of transmission block; estación terrena de barco (s), ship earth station [SES]; exchange terminal board

ETBDB ETB de datos a baja velocidad (s)

ETBEIA1 Electronics, Telecommunications and Business Equipment Industries Association (UK)

ETBPT ETB de petición (s)

ETBRP ETB de respuesta (s)

ETBT ETB télex (f)

ETC domaine d'état du canal sémaphore (f), status field [SF]; enhanced throughput cellular (AT&T); estación terrena costera (s), coast earth station [CES]; exchange terminal chain/circuit; exempt telecommunications company; extendible compiler

ETCA ETC de asignación (s)

ETCD équipement de terminaison du circuit de données (f), equipo de

terminación del circuito de datos (s), data circuit-terminating equipment [DCE]; équipement de terminaison de circuit de données (f), data terminal-circuit equipment

ETCS European Train Control System *tn*

ETCT ETC télex (s), coast earth station telex

ETD equipo terminal de datos (s), data terminal equipment [DTE]

E-TDMA extended time-division multiple access

ETEM états d'encombrement multiples (f), multiple-congestion states [MCS]

ETF electronic toll fraud

ETH équipement télégraphique à fréquences harmoniques (f), voice-frequency telegraph equipment [VFTE]

Ethernet LAN communications standard (defined in IEEE 802.3)

ETHS indication d'état 'hors service' (f), status indication 'out of service' [SIOS]

ETIC événements téléphonique transféré á l'interface entre la signalisation et la commutation (f), evento telefónico en el interfaz de conmutación (s), switching processing interface telephone events [SPITE]

ETIP indication d'état 'isolement du processeur' (f), status indication 'processor outage' [SIPO]

ETL effective testing loss; European Testing Laboratory; équipement de termination de ligne (f), equipo de terminación de línea (s), line terminating equipment [LTE]

ETM entorno del tratamiento de mensajes (s), message handling equipment [MHE]; eventos de transferencia de mensajes (s), message transfer event [MTE]

ETN electronic tandem network

ETNA Ericsson Transport Network Architecture *tn*

ETNO European Public Telecommunications Network Operators' Association

ETOC indication d'état 'O' ('occupé')
(f), status indication 'B' ('busy') [SIB]
ETOM electron-trapping optical memory
ETQR External Total Quality and
Reliability
ETR early token release; easy to reach;
estación de trabajo (s), work station
[WS]; ETSI technical report
ETS econometric time series; electronic
tandem switching; electronic translation
system; Engineering Test Satellite
(Japan); European Telecommunications
Standard
ETSACI electronic tandem switching
administration channel interface
ETSI European Telecommunication
Standards Institute, Institut européen des
normes de télécommunications (f),
Europäisches Institut für
Telekommunikationsnormen (g)
ETSI ATM ETSI Advanced testing
method (committee)
ETSSP electronic tandem switching
status panel
ETT estación terrena de tierra para los
servicios aeronáuticos (s), estación
terrena aeronáutica (situada en tierra) (s),
(aeronautical) ground earth station [GES]
ETTD équipement terminal (de
traitement) de données (f), data terminal
equipment [DTE]
ETV Educational Television Station (US)
ETX end of text
ETX/ACK end of text/acknowledgement
ETXX ETD o ETCD (s), data terminal
equipment-data circuit-terminating
equipment [DTE/DCE]
EU end user; execution unit
EUC extended UNIX code (IBM)
EUI end-user interface
EULA end-user licensing agreement
EULisp European dialect of LISP
EUM extended unsuccessful backward
set-up information message indication
EUMETSAT European Meteorological
Satellite Organisation
Eunet European UNIX Network

EUPOT end-user point of termination
EUREKA European Research
Coordination Agency
EURESCOM European Institute for
Research and Strategic Studies in
Telecommunication
euro-connector *v* SCART
EUROCONTROL European
Organization for the Safety of Air
Navigation
Eurokom teleconferencing system for
ESPRIT
Euronet European network to access
scientific and economic information
Eurovision programme exchange within
EBU
EUSR end user
EUT equipment under test
EUTELSAT European
Telecommunications Satellite
Organization, Organisation Européenne
de Télécommunications par Satellite (f)
EUUG European Unix User Group
EUV extreme ultraviolet
eV electron-volt
EVA extremos virtuales analógicos (s),
virtual analogue switching points
[VASP]
EVAX Extended VAX (original DEC
name for ALPHA multiprocessor)
EVC equilibrium vapour concentration;
essai de validation d'un circuit (f), circuit
validation test [CVT]
EVE extensible VAX editor
EVF electronic view-finder
EVGA expanded video-graphics array
EVL event log subsystem
EVM elektrónno-vychislítel'naia mashína
(Russian), electronic computing
machine; event monitor
EVO Ericsson Virtual Office *tn*
EVPNUA European Virtual Private
Network Users' Association
EVSt Endvermittlungsstelle (g), terminal
switching centre
EVUA European Virtual User
Association

EVX electronic voice exchange

EW electronic warfare

EWACS Ericsson Wide Area Coverage *tn*

EWAN emulator without a name

Ewk Einheitswecker (g), standard bell

EWMA exponentially-weighted moving average

EWOS European Workshop for Open Systems

EWS employee-written software (IBM)

EWSD Elektronisches Wählsystem digital (g), electronic digital dialling system

EX données exprés (f), expedited data [ED]

EXALI external alarm interface

EXAPT extended APT

ExCA exchangeable card architecture (Intel)

Excel Microsoft spreadsheet application *tn*

EXM enterprise messaging exchange (Lotus)

EXNOR exclusive NOR gate

EXOR exclusive OR gate

EXR indication d'extension pour les messages de réponse [EAM]

EXRANG external alarm ranging field

EXT indication d'extension pour les messages d'échec de l'établissement émis vers l'arrière [EUM]

Extranet intranet allowing access to/partial access from Internet

EXUG European X User Group (X window system)

EZ excessive zeros

ezd easy drawing (graphics server)

EZTG elektronischer Zeichentaktgeber (g), electronic clock pulse generator

F

f farad; femto 10^{-15}

F facultativo (s), optional [O]; flag, fanion (f); final bit, bit final (f); físico (s), physical; frame synchronization; (auxiliary) framing bit

F/I final inspect

F/m farad per meter

F/S fetch and send

F+L functions plus logic

f2c Fortran (77)-to-C-translator

F2F face to face

FA fax adaptor; fin de actividad (s), activity end [AE]; frame alignment, alignement de trame (f), alineación de trama; frequency-agile (modem); full allocation; functional architecture/area; fuse alarm

FA&T final assembly and test

FAA facile à atteindre (f), easy to reach [ETR]; facility-accepted message, mensaje facilidad aceptada (s)

FAB fast atom bombardment

FAC facility; file access code; final assembly code; functional array calculator

FACCH fast associated control channel

FACCH/F fast associated control channel/full rate

FACCH/H fast associated control channel/half rate

FACD foreign area customer code

face time speaking face-to-face [F2F] rather than electronically

FACS facility assignment (and) control system

FACT Federation Against Copyright Theft; fully-automated compiling technique

FACTR Fujitsu Access and Transport System (Fujitsu *tn*)

FAD facility de-activated, facilidad desactivada (s); facilidad de ampliación de dirección (s), address extension

facility [AEF]; facility access di-group; fraudulent activity detection

FADEC fully authorized digital engine control

FAFIEC Fonds d'Assurances Formation des Entreprises d'Ingénierie, d'Informatique, d'Etudes et de Conseil

FAI facility information; field-aligned irregularities; fournisseur d'accès Internet (f), Internet access provider [IAP]

FAM fast access memory; final address message; forward-address message

FAME Fujitsu ASIC management environment *tn*

FAMOS floating-gate avalanche-injection metal-oxide semiconductor

FANS future air navigation system

FANU fan unit

FAP fault analysis process; file access protocol; format and protocol; message à faire-passer (f), pass-along message [PAM]

FAQ frequently-asked questions, foire aux questions; función de adaptador a interfaz (s), Q-adaptor function [QAF]

FAQL FAQ list

FAR facility request message; false-alarm rate; Federal Acquisition Regulation (US); forward request

FARM-IT Farmers and IT Conference (UK)

FARNET Federation of American Research Networks (US)

FAS file access subsystem; frame alignment signal, signal de verrouillage de trame (f); frame alignment sequence; frequency allocation support

FASE fundamentally-analysable simplified English

FASIC function and algorithm-specific integrated circuit

FASL Fujitsu AMD Semiconductor Ltd (Japan)

FAST Federation Against Software Theft (UK); first application system test;

FORTRAN automatic symbol translator; high-speed TTL logic circuit

fastext development of teletext

FAT file allocation table (Microsoft Windows); foreign area toll

FAW frame alignment word

FAX G4 group 4 facsimile service

FAX G3 group 3 facsimile type

fax facsimile; Fernkopie (g)

FB framing bit

FBD functional block diagram

FBE framing bit error

FBGA flat BGA

FBN form Backus Naur [BNF]

FBRAM frame buffer RAM

FC factor of cooperation, facteur de coopération (f) [FOC]; fault condition; feedback control; find called party; flip chip; format coding; frame control (field); function change; functional components; fuse chamber

FC-AL fibre-channel arbitrated loop

FCAM frequency/code allocation management

FCB fan control board; file control block/box; frequency correction burst

FCC Federal Communications Commission (US); Federal Cracking Consortium; file carbon copy; flow control confirmation; forward control channel; fraud control centre

FCCH frequency correction channel

FCCSET Federal Coordinating Council for Science Engineering and Technology (US)

FCD facsimile coded data; (bloc de) fonction de communication de données (f), (bloque de) funciones de comunicaciones de datos (s), data communication function [DCF] block

FC-EL fibre channel-enhanced loop

FCF facsimile control field

FCI flux changes per inch; funciones de capa inferior (s), low-layer function [LLF]

FCIA función de control de intrusión adaptativa (s), adaptive break-in control

function [ABCF]; funciones de capa inferior adicionales (s), additional low-layer functions [ALLF]

FCIB funciones de capa inferior básicas (s), basic low-layer functions [BLLF]

FCIF full common intermediate format

FCM facilities cost model; facteur de crête multicanal (f), multi-channel peak factor [MPF]; fault control module; flow control message; functional configuration management; signalling traffic flow control messages (f)

FCO field change order; señal de fallo de continuidad (s), continuity failure signal [CCF]

FCOMB filter combiner

FCP flat concurrent Prolog

FCR fallo de una conexión de red (s), network connection failure [NCF]; FIFO control register

FCS factory control system; fast circuit switching; Federation of Communications Services (UK); fibre channel standard; file control system; first customer release; frame check sequence/sum; funciones de capa superior (s), high-layer function [HLF]; séquence de contrôle de trame (f), frame control sequence

FCSA funciones de capa superior adicionales (s), additional high-layer functions [AHLF]

FCSB funciones de capa superior básicas (s), basic high-layer functions [BHLF]

FC-SCSI fibre channel to SCSI

FCT fixed cellular terminal

FCVC flow-controlled virtual circuit

fd leak file-descriptor leak

FD fibre duct; fin de introducción (s), end of introduction [EOI]; finished dialling; floppy disk; frequency distance; full duplex; functional description

FD n do-not-overflow filtering to the order n

FD/PSK frequency-differential/PSK

FDA fácil de alcanzar (s), easy to reach [ETR]; field despatcher application

FDB functional description block

FDBF fin de bloc facsimilé (s), end of facsimile block [EOFB]

FDC factor de cooperación (s), factor of cooperation [FOC]; fault detection and classification; final-de-contenido (s), end of contents [EOC]; floppy disk controller; frame delete compression

FDCC fixed direct control channel

FDD floppy disk drive; frequency division duplex

FDDI fibre digital device interface; fibre-distributed data interface

FDE frequency domain experiments

FDFS file descriptor file system

FDGW Ferndurchgangsgruppenwähler (g), national transit group selector

FDHP full-duplex handshaking protocol

FDI feeder-distribution interface

FDISK fixed disk; format (hard) disk utility (Microsoft Windows)

F-display type of radar display (used in aiming)

FDIV flaw in the division of rare number pairs (Intel Pentium bug)

FDL facility data link; fin de línea (s), end of line [EOL]

FDM (señal de) fin de mensaje (s), end of message [EOM]; frequency division multiplex/multiplexing

FDMA frequency-division multiple access

FDMH fixed daily measurement hour

FDMP fixed daily measurement period

FDN fixed dialling number

FDP fibre distributed/distribution panel; fin de procedimiento (s), end of procedure [EOP]; fixed-disk parameter table; flight data processing, traitement automatique des données de vol (f)

fdp fonction densité de probabilité (f), probability density function [PDF]

FDPE señal de fin de petición de estado (s), end of status request signal [EOSR]

FDPS flight data processing system, système de traitement automatique des données de vol (f)

FDR fin de retransmisión (s), end of retransmission [EOR]; frequency dependent rejection

FDS (carácter de) fin de selección (s), end of selection [EOS]

FDSE full-duplex switched Ethernet

FDT formal description technique

FDX full duplex

Fe iron

FE fetch-execute (instruction) cycle; forward error-correction; framing error; functional entity; functional element

FEA finite element analysis; functional entity action

FEAD Fernsprech-Auftragsdienst (s), operator services for customers

FEADS FEAD-Satz (g), FEAD unit

FeAp Fernsprechapparat (g), telephone set

FEBE far-end block error

FEC fabrication evaluation chip; fetch-execute (instruction) cycle; forward error-correction, corrección de errores sin canal de retorno (s); forward error compilation/control; filtro de espejo en cuadratura (s), quadrature mirror filters [QMF]

FECN forward explicit congestion notification

FED far end data; Fernsprech-Auftragsdienst (g), operator services for customers; field-emission display

FEDS fixed and exchangeable disk storage

FEE far-end block error; far-end error

Feel free and eventually Eulisp

Feevee pay cable TV in US

FEFO first ended, first out

FeH Fernsprechhäuschen (g), telephone kiosk

FeHAs Fernsprech-Hauptanschluß (g), telephone exchange line

FeHb Fernsprechhaube (g), telephone hood

FEI Federation of the Electronic Industry (UK)

FeK Fernsprechkarte (g), telephone PCB

FEL function equation language

FEM finite element model; función de entidad de mantenimiento (s), maintenance entity function [MEF]

FEMF foreign electromotive force

FeN Fernsprechnetz (g), telephone network

FeNStAnl Fernsprechnebenstellenanlage (g), private branch exchange [PBX]

FEO fracción de exclusión por ocupación (s), freeze-out fraction

FEOL front-end of line

FEP Fernsprechapparat (g), telephone set; front-end processor (IBM)

FEPI front-end programming interface

FEPROM flash erasable PROM

FEPS fast Ethernet parallel port SCSI (Sun Microsystems *tn*)

FER frame erasure/error rate; funciones de elemento de red (s), network element function [NEF]

FeRAM ferro-electric RAM

FERF far-end receive failure

FES función de entidad de soporte (s), support entity function [SEF]

FESDK Far East Software Development Kit (Microsoft)

FESEM field-emission scanning electron microscope/microscopy

FET field-effect transistor

FETR funciones de estación de trabajo (s), workstation functions [WSF]

FEV far-end voice

FEXT far-end crosstalk, télédiaphonie (f), telediafonía (s)

FF flicker-free (monitor display); flip-flop; form feed

FFB fan filter board

ffccc floppy Fortran coding convention checker

FFDC first failure data capture (IBM)

FFDI fast-fibre data interface

FFIF file format for internet fax [RFC 2301]

FFP formal functional programming

FFR freeze frame request

FFS fast file system; for further study

FFSK fast frequency shift keying

FFST first failure support technology (IBM)

FFT fast Fourier transform (the 'Butterfly')

FFTDCA final-form-text data communications architecture

FG finished goods; floating gate; frame ground; función global (s), global function [GF]; Functional Group (ATM Forum)

FGA funciones globales adicionales (s), additional global functions [AGF]

FGB funciones globales básicas (s), basic global function [BGF]

FGDC Federal Geographic Data Committee (US)

FGL first generation language; flow-graph Lisp; function graph language

FGNr Fernmeldegebühren-Nummer (g), telecommunication charge number

FGRAAL Fortran extended graph algorithmic language

FGREP fixed global regular expression print (Unix)

FGW Ferngruppenwähler (g), trunk group selector

FH faisceau hertzien (f), microwave link; frame handler; frequency hopping

FHD fixed-head disk

FHS frequency-hopping set

FHSS frequency-hopping spread-spectrum

FHT fast Hartley transform

FI comunicación entre capa enlace de datos y capa física (s), packet handler/handling [PH]; format identifier; fréquence intermédiaire (f), intermediate frequency [IF]; functional interface

FIA Fibre-optic Industry Association; signal d'accusé de réception de désinhibition de faisceau (f), link uninhibit acknowledgement signal [LUA]

FIB focussed ion beam; forward identity/indicator bit

FIBR FIB received

FIBT FIB transmitted

FIBX FIB expected

FIC signal de désinhibition forcée de faisceau (f), link forced uninhibit signal [LFU]

FICS file and information converting system

FID flame ionization detector; formato de intercambio de documento (s), document interchange format [DIF]

FIDA formato de intercambio de documento abierto (s), open document interchange format [ODIF]

FIDDI fibre-distributed data interchange

FIDO finite domains; spontaneous name for bulletin board (1983)

FidoNet cooperative network between PCs in US (*also* Fidonet)

field mouse mouse without cable connector

Fieldbus data transmission between robots (EUREKA programme)

FIF facsimile information field; fonctions d'interfonctionnement (f), función de interfuncionamiento (s), interworking function [IWF]; fractal image file (format)

FIFO first-in first-out

FIFO list a queue

FIFTP función de interfuncionamiento télex paquete (s)

FII failure indication information

FILS Federal Information Locator System (US)

FILT filter unit

FIM feature interaction manager

FIMAS financial institution message authentication standard (ANSI)

FIMS form interface management system

FIN facility information; font identification number; signal de fin (f), clear-forward signal [CLF]; señal de fin (desconexión) (s), clear-forward signal [CLF]

FIN de TSDU fin de marque UDST (f), end of transaction transport service data unit [EOT]

FIND Financial Information Net Directory

Finder GUI on Apple Macintosh *tn*

FIO signal de désinhibition de faisceau (f), link uninhibit signal [LUN]

FIOC frame input-output controller

FIOL file transfer on-line

FIOLOG file optimization log (Microsoft Windows 98)

FIP facility interface processor; factory information protocol; file processor buffering; International Federation for Information Processing

FIPR Foundation for Information Policy Research (UK)

FIPS Federal Information Processing Standards (maintained by NIST)

FIR fast infrared (port); finite impulse response

Firefly software company working with W3C on P3P

FiRM Five-O is Rigor Mortis

Firmware software or data hardwired/embedded onto ROM, PROM etc

firmy old usage for 3.5 inch 'floppy' disk (*aka* stiffy)

FIRP Federal Interworking Requirements Panel (US)

FIRS Fax Information Retrieval System (BT)

FIRST Forum of Incident Response and Security Teams

FIS functional interface specification

FISH first-in still-here

fish-tank cabinet for microfiche files

FISU fill-in signal unit

FIT failure unit; failure in time; fingerprint identification technology (Compaq *tn*); formato de intercambio de texto (s), text interchange format [TIF]

FITCCD frame interline transfer CCD

FITE forward interworking telephony event

FITL fibre-in-the-loop

FITLM facilidad de interfuncionamiento

telemático (s), telematic interworking function [TIF]

FITNR fixed in the next release (optimistic comment used ironically)

FITS flexible image transport system

FIU fan interface unit; fingerprint identification unit (Sony)

Five-O cracking group – formerly Imperial Warlords

FIX Federal Information Exchange (US); Federal Internet Exchange (US)

FKTO Fernmelde-Konto (g), telecommunication account

FL fin de libération (f), release complete [RLC]; fuzzy logic; function level (language)

FLAG Fibre-optic Link Around the Globe (17,500 miles: UK-SE Asia)

flamage combination of *flaming* and *verbiage*

flame mail abusive e-mail

flame to criticise/abuse a person severely, publicly on BBS

FLC ferro-electric liquid crystal; function level code; fuzzy logic control

FLEFO Foreign Language Forum (CompuServe)

FLEX faster LEX

FLIC functional language intermediate code

FLIH first-level interrupt handler

FLINK flash-wink signal

FLIR forward-looking infrared (sensor)

FLL FoxPro link library (Microsoft/FoxPro)

FLOODS Florida Object-Oriented Device Simulator

FLOOPS Florida Object-Oriented Process Simulator

FLOPC floating-point operations per cycle

flops floating-point operations per second

floptical combination of floppy and optical (disk)

FLOTOX floating-gate tunnel oxide

FLPL FORTRAN list processing language

FLR frame loss ratio

FLRT factory layout/re-layout tool

FLS first-line support; free-line signal

flush to halt an operation; to delete something

flyspeck 3 unreadably-small font

FM fault management/monitor; foreign material; frequency modulation, Frequenzmodulation (g); funciones de mediación (s), mediation functions [MF]; function management; functional model

FM1 frame mode 1

FMA fault management application

FMAC facility maintenance and control

FMAP field manufacturing automated process

FMAS facility management application area in TMOS

FMAU fault management application unit

FMB fixed multi-beam

FMC fixed message cycle fixed-mobile convergence

FMDU FM-Datenumsetzer (g), FM data modems

FMEA failure modes and effects analysis, análisis de modos de avería y de sus efectos (s)

FMECA fault modes effects and criticality analysis, análisis de los modos de avería, sus efectos y su criticidad (s)

FMFB frequency-modulation feedback

FMH function management header

FMU fin de marca de UDST (s) [TSDU end mark]

FMI fixed-mobile integration; Fujitsu Microelectronics Inc

FMMC factory material movement component

FMPL Frobozz Magic Programming Language

FMR frequency modulation receiver

FMS facility management system; file management subsystem; flexible manufacturing system (computer-controlled machines); forms management system

FMT format; frequency-modulation transmitter

FMTP file management transaction processor (Bank of America)

FMV full-motion video

FMVFT frequency-modulated voice-frequency telegraph

FN finish; frame number

FNAL Fermi National Accelerator Laboratory (US)

FNC Federal Networking Council (US)

FND find

FNI first network implementation

FNN facilidades no normalizadas (s), non-standard facilities [NSF]; feed-forward neural network

FNP front-end network processor; front-end processor

FNPA foreign numbering plan area

FNR flexible number/numbering register

FNRC Financial Networks Readiness Consortium

FNT font

FO Fernsprechordnung (g), official telephone regulations; flash override; formatting and output

FOA fibre-optic adaptor; first office application

FOAD fuck off and die, va te faire foutre et crève (f)

FOAF friend of a friend

FOBus fibre-optic bus

FOC factor of cooperation; fibre-optic communications

FOCAL formal calculator

FOCS fibre-optic chemical sensors; Focus on Computing in US (joint committee of ACM/IEEE); Forum of Control Data Users (US); forward control channel

FOD finger of death

FODgTVSt Fern- und Ortsdurchgangs- und Teilnehmervermittlungsstelle (g), trunk long distance-local transit and subscriber switching centre

FODgVSt Fern- und Ortsdurchgangs-Vermittlungsstelle (g), trunk and local transit switching centre

FOE festering operator error

FOF freeze-out fraction

FOIL file-oriented interpretive language

FOIRL fibre-optic inter-repeater link

FOLDOC Free On-line Dictionary of Computing

FON fibre-optic network

FOOL Fool's Lisp

FOOS force out of service

FORMAC formula manipulation compiler

FORMAT-FORTRAN FORTRAN matrix abstraction technique

FORML formal object role-modelling language,

FORTH Foundation for Research and Technology in Heraklion; programming language using 'reverse Polish notation'

FORTRAN formula translation (programming language)

FORTRANSIT FORTRAN internal translator

FORTWIHR Bavarian Consortium for High-Performance Scientific Computing

FORWISS Bayerische Forschungszentrum für Wissensbasierte Systeme (g)

FOSDIC film optical-scanning device for input into computers

FOSE Federal Office Systems Exposition (US)

FOSI formatting output specification instance

FOSIL Fredette's OS interface language

FOSSIL Fido/Opus/Seadog standard interface level

FOT fibre-optic terminal/transceiver; forward transfer message/signal, signal d'intervention (f); optimum traffic frequency

FOTS fibre-optic transmission system

four Russians an algorithm

fourcc four-character code

FOV field of view

FOX field oxide; field operational X.500 project

FP flash point; floating point; functional processor; functional programming

FP2 functional parallel programming

FPA floating point accelerator/adaptor; function point analysis

FPB function progammable type B, Funktionsprogrammierbarer Type B (g)

FPC floating point calculation; functional progression chart

FPCP Funky Pack of CyberPunks

FPD flat panel display; floating point divide (Intel Pentium bug); focal-plane deviation; full-page display

FPF facility parameter field

FPGA field-programmable gate array

FPH freephone

FPIS forward propagation ionospheric scatter

FPLA field-programmable logic array

FPLF field-programmable logic family

FPLMTS future public land mobile telecommunications system

FPLS field-programmable logic switch

FPM flat page mode; flexual plate mode

FPMDRAM fast page mode DRAM

FPMH failures per million hours

FPMRam fast page mode RAM

FPMS Factory Performance Modeling Software

FP n partial filtering to the order n

FPNW file and print service for NetWare

FPP fixed path protocol; floating-point processor

FPR flat plate radiometer; floating-point register

FPRD formato de publicación de referencia decimal (s) [DRPF]

FPRH formato de publicación de referencia hexadecimal (s) [HRPF]

FPROM field-programmable ROM; fusible-link programmable ROM

FPS framing-pattern sequence

FPT forced perfect/point termination

FPTS forward propagation tropospheric scatter

FPU floating-point unit

FPY failures per year

FQA fixed quality area

FQDN fully-qualified domain name

FQPCID fully-qualified procedure correlation identifier (*v* APPN)

FR frame relay (interface between LAN and WAN); frame reject; frame reset; force release, full rate

FRA frequency allocation/re-allocation

FRAD frame relay access/asynchronous device; frame relay assembler/dis-assembler

FRAM ferric oxide RAM (*ie* stored on Fe$_2$O$_3$ coated magnetic film)

FranzLisp dialect of LISP

FRB (International) Frequency Registration Board

FRBS frame relay bearer service

FRBSI frame relay bearer service interworking

FRC fault reporting centre; final routing centre; funciones relacionadas con la conexión (s), connection-related functions [CRF]; functional redundancy checking

FRED fallo de recepción en el extremo distante (s), far-end receive failure [FERF]; frame editor

Free BSD version of Unix OS (BSD)

FREL frame relay

freq-mult frequency multiplier

FRF Frame Relay Forum

FRFTC Frame Relay Forum Technical Committee

FRICC Federal Research Internet Coordinating Committee (US)

friode fried diode (*ie* a fused/blown diode)

FRJ facility reject message

FRL frame representation language

FRM fault-reporting module; functional reference model

FRMB fast-ramp mini batch

FRMBE framing bit errors

FRMR frame reject, rechazo de trama (s)

FRP fault-report point; field repetition; fragmentation protocol

FRPI flux reversals per inch

FRS forward ready signal; frame relay service/switch; freely redistributable software; fundamental reference system

FRSSCS frame relay service-specific convergence sublayer

FRT front

FRTT fixed round-trip time

FRU field-replaceable unit

FRW Fernrichtungswähler (g), trunk route-selector

FRXD fully automatic reperforator-transmitter distributor

fry (to) fail

fs femto second, 10^{-15}

Fs signal/signalling framing (bits)

FS fast store; figure shift; file separator (ASCII 28); file services; final splice (of cable); frame status; frame sync; frequency shift; functional specification; further study

FSA finite-state automaton (*ie* a machine)

FSB functional specification block

FSC field support centre; frequency of sub-carrier

FSCA fuente de señales cuasialeatorias (s), quasi-random signal source [QRSS]

fsck file system check (Unix)

FSD file system driver (OS/2)

FSE fonction de système d'exploitation (f), operating system function [OSF]

FSF fading safety factor; Free Software Foundation (US)

FSK frequency-shift keying

FSL flexible service logic; formal semantics language

FSM FDDI switching module; finite state machine; forward set-up message; Freiwillige Selbstkontrolle Multimedia-Diensteanbieter (g), Voluntary Self-regulation Multimedia Service Providers' Association

FSML financial services mark-up language

FSN forward sequence number; full service network

FSNC FSN of last MSU accepted (by remote level 2)

FSNF FSN of the oldest MSU in the RTB

FSNL FSN of the last MSU in the RTB

FSNR FSN received

FSNT FSN of the last MSU transmitted

FSNX FSN expected

FSO funciones de sistema de operaciones (s), operating system function [OSF]

FSP fault servicing process; file service protocol; flaky stream protocol; flexible service profile; frequency shift pulsing

FspWNBw Fernsprechwählnetz der Bundeswehr (g), German Armed Forces telephone network

FSS Fixed Satellite Service; formulario de solicitud de servicio (s), service order form [SOF]

FST field start code; flat square tube/flatter squarer tube (for CRT monitor); formato de supertrama (s), superframe format [SF]

FSTA formato de supertrama ampliado (s), extended superframe format [ESF]

FSU final signal unit

FSvAnl Fernsprech-Stromverteilungsanlage (g), current distribution for telephony

FSWI frame switching

Ft terminal framing (bits)

FT fault tolerant; final test; forward transfer; France Télécom (f); functional test; total filtering (f)

FT1 fractional T1 (*aka* channelized T1)

FT CCD frame transfer CCD

FTA fault tree analysis, análisis en árbol de averías (s); field to advise; final/full type approval

FTAB Focus Technical Advisory Board

FTAM file transfer access and management (ISO protocol)

FTC Federal Trade Commission (US)

FTD fichier de transfert direct (f), fichero

de transferencia directa (s), direct transfer file [DTF]; frame transfer delay

FTFL code fixed-to-fixed-length code (*aka* block code)

FTG final trunk group

FTI France Télécom Interactive (f)

FTIR Fourier transform infrared

FTM file transfer manager; flat tension mask (Zenith)

ftn forwarded-to number

FTP file transfer protocol (RFC 859); filtre de groupe secondaire (f), filtro de grupo secundario (s), through-supergroup filter [TSF]; foiled twisted pair

FTPI flux transitions per inch

FTR functional throughput rate

FTS Federal Telecommunication System (US); fixed telephone service with radio accessory; fixed telephone subscriber; full-text search

FTT failure to train

FTTB fibre to the building

FTTC fibre to the curb

FTTH fibre to the home

FTTK fibre to the kerb

FTTM file transfer and access method

FTU fixed test unit

FTVL fixed-to-variable (length) code

FTVSt Fern- und Teilnehmervermittlungsstelle (g), trunk and subscriber switching centre

FTX fault tolerant UNIX

FTZ Fernmeldetechnisches Zentralamt (Darmstadt) (g), Central Telecommunications Authority

FUBAR failed UniBus address register (in a VAX); fouled up beyond all recognition/repair

FUD fear - uncertainty - doubt

FuFSt Funkfestation (g), fixed radio station

FUI file update information

FuKo Funkkonzentrator (g), radio line concentrator

FUNET Finnish University and Research Network

FUNI frame-based user-to-network interface (ATM Forum)

FUR fast update request

fureteur logiciel de navigation sur le Web (f) (browser)

futz to re-arrange Windows desktop icons

FuVB Funkvermittlungsbereich (g), radio switching centre area

FUZ Funkzelle (g), radio cell

FV channel filtering; frecuencias vocales (s), voice frequencies [VF]; forward voice channel

FVIPS first virtual Internet payment system

FVSt Fernvermittlungstelle (g), radio switching centre

FVT full video translation

FVW forward volume wave

FW firmware; full wave

FWB Fahrenheit wet bulb

FWHM full-width – half-maximum

FWIW for what it's worth

FWS filter wedge spectrometer

FX effects (as in 3D FX)

FXC FerroxCube

FXO foreign exchange office

FXRD fully automatic reperforator-transmitter distributor

FXS foreign exchange station

FYI for your information

FZ float zone

FZA Fernmeldezeugamt (g), telecommunications procurement office

FZG Folo Zone Gang

G

G giga, 10^9, one thousand million; profile parameter; symbol for conductance; symbol for electrical conductivity

G/T antenna gain-to-noise temperature (satellites)

G1 extension-band for GSM 900

G2 system from Gensym Corp

G3 Generation 3 IBM *tn*; third generation Power PC 750 processor (Apple Macintosh)

G3/G4 group 3/4 facsimile (type)

Ga gallium (used in semiconductors and LEDs)

GA general aspects; general availability; generic algorithm

GaAs gallium arsenide (alternative to Si as semiconductor material)

GaAsP gallium arsenide phosphide

GABC attenuation limiting curve

GAC granular activated carbon

GAI general alarm interface

GAIA GUI application inter-operability architecture (OSF)

GAK government access to key

GAL genetic array logic

GAMM Gessellschaft für Angewante Mathematik und Mechanik (g)

GAMS global account management (system); guide to available mathematical software

GaN gallium nitride; generating and analysing networks; global area network (Internet)

GAO Government Accounting Office (US)

GAP generic access protocol/(profile); groupe analyses et prévision (f); group algorithms and programming

GAPI gateway application programming interface

GAPLog general amalgamated programming with logic

GARP Global Atmospheric Research Program

GASAD gate and source and drain

GASP graph algorithm and software package

GAT generalized algebraic translator; group audio terminal

GaVr Gabelverstärker (g), hybrid bridge amplifier

GAWK Gnu AWK

Gb gigabit (one thousand million bits)

GB base antenna gain; gigabyte (one thousand million bytes); grid bias

GbAnz Gebührenanzeiger (g), call charge indicator

GBH group busy hour

GBML genetics-based machine learning

GBN Global Business network

Gbps gigabits per second

GBps gigabytes per second

GBS Gragg-Burlisch-Stoer (extrapolation) method

GBSC group of blocks start code

GBSVC general broadcast signalling virtual channel (B-ISDN)

Gbyte gigabyte

GC gain de concentration (f), interpolation gain [IG]; garbage collection; gas chromatography; global control; group command; group connector; gravimetric calibrator

GCA ground-controlled approach (of aircraft by radar)

GCAC activation des canaux sémaphores (f), signalling link activation [LSLA]; generic connection admission control

GCAL affectation des liaisons sémaphores de données (f), signalling data link allocation [LSDA]

GCAT affectation des terminaux sémaphores (f) [LSTA]

GCC generic cell controller; generic conference control; GNU C compiler; graphic cell configuration; graphic character composition; The Telecommunications Bureau of the

Cooperation Council for the Arab States of the Gulf

GCD gas chromatography distillation; graphical cell display; greatest common divisor

GCDA désactivation des canaux sémaphores (f), signalling link de-activation [LSLD]

G-CDR GGSN call detail record

GCF generation control function

GCIS (UK) Government Centre for Information Systems

GCL generic control language

GCLR sélection des liaisons sémaphores de réserve (f), stand-by data link selection [LSDS]

GCMS gas chromatography mass spectroscopy

G-code language used by G-machine

GCOS general comprehensive operating system (Bull); God's chosen operating system

GCR grey component replacement; group cell register; group code record/recording

GCRA generic cell-rate algorithm

GCRE rétablissement des canaux sémaphores (f), signalling link restoration [LSLR]

GCS gestion des canaux sémaphores (f), signalling link management [SLM]

GCSA supervision de l'activité des canaux sémaphores (f), signalling link activity control [LSAC]

GCSF supervision des faisceaux de canaux sémaphores (f), link set control [LLSC]

GCT Greenwich Civil Time

GCU grupo cerrado de usuario (s), closed user group [CUG]

GCUB grupo cerrado de usuarios bilateral (s), bilateral closed user group [BCUG]

Gd gadolinium (used in electronic components)

GD diversity gain

GD&R grinning ducking and running

GDA global data area

GDB GNU debugger

GDBPSK Gaussian differential binary PSK

GDCI general data communications interface

GDD group delay distortion

GDDM graphics data display manager

GDF group distribution frame

GDFI Groupement de Freelance en Informatique (f)

GDG Generation Data Group (IBM)

GDI graphical device interface

GDLC generic data link control (IBM)

GDM generic device magazine; global data module

GDMO Guidelines for the Definition of Managed Objects (ISO/IEC 10165-4)

GDN Gleichstrom-Datenübertragung Niederpegel (g), low-level DC data transmission

GDNDU GDN-Datenumsetzer (g), GDN data converter

GDPL generalized distributed programming language

GDPS Global Data Processing System (World Meteorological Organization)

GDS Generation Dataset; grado de servicio (s), grade of service [GOS]; graphical design software/system

GDT geometric dimensioning and tolerancing; global descriptor table

GDU graphical display unit; Grave Dancers' Union

Ge germanium (much used in semiconductors)

GE Gebühreneinheit (g), call charge unit; General Electric Corp; generic element; gestionnaire équipement (f), equipment manager

GEA Gigabit-Ethernet Alliance; graph-extended Algol

GEC General Electric Co (UK)

GECOS General Electric comprehensive operating system *tn*

GED comunicación entre entidad de gestión y capa enlace de datos (s), communication between management entity and data link layer [MDL]; gestion électronique de documents (f), electronic document management

GEDAN Gerät zur dezentralisierten Anrufweiterschaltung (g), equipment for decentralized call forwarding

Gedcom genealogical data communications

GEDI Group on Electronic Document Interchange

GEIDE gestion électronique de l'information et des documents existants (f)

GEISCO General Electric Information Service Company

GEM generic equipment model; Graphics Environment Manager (Digital Research Inc *tn*)

GEMM generic equipment module magazine

GEMVS GEM verification system

gen generate/generator

GEN General European Network (CEPT)

GEnie General Electric network for information exchange

GENLOCK generator lock

gensym contraction-generated symbol

GEO geostationary/synchronous (earth) orbit

GEOS geosynchronous (earth) orbit satellite; Graphic Environment OS (Geoworks *tn*)

GEPC gestión de la PCCS (f), SCCP management

GES gestión de enlaces de señalización (s), signalling link management [SLM]; generic equipment simulator; (aeronautical) ground earth station; système de gestion (f), management system [MGMT]

GEST generic expert system tool

GESTE Groupement des editeurs de service en ligne (f), on-line publishers' service group

GETS Government Emergency Telecommunications Service (US)

GF global functions

GFAAS graphite furnace atomic absorption spectrometry/spectroscopy

GFC gas-filter correlation; generic flow control

GFCI ground-fault circuit interrupter

GFE government-furnished equipment

GFI general format identifier (X.25)

Gflops one billion (10^9) floating-point operations per second

GFP global functional plane

GFR Grim File Reaper

GFS gehender Fangsatz (g), switched connection hold circuit outgoing

GFSK Gaussian frequency shift keying

GFU groupe fermé d'usager (f), closed user group [CUG]

GFUB groupe fermé d'usagers bilatéral (f), bilateral closed user group [BCUG]

GFUBAS groupe fermé d'usagers bilatéral avec accès sortant (f), bilateral closed user group outgoing [BCUGO]

GGCA geometric graphics content architecture

GGCL government-to-government communications link

GGG gadolinium gallium arsenide

GGMV group of blocks global motion vector

GGP gateway-gateway protocol

GGSN gateway GPRS support node

GH gain hit

GHC Glasgow Haskell compiler (Glasgow University); guarded horn class

GHOST general hardware-oriented software transfer

GHTh gain hit threshold

GHz gigahertz

GI group identifier, identification du groupe (f); ganancia de interpolación (s), interpolation gain [IG]; group identity; señales de acondicionamiento de línea (s), line conditioning signal [LCS]

gid global index file extension; group identifier

GIF global index files (Microsoft

Windows); graphic interchange format, formato de intercambio de gráficos (s)

Gigaflops one billion (10^9) floating-point operations per second

GIGO garbage in – garbage out, déchets en entrée, déchets en sortie (f)

GII Global Information Infrastructure

GILC Global Internet Liberty Campaign

GILD gas-immersion laser doping

GIM 1 generalized information management 1

Gimp GNU image manipulation program

GIMPS Great Internet Mersenne Prime (number) Search

GINA generic interactive application

GINO graphical input-output

GINO-F graphical input-output (Fortran sub-routines)

GIO generic interface for operations

GIOP general inter-ORB protocol

GIP general interpretive programme; GSM interworking profile

GIPS giga (10^9) instructions per second

GIRL generalized information retrieval language; graph information retrieval language

GIS gas-insulated switchgear; Geographical Information System (geomatique in Canada); global information solutions (AT&T)

GIT groupe d'identification de télégramme (f), grupo de identificación telegrama (s), telegram identification group [TIG]

GITSO garbage in, toxic sludge out

GIWU GSM interworking unit

GIX Global Internet eXchange

GKS graphical kernel system (ISO 7942)

GKS-3D GKS for three dimensions (ISO 8805)

GKS-94 major revision of GKS

GKT grosser Konzentrator (g), long line concentrator

GL graphics language; ground line; group length

GLASS general language for system semantics

GLB greatest lower bound; Microsoft Mail global system file *tn*

GLC gas liquid chromatography

GLE gestionnaire local d'équipement (f)

glibc GNU C library

GLIS Global Land Information System (US Geological Survey)

Glisp generalized LISP

GLM generalized linear model

GLONASS Russian military satellite constellations

GLOS graphics language object system

GM general MIDI; mobile antenna gain

GMAP GCOS macro-assembler program

GMC guide monomode câblé (f)

GMCD ganancia del EMCD (s), digital circuit multiplicated equipment gain [DCMG]

GMD Gesellschaft für Mathematik und Datenverarbeitung mbH (g), Society for Mathematics and Data Processing; GMD-Forschungszentrum Informationstechnik GmbH (g), GMD-National Research Centre for Computer Science

GMDSS global maritime distress and safety system (Royal Navy)

GME generic macro-expander; group-modulating equipment

GMM global multimedia mobility

GMP global mobile professional

GMR giant magneto-resistive (IBM hard drive *tn*)

GMS generic maintenance system; Geostationary Meteorological Satellite (Japan); global management system; global messaging service (Novell)

GMSC gateway mobile services switching centre [MSC]

GMSK Gaussian minimum shift keying (modulation)

GMT Greenwich Mean Time

gn green; grün (g)

GN group number

gnd ground, earth

GNIP graphical network information presentation

GNN Global Network Navigator *tn*
GNS geographical network surveillance; Global Network System; green number service (freephone)
GNSS Global Navigation Satellite System
GNU GNU's not UNIX (FSF)
GNU DC GNU desktop calculator
GNU sed GNU standard stream editor
GNUS GNU news
Go giga-octet [GB]
GOD Givers of Destruction; global on-line directory
GOES Geosynchronous Operational Environmental Satellite
GOI gate-oxide integrity
GOL general operating language
GOOD graph-oriented object database
GOP Grand Old Pirates; group of pictures
GOPHER Internet browser (University of Minnesota; RFC 1436)
GOS grade of service
GOSIP Government Open Systems Interconnection [OSI] Profile (US)
GOSPL graphics-oriented signal processing language
GOSUB go to sub-routine (instruction)
GOTO jump instruction
GOVSt gesteierte OVSt (g), controlled local switching centre [OVSt]
GP general purpose; genetic programming; grupo de parámetro (s), parameter group [PG]; guard period
GPA gas pressure alarm; GSM PLMN Area
GPAO gestion de production assistée par ordinateur (f), computer-assisted management
GPC general-purpose computer
GPD gas plasma display (monitor)
gpDm geopotential decameter
GPF general protection fault (illegal operation in Microsoft Windows)
GPI graphics programming interface
GPIB general-purpose interface bus (HP-IB; IEEE standard)

GPIM generic prediction interference module
GPL general public licence (for GNU); general purpose language (ALGOL variant); general purpose library; Genken Programming Language; graphical procedures language; group processing logic
GPM general purpose macro-generator
GPO group policy objects (Microsoft Windows NT)
GPR general purpose register (IBM); ground-probing radar
GPRS general packet radio service
GPS general problem solver; generic processing system; Global Positioning System (of satellite constellations)
GPSS general-purpose systems simulator/simulation system
GPT GEC-Plessey Telecom (UK)
GPV general public virus (derogatory reference to GPL)
GPX graphics processor accelerator
GR gestión de red (s), network management [NM]; graphic representation; Gruppenrahmen (g), group frame–construction
gr&d grinning running and ducking
GRA circuit group reset acknowledgement message
GRAAL general recursive applicative and algorithmic language
GRADD graphics adaptor device driver (IBM)
GRAF graphic additions to FORTRAN
grafPort graphics environment in Apple Mac systems
graftal geometric form similar to while easier to compute than fractals
GRAIL graphical input language
GRAN generic radio access network
GRAPPLE graph processing language
GRASP graphical representation of algorithms structures and processes
Gray Book NCSC publication (security aspects of subsystems)
gray code binary block code

GRC grupo de mensajes de gestión de red/circuito (s), circuit network management message group [CNM]

GRE generic routing encapsulation (RFC 1701); graphics engine

GRENDL Group for Research in Electronically-Networked Digital Libraries

grep global-search for regular expression and print (Unix command)

GRIAP gateway roaming interrogation application part

GRIB grid in binary (World Meteorological Organization)

GRIN graded-index fibre

GRIND graphical interpretive display

GRINSCH-MQW graded refractive-index separate-confinement heterostructure - multi-quantum well

GRIP graph reduction in parallel

Grl Gruppenvermittlungsleitung (g), line of group switching centre

GRM circuit group supervision message(s)

groff GNU roff *qv*

GRP MOD group modulator

GRP group reference pilot

GRQ general request message

GRR gateway roaming re-routing

GRS gestion des routes sémaphores (f), gestión de rutas de señalización (s), signalling route management [SRM]; circuit group reset message

GRT gestión de los recursos de transmisión (s), message transfer [TRM]

GRTA contrôle de transfert autorisé (f), transfer-allowed protocol [RTAC]

GRTC commande de transfert sous contrôle (f), transfer-allowed control [RTCC]

GRTE gestion de tests d'encombrement de faisceau de routes sémaphores (f), signalling route set congestion test control [RCAT]

GRTF commande des tests de faisceau de routes sémaphores (f), signalling route set test control [RSRT]

GRTI contrôle de transfert interdit (f), transfer-prohibited control [RTPC]

GRTR contrôle de transfert restreint (f), transfer-restricted control [RTRC]

GrVSt Gruppen-vermittlungsstelle (g), group switching centre

GS Games Society; gehender Satz (g), unit outgoing; group separator (ASCII 29); group switch

GSA General Services Administration (US); generic service adaptor; group-switching architecture; GSM system area

GSAT General Telephone and Electronic Satellite Corp

GSC global support centre; Group Switching Centre (UK); GSM speech coder

GSCE gas source control equipment

GSD generic structure diagram; group switching device

GSEP generic session end-point (TINA)

GSGF Government Systems Group

GSI general server interface; Gensym standard interface; Government Secure Intranet (UK)

GSL global service logic; Grenoble System Language

GSLB Groupe spéciale large bande (f) (CEPT group)

gsm generalized sequential machine (mapping)

GSM general forward set-up information message; global shared memory; Global System for Mobile (Telecommunications); graphic size modification; Groupe Spéciale Mobile (f) (CEPT group original designation)

GSM BC GSM bearer capability

GSM mapping generalized sequential machine mapping

GSM MS GSM mobile station

GSM PLMN GSM public land mobile network

gsm SSF GSM service switching function

GSMBE gas source molecular beam epitaxy

GSMC Global System for Mobile Communications

GSN group switching nodes

GSNW gateway service for NetWare (Microsoft)

GSO GNU super-optimizer

GSP graphics system processor

GSPID gain-scheduled proportional integro-differential

GSPL Greenberg's system programming language

GSR gestión de rutas de señalización (s), signalling route management [SRM]

GSS group-sweeping scheduling; group switching subsystem

GSS-API generic security service API (RFC 1508/9)

GSSP general systems security principles

GSTN general switched telephone networks

GT gain de transcodage (f), ganancia de transcodificación (s), transcoding gain [TG]; gestionnaire de transactions (f), transaction capabilities [TC]; give tokens; global title; Guernsey Telecoms

GT/SQL Greystone Technologies SQL *tn*

GTA give tokens acknowledgement; grading terminal assembly; Guam Telephone Authority

GTAC commande des acheminements sémaphores (f), signalling routing control [TSRC]

GTC General Telephone Company (US); give tokens confirm

GTCN commande de retour sur canal sémaphore normal (f), change-back control [TCBC]

GTCS commande de passage sur canal sémaphore de secours (f), change-over control [TCOC]

GTE General Telephone and Electronics Corp; group-translating equipment

GTEI group terminal end-point identifier

GTEM gigahertz transverse electromagnetic cell

GTEP general telephone and electronics practice

GTER signalling route set congestion control (f) [TRCC]

GTFX contrôle de flux de trafic sémaphore (f), signalling traffic flow control [TSFC]

GTI generic text interface; global title indicator

GTK+ Gimp tool kit +

GTL gunning transceiver logic

GTLD generic top-level domain

GTM general teleprocessing monitor (IBM for OSI)

GTN Government Telecommunications Network (UK)

GTO gate turn-off (thyristor); guide to operations

GTP Government Telecommunications Program (US); GPRS tunnelling protocol

GTP-id GTP identity

GTPNet Global Trade Point Network

GTRN commande de retour sous contrôle sur route normale (f), controlled re-routing control [TCRC]

GTRS commande de passage sous contrainte sur route de secours (f), forced re-routing control [TFRC]

GTS gamma transfer service; GEM test system; gestion du trafic sémaphore (f), gestión del tráfico de señalización (s), signalling traffic management [STM]; Global Telecommunications System (World Meteorological Organization); Government Telecommunications System (US)

GTSD supervision de la disponibilité des canaux sémaphores (f), link availability control [TLAC]

GTT global title transmission/translation/transaction

GTU group terminal unit

GU Gabelumschalter (g), hook-switch on handset; generic unit

GUARDSMAN guidelines and rules for data systems management

GUe Gleichstromübertragung (g), DC signal transmission unit

GUG GVNS user group

GUI graphical user interface, interface utilisateur graphique (f)

GUID globally unique identifier (VB5 control)

GUIDE GUI Development Environment (Sun Microsystems *tn*)

Guidgen globally-unique identifier generator

GUIS graphics user interface system

GURPS generic universal role-playing systems

GVB Global Village Bank (cyber barter/business web site)

GVHRR geosynchronous very-high resolution radiometer

GVL graphical view language

GVNS global virtual network service

GVPN Global Virtual Private Network (UK)

GVT global virtual time

GW gateway; Gruppenwähler (g), group selector

GWA gateway administration

GW-Basic Gee Whizz BASIC (early Microsoft version of BASIC)

GWE global write enable

GWIM global warming impact mode

GWM generic window manager

GWT gee-whiz technology

GXC Global Exchange Carrier

gzip GNU zip (compression utility)

H

H henry (unit of inductance); hydrogen; hybrid, hybride (f), hibrido (s)

H&J hyphenation and justification

H/m henry per meter

H/PC hand-held PC

H1 code d'en-tête 1 (f), heading code 1

HA Handapparat (g), handset

HACK hacking and computer krashing

HACMP high availability cluster multi-processing (IBM)

HAD half-amplitude modulation; head disk assembly (mechanical parts of disk drive)

HAGS Herbally Aroused Gynaecological Squad

HAKMEM hack's memo

HAL hard array logic; heuristic algorithm; heuristically-programmed algorithm; name of computer in 2001 (letters *I B M* shifted left); hardware abstraction layer (Microsoft Windows NT *tn*)

HALE Hackers Against Law Enforcement

HAN home area network

HAND have a nice day

HANDO handover

HAP hazardous air pollutant; host access protocol

Harvard M1 Harvard Mark 1

HAs Hauptanschluß (g), direct PSTN subscriber

HAS mise hors service d'un sous-système acceptée (f), subsystem out-of-service grant [SOG]

HAsl Hauptanschlußleitung (g), direct PSTN subscriber line

HASP Houston automatic spooling priority/(program)

HAST highly-accelerated stress testing

HATE Highly Artistic Talented Enterprises

HATS head and torso simulator

HAUTHD home authentication data

HAZCOM Hazard Communication Standard

HAZOP hazard and operability study

HB home-base; horizontal Bridgeman crystal

HBA hardware failure-oriented circuit group blocking acknowledgement message; host bus adaptor

HBD Honeywell brain damage

HBFG Host Behavior Functional Group (ATM Forum)

HBS home base station

HBT hetero-junction bi-polar transistor

HBUR hardware failure-oriented circuit group blocking and unblocking receipt

HBUS hardware failure-oriented circuit group blocking and unblocking sending

HC handover control/request channel; handover criteria; horizontal cross-connect (DEC)

HCA handover criteria adjustment

HCDS high-capacity digital services (*v* HiCap)

HCF halt and catch fire; highest common factor; host command facility (IBM)

HCGS congestión del enlace de señalización (s), signalling link congestion [HMCG]

HCI hot-carrier injection; human-computer [man-machine] interaction/interface

HCLP hierarchical CLP

HCM high-capacity multiplexing; human capital and mobility programme (CEC DG XII)

HCMOS high-density complementary metal-oxide semiconductor

HCMR hora cargada media repetitiva (s), average daily peak hour [ADPH]

HCMTS high-capacity mobile telecommunications system

HCNAN home C-number analysis

HCNBA HLR C-number administration

HCOMB hybrid combiner

HCS hard clad silica (fibre), header

check sequence; heterogeneous computer system, hot-carrier suppressed

HCSDS high-capacity satellite digital service

HCSPR hundred call seconds per hour

HCSS high-capacity storage system

HCTDS high-capacity terrestrial digital service (AT&T)

HCU handover (request) channel unit; home computer user

HCV high-capacity voice

HD hard disk, disque dur (f); half duplex; handover decision; high density (of floppy disk)

HD0 high-definition TV level-0 format (MS/Compaq/Intel specification)

HDA head-disk assembly

HDB high-density binary

HDB2 código bipolar de alta densidad de orden 2 (s), high-density bipolar code 2nd order

HDB3 high-density bipolar code 3rd order (European ISDN protocol)

HDCD high-definition compatible digital

HDCM discriminación de mensajes (s), message discrimination [HMDC]

HDD hard disk drive

HDF hierarchical data format; horizontal distribution frame

HDH HDLC distant host

HDI head-to-disk interference; high-definition interlaced (scanned image)

HDK Hippy Dippy Klub

HDL hardware description language

HDLC high-level data link control, control de alto nivel para enlaces de datos (s) [HDLC]

HDLC-ISO high-level data link control (ISO protocol)

HD-MAC (wide-screen) high-definition multiplexed analogue components (TV)

HDML hand-held device mark-up language

HDND home digital network device (interface specification)

HDP high-density plasma; high-definition progressively (scanned image)

HDPE high-density polyethylene

HDR header

HDRC hypothetical digital reference connection

HDRSS high-data rate storage system

HDSC high-density signal carrier (DEC)

HDSL high data-rate digital subscriber line (2 Mbit/s on copper pairs)

HDSS holographic data storage system

HDT host digital terminal

HDTM half-duplex transmission module (X.25); distribución de mensajes (s), message distribution [HMDT]

HDTP hand-held device transport protocol

HDTV high-definition TV

HDVD high-definition volumetric display

HDVTR high-definition video tape recorder

HDW Hebdrehwähler (g), two-motion selector

HDWDM high-density wavelength-division multiplex

HDX half-duplex

HE header extension

HE$_{11}$mode fundamental hybrid mode (of optical fibre)

HEAnet Higher Education Authority Network (Ireland)

HEAP Home Energy Assistance Program (US)

HEC header error control; Higher Education Communication (UK)

HECI human-interface equipment catalogue item

HECIG human-interface equipment catalogue item group

HEHO high-end hop-off

heisenbug random bug that disappears when examined

HEL header extension length

Helen Keller mode a system which neither accepts nor outputs data

Helios French photographic satellite

Helix hardware description language

HEM hybrid electromagnetic wave

HEMP high-altitude electromagnetic pulse

HEMT high-electron mobility transistor

HENM encaminamiento de mensajes (s), message routing [HMRT]

HENSA Higher Education National Software Archive (UK)

HEOS highly-elliptical/(-eccentric) orbit satellite

HEP high-energy (particle) physics

HEPA high-efficiency particulate air

HEPDB HEP database (management system)

HEPix HEP-compatible Unix

HEPNET high-energy physics network

HERF hazards of electromagnetic radiation to fuel

HERP hazards of electromagnetic radiation to personnel

hex hexadecimal

HF hands-free; high frequency, Hochfrequenz (g); human factors

HFC hybrid-fibre co-axial

HFMD hora fija de medidas diarias (s), fixed daily measurement hour [FDMH]

HFS hierarchical file system (Apple); hopping-frequency set

HFT hands-free telephone; high-function terminal (IBM)

HGA Hercules graphics adaptor *tn*

HGB hardware failure-oriented group blocking message

HGC Hercules graphics card *tn*

HGCP Hercules graphics card plus *tn*

HGS high-capacity group switch

HGU hardware failure-oriented group unblocking message

HH hanging handset

HHC hand-held computer, ordenador portátil (s)

HHCP host-host copy

HHOJ ha, ha, only joking

HHOK ha, ha, only kidding

HHOS ha, ha, only serious

HI handover initiation

HIBS heavy ion back-scattering spectrometry

HIC hybrid integrated circuit

HiCap high-capacity digital services

HICL High-Integrity Computing Laboratory (IBM)

HID handover initiation and decision; human-interface device

HIDF horizontal side of intermediate distribution frame

HIF hyper-G interchange format

HIFD high-density floppy disk

High Sierra early CD-ROM standard specification

highway obsolete UK term for *bus* (*nb* not the IT super-highway)

HIGZ high level interface to graphics and zebra

HIL human-interface link (Hewlett Packard)

HILI higher layers and interworking (IEEE 802); higher-level interface

HiLog higher-order logic

HIMEM high memory (memory manager loaded at DOS system boot)

HIN hybrid integrated network

Hiperlan high-performance radio LAN

HIPERLAN high-performance LAN

HiPOx high-pressure oxygen

HIPPI high-performance parallel interface

hi-res high resolution (graphics)

HIRS high-resolution infrared sounder

HIT Hackers in Touch

HITECH Hartford Integrated Technologies

HITL Human Interface Technology Laboratory (University of Washington)

HITS hyperlink-associated topic search engine (Cornell University)

Hivol (heat-dissipating) silicon carbide and aluminium composite

HKCC Hong Kong Cable Communications

HKT Hong Kong Telecom

HL hearing loss

HLA home location area

HLC high-layer compatibility

HLCMAP home location cancellation mobile application part
HLCO high-low close-open
HLD HLR location data
HLDA HLR subscriber location administration
HLDL high-level data link
HLDLC high-level data link control
HLF higher-layer function; high-level formatting; horizontal laminar flow
Hlg Hauptvermittlungsleitung gehend (g), PSTN switching centre line outgoing
HLHSR hidden-line hidden-surface removal
HLI high-speed LAN interconnect
Hlk Hauptvermittlungsleitung kommend (g), PSTN switching centre line incoming
HLL half-loop loss; high-level language
HLLAPI high-level language application programming interface
HLM heterogeneous LAN manager
HLNA HLR local analysis administration
HLNAN HLR location analysis
HLPI higher layer protocol identifier
HLQ high level qualifier
HLR home location register
HLRG home location register gateway
HLRH home location register home
HLRV home location register visitor
HLS hue - lightness - saturation (colour model)
HLT heterodyne look-through
HLU high-gain line unit
HLUAP home location updating mobile application part
HMA high-memory area (*cf* HIMEM); hub management architecture
HMAC keyed-hashing message authentication
HMAPTC home mobile application part incoming transaction coordinator
HMCG signalling link congestion
HMD head-mounted display
HMDC message discrimination
HMDF horizontal side main distribution frame

HMDT message distribution
HMG hardware message generator
HMI human-machine interface
HML human-machine language (ITU-T)
HMMP hazardous materials management plan; hyper-media management protocol
HMOS high-density MOS; high-performance MOS; high-speed MOS (Intel Corp)
HMP host monitoring protocol; hybrid multiprocessing
HMRT message routing
HMS hand-held mobile station; head-mounted screen (*cf* HMD); casque de vision tridimensionnelle (f) [HMD]; casque vidéo stéréoscopique (f) [HMD]; hipsómetro selectivo (s) [SLM]
HMSC home mobile service switching centre
HMZ Hauptanschluß mit Mehrfachzugriff (g), multiple access PSTN subscriber line
HNF head normal form
HNIL high noise-immunity logic
HNN Hackers' News Network
HNPA home numbering plan area
HØ code d'en-tête O (f), head code Ø; handover
HOB head of bus
hobbit high-order bit
HOC handover control
HOCM handover configuration management
HOF higher-order function
HOFM handover fault management
HOL higher order logic
HOLD call HOLD (supplementary service)
HOLMES Home Office Large Major Enquiry System (UK)
holy war extended flame war
HomePNA Home Phoneline Networking Alliance (US)
HOMER hazardous organic mass emission rate
HON handover number
HONE hands-on network environment

HOOD hierarchical object-oriented design

HOPE early functional language; Hackers on Planet Earth

HOPG highly-oriented pyrolitic graphite

HOPM handover performance management

HOR execution at a specified time

hors connexion off-line [OL]

hors ligne off-line [OL]

HOS-STPL Hospital operating system-structured programming language (US)

HOSTID host identifier

HOT Home Order Television (Germany)

HOTT Hot Off The Tree (Internet magazine)

HO-UDM handover user data management

HOV handover

HOYEW hanging on your every word

HP Hewlett Packard; high pass; high purity

HP VEE Hewlett Packard Visual Engineering Environment *tn*

HP/UX Hewlett Packard's version of UNIX operating system

HPA Hackers-Phreakers Association

HPAD host packet assembler/disassembler

HPAS high-performance application software

HPB character position backward

HPBW half-power bandwidth

HPC hand-held PC; high-performance computing; high probability of completion

HPCC high-performance computing and communications; High Performance Computing Centre (UK)

HPCLD high-capacity programmable logic device

HPCN high-performance computing and networking, calcul et réseaux à haute performance (f), informatique distribuée à haute performance (f), informática y redes de alto rendimiento (s), Hochleistungsrechnentechnik und-Netze

HPD hybrid passive display

HPDJ Hewlett Packard Desk Jet (colour printer) *tn*

HPEM hybrid plasma equipment model

HPF high-pass filter; high-performance FORTRAN; highest priority first

HPFS high-performance filing system (OS/2)

HPG Hewlett Packard graphics

HPGL Hewlett-Packard Graphics Language *tn*

HPI high-pressure isolation

HPIB Hewlett-Packard interface bus (*aka* GPIB) *tn*

HPL high-performance logic

HPLC high-performance liquid chromatography

HPLJ Hewlett-Packard Laser Jet (laser printer) *tn*

HPLMN home public land mobile network

HPM hazardous production materials, high-purity metal

HP-MPE Hewlett Packard Multi-Processing Executive *tn*

HPN high-penetration notification

HPN-SC HPN service centre

HPO high-performance option

HP-PA Hewlett-Packard Precision Architecture *tn*

HP-PCL Hewlett-Packard Printer Control Language *tn*

HPPI high-performance parallel interface

HPR character position relative; high-performance routing

HPS high-performance switch/(systems)

HPTC high-performance technical computing

HPU hand/hand-held portable unit

HP-UX Unix running on Hewlett Packard workstations *tn*

HPV high-pressure vent

HPW high-performance workstation (Sun)

HQ high quality (upgrade of VHS specification)

HR half rate; hypothetical reference

HRA human reliability analysis
HRC hypothetical reference circuit
HRD HLR roamer routing determination
HRD high-resolution diagnostic disk(-ette)
HRDL hypothetical reference digital link
HRDP hypothetical reference digital path
HRDS hypothetical reference digital section
HRDTAP home-routing determination telephony mobile application part
HRFAX high-resolution fax
HRG high-resolution graphics
HRIR high-resolution infrared scanning system
HRIS human resource information system
HRMS human resource management system
HRN handover reference number
HRPF hexadecimal reference publication format
HRS home (location) register subsystem
HRSC high-resolution stereo camera; HLR subscriber service call
HRT high-rate telemetry; Hraviti Radio Televisija
HRTF head-related transfer function
HRU Hackers R Us; HLR roamer data updating
HRW Hauptrichtungswähler (g), main route selector
HRX hypothetical reference connection
HS hard-sectored; high speed
HSB hue – saturation – brightness (colour model)
HSC hierarchical storage controller; high-speed channel; High Speed Contact (HP *tn*)
HSCP high-speed card punch
HSCSD high-speed circuit-switched data
HSCT high-speed compound terminal
HSD HLR subscriber data; honestly significant difference
HSDA HLR subscriber data administration
HSE handover security entity

HSF High Sierra format (ISO 9660 for CD-ROMs); hot-spot finder
HSFS High Sierra file system [HSF]
HSGA HLR subscriber class group administration
HSI human-system interface
HSL high-speed link
HSL-FX hierarchical specification language function extension
HSM handover state model; hierarchical storage management
HSMS high-speed message services
HSN Home Shopping Network; hopping sequence number
HSNA HLR subscriber number data administration
HSNAN HLR subscriber number analysis
HSP high-speed printer
HSPAP home subscriber procedure – application part
HSPSD high-speed packet switched data
HSR horizontal scan rate/(route)
HSRC hypothetical signalling reference connection
HSRP hot-swappable routing protocol
HSSDS high-speed switched digital service
HSSI high-speed serial interface (ANSI)
HST High Speed Technology (US Robotics/3Com *tn*); hyper-script tool
HSUPA HLR subscriber supplementary administration
HSV hue – saturation – value (colour model)
HT high tension; holding time; horizontal; horizontal tabulation
HTH head-to-head; hope this helps
HTL high-threshold logic; hotel call (with time and charges)
HTM half-transponder mode
HTML hypertext mark-up language
HTO high-temperature oxidation
HTR hard-to-reach
HTRAN HLR register translation functions
HTRB high-temperature reverse bias

HTTL high-speed transistor-transistor logic

HTTP hypertext transfer protocol (RFC 2068)

HTTPd HTTP daemon

HTTP-NG HTTP next generation

HTTPS HTTP secure (Netscape)

HTU height of transfer unit

HU hang-up; height unit; high usage; home unit; housing unit

HUA hardware-failure oriented group unblocking acknowledgement message

HUD head-up display

HUGS Haskell User's Gofer System

HULD HLR unreliable location data

HUMBUL Humanities Bulletin Board (UK)

Hungarian naming convention used in programming

HUP handover user profile

HUPN handover user profile – network

HUPN/U handover user profile network/user

HUPU handover user profile – user

HUPW hot ultra-pure water

HUT Hopkins ultra-violet telescope

HUTG high-usage trunk group

HVAC heating, ventilating and air-conditioning

HVLRAD home visitor location register address

HVQ hierarchical vector quantisation

HVSt Hauptvermittlungsstelle (g), main switching centre

HVt Hauptverteiler (g), main distribution frame [MDF]

HW handset on wall; hardware; HRSC/WAOSS

HWCP hardware code page

HWY highway

HY hybrid

Hyb hybrid

hybrid ring microwave junction (*aka* rat race)

HYPS hypsometer (level-measuring set)

HYTELNT hypertext browser for Telnet accessible sites

Hz Hertz (cycles per second)

I

i instantaneous current

I (tiempo de) inactividad (s); incorrecto (s) [NOK]; information, información (s); iodine; current immediate (precedence); intra-pictures (MPEG)

I&B information and business

I&C installation and checkout

I&D integrate and dump detection

I&M installation and maintenance

i,k-display radar display

i.Link IEEE 1394 port (specification *aka* Firewire; Sony *tn*)

I/C incoming; kommend

I/F interface

I/G individual/group

I/O input/output

I/P input

I/Q in phase/quadrature phase

I/R bit de instrucción/respuesta (s), bit de campo de instrucción/respuesta (s), command/response (field) bit [C/R]

I2 Internet 2

I^{2}C inter-integrated circuit

I^{2}ICE integrated instrumentation and in-circuit emulator (Intel *tn*)

I^{2}L integrated injection logic

I2O intelligent input-output

I300I International 300 mm Initiative

I$_A$ articulation impairment values

IA incoming access; information appliance; intelligence artificielle (f) [AI]; instalación de alquilador (s), renter's premises [RP]; International Alphabet, alphabet international (f) (CCITT); International Ångström; interrupción de actividad (s), activity interruption [AI]

IA5 International Alphabet N° 5 (ITU-T/CCITT Rec T.50)

IA-64 Intel architecture 64-bit processor *tn*

IAAB Inter-American Association of Broadcasting

IAB Internet Activities Bureau; Internet Architecture Board

IAB-OPS Internet Architecture Board – Official Protocol Standards (RFC 1140)

IAC initial-alignment control; inter-application communication (Apple Macintosh); International Accounting Center

IACB Inter-Agency Consultative Board; international automatic call-back

IACK service acknowledgement signal

IACS integrated access and cross-connect system

IAD indexed attribute directory; indicación de alarma distante (s), remote alarm indication [RAI]; information à la demande (f), information on demand; Internet addiction disorder

IADMT far-end fault indication (f)

IAE indicación de estado de 'alineación de emergencias' (s), status indication E 'emergency alignments' [SIE]

IAEA International Atomic Energy Agency

IAF identificador de autoridad y de formato (s), authority and format identifier [AFI]; intelligent access function; International Astronautical Federation

IAFIS integrated automated fingerprint identification system

IAG instruction address generation

IAGC instantaneous automatic gain control

IAHC Internet Ad Hoc Committee

IAI initial address information; initial address message with additional information, message initial d'adresse avec informations supplémentaires (f)

IAK Internet access kit (IBM)

IAL extended unsuccessful backward set-up information message indication; international algorithmic language

IALD identificación de abonado llamado (s), called subscriber identification [CSI]

IALT identificación del abonado llamante (s), calling subscriber information [CIG]

IAM identification d'appel malveillant (f), malicious call identification/trace [MCI]; initial address message, message d'adresse initial (f), mensaje inicial de dirección (s); interactive algebraic manipulation; intermediate access memory

IAN indicación de estado de 'alineación normal' (s), status indication 'normal terminal status' [SIN]

IANA Internet Assigned Names Authority

IANAL I am not a lawyer; je ne suis pas un juriste (f)

IAP interfuncionamiento mediante acceso por puerto (s), interworking by port access [IPA]; Internet Access Provider

IAPS I am pretty sure

IAR instruction address register (IBM)

IARU International Amateur Radio Union

IAS immediate access store; integrated access system; instrucción de aborto de sesión (s)

IAS comp Institute for Advanced Study Computer

IASG interwork Address Sub-Group

IAT identificación del abonado que transmite (s), transmitting subscriber information [TSI]; import address table; International Atomic Time

IATA International Air Transport Association

IAW inactive window

IAWC in a while crocodile

IB in-band; indicator-bit; input/instruction buffer

IBAG INFOSEC Business Advisory Group

IBC instrument bus computer; integrated broadband communications, integrierte Breitbandkommunikation (g)

IBCN Integrated Broadband Communication Network

IBCS Intel binary compatibility specification

IBDN Integrated Building Distribution Network

I-beam 'I' shaped cursor (*ie* GUI marker)

IBEW International Brotherhood of Electrical Workers

IBG interblock gap/group

IBM insidious black magic; International Business Machines Corp; it's been malfunctioning; it's better manually

IBM-GL IBM graphics language

IBOC in-band on-channel

Ibpag2 Icon-based parser generation – system 2

IBS Intelsat Business Services

IBSG Interwork Broadcast Sub-Group

IBT intrinsic burst tolerance

IBTO International Broadcasting and Television Organization

IBUFG Internetwork Broadcast/Unknown Functional Group

IBX integrated business exchange

IC identificateur de commande (f), command identifier [CI]; inactivity counter; incoming call; independent carrier; indicación de concatenación (s), concatenation indication [CI]; índice de calidad (s), performance index [PI]; interlock code (CUG supplementary service); integrated circuit; interexchange carrier; interface cancellation; Investment Council; ion chromatography

IC circuit inductor-capacitor circuit

IC(pref) interlock code of the preferential CUG

ICA independent computing architecture (Citrix Systems Inc *tn*); índices por categorias absolutas (s), absolute category rating [ACR]; International Communications Association

ICAM integrated CAM

ICAN individual circuit analysis

ICANN Internet Corporation for Assigned Names and Numbers

ICAO International Civil Aviation Organization

ICAP inductively-coupled argon-plasma

spectrometry; Internet calendar access protocol (Lotus *tn*)

ICAS Intel communicating applications specifications

I-CASE integrated computer-aided software engineering

ICB incoming calls barred (within the CUG); intercommunication (computer) bus; Internet citizens' band; Inter-project Coordination Board

ICC incoming trunk circuit; integrated circuit card; International Conference on Communications; International Color Consortium; Interstate Commerce Commission (US)

ICCB Internet Control and Configuration Board

ICCG incomplete conjugate gradient

ICCL I couldn't care less

ICCM Interworking by call-control mapping, interfoncionnement par mappage de commande d'appel (f), interfuncionamiento mediante correspondencia de control de la llamada (s)

ICCP Institute for the Certification of Computing Professionals

ICD índices por categoria de degradación (s), degradation category rating [DCR]; Informatik Centrum Dortmund; installable client driver (OpenGL *tn*); interactive call distribution; international code designator (ISO)

ICDR inward call detail recording

ICE in-circuit emulator (Intel *tn*); information et communications de l'entreprise (f), initial address message with additional information [MIS]; Inner Circle Elite; Insane Creator Enterprises; insertion communication equipment (BBC *tn*); intelligent concept extraction (search engine technique); International Communications Echomail; intrusion countermeasure electronics

ICEA Insulated Cable Engineers Association

ICECAN Iceland-Canada submarine cable system

ICED identificador de conexión de enlace de datos (s), data link connection identifier [DLCI]

ICER Industry Council for Electronic Recycling

ICES Integrated Civil Engineering System; interference-causing equipment standard

ICETRAN component of ICES and extension of FORTRAN

ICFG IASG Coordination Function Group

ICG índice de calidad global (s), overall performance index [OPI]

ICH international call handling

ICI image component information; immediate call itemization; incoming call identification; interactive C interpreter; inter-carrier interface; inter-exchange carrier interface (Bellcore); interface control information (OSI); Internet Computer Index

ICID inter-cell identity

ICJ incoming junction

ICL inserted connection loss; interface clear; International Computers Limited (formerly ICT)

ICLID incoming-call line identification

ICM image colour matching (Kodak *tn*); in-call modification; integrated computer module

ICMP internet control message protocol (RFC 792)

ICMS integrated circuit measurement system

ICN integrated control node; interconnect node; International Computers Nederland; international CUG number

ICND instrucción de continuación de documento (s)

ICNI integrated communications navigation and identification

ICO message d'incohérence (f), confusion message [CFN]

ICOA image communication open architecture

iCOMP Intel Comparative Microprocessor Performance index

ICON programming language – successor to SNOBOL

ICOT Institute for New Generation Computer Technology (Japan)

ICP indicateur de contrôle de protocole (f), indicador de control de protocolo (s), información de control de protocolo (s), protocol control indicator [PCI]; inductively-coupled plasma; integrated channel processor; intelligent communications processor; Internet control protocol

ICPA información de control del protocolo de aplicación (s), application protocol control information [APCI]

ICP-AES inductively-coupled plasma atomic emission spectroscopy

ICP-MS inductively-coupled plasma mass spectrometry

ICPOT InterExchange carrier point of termination

ICPP información de control del protocolo de presentación (s), presentation protocol control information [PPCI]

ICPR information de contrôle du protocole de réseau (f), información de control del protocolo de red (s), network protocol control information [NPCI]

IC-Prolog Imperial College Prolog

ICQ I seek you (chat system program) *tn*

ICR inductance – capacitance – resistance; initial cell rate; intelligent character recognition

ICRA Internet Content Rating Association

ICRIS intelligent customer record information system

ICS IBM cabling system; identification of character set; interactive collaborative systems; interception service; interference cancellation system; internet connection sharing (Microsoft Windows 98); intuitive command structure; Italian Cracking Service

ICSA International Computer Security Association

ICSC Interim Communications Satellite Committee (Intelsat)

ICSI international charged subscriber identifier; International Computer Science Institute (Berkeley)

ICST Institute for Computer Science and Technology

ICSU International Council of Scientific Unions

ICT ideal cycle time; in-circuit test/tester; information communication technology; interfaz de codificación de telescritura (s), telewriting code interface [TCI]; International Cablecasting Technology (US); International Computers and Tabulators (*now* ICL)

ICTF (German) Federal Associations Of Content Providers

ICTI International C-MOS Technology Inc

ICU instruction-cache unit; Intel configuration utility; international communication unit; interrupt control unit; ISA configuration utility

ICUG international closed user groups

ICUP individual circuit usage and peg count

ICV identificación de canal virtual (s), virtual channel identification [VCI]

ICVT incoming verification trunk

ICW Interactive CourseWare *tn*; interrupted continuous wave

ICWS International Core War Society

ICX Internet College Exchange (US)

ICZD instrucción de comienzo de documento (s) (document start command)

Id Irvine dataflow

ID identifier, identity, identification, identificateur, identité, identification (f), identificador, identificación (s); instruction decoder (Intel); Internet draft; (señalización por) impulsos decádicos (s), decadic pulsing (signalling) [DP]; non-urgent alarm

ID de TSAP identificateur de point d'accès au service de transport (f), transport service access point identifier [TSAP-ID]

IDA indication de défaillance en amont (f), upstream failure indication [UFI]; integrated digital access; intelligent disk (drive) array; Telematic Interchange of Data between Administrations in the EU community, échange (télématique) de données entre administrations (f), intercambio de datos entre las administraciones (s), Datenaustausch zwischen Verwaltungen (g)

IDB installation database

IDC image dissector camera; índice de cooperación (s), index of cooperation [IOC]; information distribution companies; insulation displacement connection; integrated database/desktop connector; intelligent data connector; Internet database connector (Microsoft *tn*); interpolación digital de conversación (s), digital speech interpolation [DSI]

IDCC international data coordinating centre

IDCMA Independent Data Communications Manufacturers Association

IDCP international data-collecting platform

IDCT inverse discrete cosine transform

IDD instrucción de descarte de documento (s); international direct dialling; signal de demande d'identité de la ligne appelante (s), calling line identity request signal [CIR]

IDDD international direct distance dialling

IDDE integrated development and debugging environment (Symantec *tn*)

IDDF intermediate digital distribution frame

IDDQ indirect drain quiescent current

IDDS International Digital Data Service

IDE interactive design and engineering; interface design enhancement; integrated/(interactive) development environment; integrated drive/(device) electronics; intelligent drive/(device) electronics

IDEA interactive data entry/access; interactive digital electronic appliance (US); International Data Encryption Algorithms

IDEAL Ideal deductive applicative language; initiating – diagnosing – establishing – acting – leveraging

IDEALS applications télématiques (f)

IDECUS Internal DEC Users' Society

IDEF ICAM definition; non-proprietary form of SADT (USAF)

IDF intermediate distribution frame

IDG identificación de grupo (s), group identification [GI]

IDH Inpherno Data Heaven

IDI initial domain identifier (ISO); Intergalactic Dismantling Incorporated

IDIV integer divide

IDK I don't know

IDL Interactive Data Analysis Language (Xerox); Interface Definition Language (SunSoft); message d'identité de la ligne du demandeur (f), calling line identity [CLI]

IDLC integrated digital loop carrier

IDLH immediately dangerous to life or health

IDLR infrastructure desservant les logements raccordables (f)

IDM internal distributed module

IDMR inter-domain multicast routing

IDMS integrated database management system (based on CODASYL)

IDMSX IDMS extended

IDN integrated text and data network, Integriertes Text und Datennetz (g); integrated digital network; signal d'identité non disponible de la ligne appelante (f), calling line identity unavailable signal [CLU]

IDNS Internet Domain Name System

IDNX integrated digital network exchange (IBM)

IDOL icon-derived object language

IDP identificación de protocolo (s), protocol identification [PI]; implementation project definition; initial domain part; initial protection point; Institute of Data Processing (UK); integrated data processing, traitement intégré de l'information (f), proceso de datos integrados (s); inter-digit pause; Internet datagram protocol; interpolación digital de la palabra (s), digital speech interpolation [DSI]

ID-PAST identificador de punto de acceso al servicio de transporte (s), transport service access point identifier [TSAP ID]

IDPE identificador de punto extremo (s), endpoint identifier [EID]

IDPM Institute of Data Processing Management (UK)

IDPR información de dirección (direccionamiento) de protocolo de red (s) [NPAI], network protocol addressing information

IDPS identificador de perfil de servicio (s), service profile identifier [SPID]

IDR identification request; información digital con restricciones (s), restricted digital information [RDI]; intelligent document recognition; intermediate data rate

IDRP interdomain routing protocol (ISO)

IDS information display system (BISYNC); interactive diagnostic system; interworking data syntax; identificación de sesión (s)

IDSCP Initial Defense Satellite Communications Project (US)

IDSE international data switching exchanges

IDSL ISDN over DSL

IDSR información digital sin restricciones (s), unrestricted digital information [UDI]

IDSS intelligent decision support systems

IDSTN integrated digital switching and transmission network

IDSU identificador de servicio de usuario (s), user service identifier [USID]

IDT identificador de terminal (s), terminal identifier [TID]; Integrated Device Technology (US corporation); integrated digital terminal; integrated document tool; inter-digital transducer; interface design tool; interrupt descriptor table; servicio suplementario de información de tarificación (s)

IDTV improved-definition television

IDU Interface Data Unit (UNI 3.0); método de prueba insertada distribuida en una sola capa (s), distributed single-layer embedded test method [DSE]

IDV independent software vendor

IE inference engine (AI expert systems); information element (ISDN); informe de estado (s), status report [SRPT]; informe de excepción (s), exception report [ER]; installation engineer/engineering; Internet Explorer (Microsoft *tn*)

IEA Integrated EDACS Alarm system (Ericsson *tn*); ion energy analysis

IEC identificador de punto extremo de conexión (s), connection endpoint identifier [CEI]; infused emitter coupling; International Electrotechnical Commission (standards); inter-exchange carrier

IEC 559 IEEE floating point standard

IED Information Engineering Directorate (UK; formerly ALVEY); International Engineering Development

IEDF ion energy distribution function

IEE Institute of Electrical Engineers

IEEE Institute of Electrical and Electronics Engineers

IEEE-CS IEEE Computer Society

IEEL índice de enmascaramiento para el efecto local (s), sidetone masking rating [STMR]

IEF Information Engineering Facility (Texas Instruments *tn*)

IEI information element identifier

IEL identificador de enlace lógico (s), logical link identifier [LLI]

IELD identificación de la estación llamada (s), called station identification [CED]

IELO índice de efecto local para el oyente (s), listener sidetone rating [LSTR]

IEM identidad de la estación móvil (s), mobile station identity [MSIN]; information sur les événements de maintenance (f), información de evento de mantenimiento (s), maintenance event information [MEI]; interferencia electromagnética (s), electromagnetic interference [EM]

IEMI identidades de los equipos móviles (s), international mobile equipment identity [IMEI]

IEN individual electronic newspapers; Internet Experiment Note

IEPG Internet Engineering and Planning Group

IERE Institution of Electronics and Radio Engineers (UK)

IERN internal-external recurrent neural network

IES incoming echo suppressor

IESG Internet Engineering Steering Group

IET identificador de punto extremo terminal (s), terminal endpoint identifier [TEI]

IETF Internet Engineering Task Force

IETG identificador de punto extremo terminal de grupo (s), group terminal endpoint identifier [GTEI]

IETS Interim European Telecommunications Standards

IEV international electrotechnical vocabulary

IF identificador de formato (s), format identifier [FI]; information flow; interface; interfrequency; intermediate frequency (300 Hz-3 kHz), Zwischenfrequenz (g)

IF strip intermediate frequency amplifier

IFA indicación de estado de 'fuera de alineación' (s), status indication 'out of alignment' [SIO]; indicación de fallo atrás (s), upstream failure indication [UFI]

IFAC International Federation of Automatic Control

IFAD Institute of Applied Computer Science Denmark

IFAM initial final address message

IFAX international facsimile service

IFC installed first cost; interface clear; Internet Foundation Classes

IFD image file directory; instrucción de fin de documento (s); International Federation for Documentation

IFDL independent form description language (DEC)

IFE intelligent front end

I-Feld Informationsfeld – Daten (g), information field – data

IFEN inter-company file exchange network

iff if and only if

IFF identification/identity friend or foe; intensity fluctuation factor; interchange/image file format (Amiga); international file format

IFG incoming fax gateway

IFHO inter-frequency handover

I-field information field

IFL integrated fuse logic; international frequency list

IFM interface module; intraframe prediction mode

IFMP Ipsilon flow management protocol [RFC 1953]

IFNN instrucción de facilidades no normalizadas (s), non-standard facilities command [NSC]

IFOV instantaneous field of view

IFP Illinois file protocol; instruction fetch pipeline

IFPI In-Flight Phone International (US corp)

I-frame information frame

IFRB International Frequency Registration Board (of ITU)

IFS indicación de estado 'fuera de servicio' (s), status indication 'out of service' [SIO]; installable file system (Microsoft Windows 95); instrucción de fin de sesión (s), end of session instruction; interactive flow simulator; international freephone service; ionospheric forward scatter; iterated function system

IFSM information systems management

IFU instruction fetch unit (Intel)

IFW inverted frame word

IG identificador de grupo (s), group identifier [GI]; internet Gopher (RFC 1436); interpolation gain; message d'inhibition de gestion (s), management inhibit message [MIM]

IGA integrated graphics array

IGBT insulated-gate bipolar transistor

IGC Institute for Global Communications (US); integrated graphics controller; mensaje de indagación sobre grupo de circuitos (s), circuit group query message [CQM]

IGD message d'interrogation de groupe de circuits (f), circuit group query message [CQM]

IGES initial graphics exchange specification (ASME/ANSI standard)

IGF identificateur général de format (f), identificador general de formato (s), general format identifier [GFI]

IGFET insulated gate field-effect transistor

IGMP Internet Group Management Protocol

IGOSS Industry/Government open systems specification; integrated global ocean station system

IGP identificateur de groupe (de) paramètres (f), identificador de grupo parámetros (s), parameter group identifier [PGI]; interior gateway protocol (IETF)

IGPL Interest group in pure and applied logics

IGR message de réponse à une interrogation de groupe de circuit (f), circuit group query response message [CQR]

IGRP interior gateway routing protocol

IGS identity graphic sub-repertoire; Internet Go server (*ie* the Japanese board game, Go)

IH interrupt handler

IHA International Hackers Association

IHD integrated help desk (IBM *tn*)

IHL internet header length (IP)

IHS integrated home system; intelligent home systems (EUREKA project)

IHT ideal handler time

IHTFP I have truly found paradise; I hate this fucking place

IHV independent hardware vendor

II identificador de instrucción (s), command identifier [CI]; indicatif interurbain (f), indicativo interurbano (s); trunk code; information indicator; ion implant (*aka* I2)

IICD International Institute for Communication and Development

IICM Institute for Information Processing and Computer-supported new Media (Graz University)

IICN integrated intelligent communications network

IIEM identidad internacional de estación móvil (s), international mobile station identity [IMSI]

IIF image interchange facility (part of IPI standard); immediate interface; información de indicación de fallo (s), fault information indication [FII]

IIG International Internet Gateway

III interstate identification index

IIJ Internet Initiative Japan

IIL integrated injection logic

IIMC indisponibilité intrinsèque moyenne cumulative (f), mean accumulated intrinsic down-time [MAIDT]

IIN integrated information network

IINREN Interagency Interim National Research and Education Network

IIOP Internet inter-operability protocol; Internet inter-object request broker [ORB] protocol

IIP indicación de estado interrupción del procesador (s), station indication processor outage [SIPO]

IIR infinite impulse response

IIRC if I remember correctly

IIRG International Information retrieval Guild

IIS idealized instruction set; Internet Information Server (Microsoft *tn*)

IISF International Information Security Foundation

IISM identité internationale de la station mobile (f), international mobile station identity [IMSI]

IISP interim inter-switch signalling protocol (ATM Forum)

IISTO Information Science and Technology Office (US DoD)

IIT Integrated Information Technology (US corp); integrated information transport; Institution Internationale de Télématique

IITF Information Infrastructure Task Force (US Govt)

IIUD instrucción de información de usuario de documento (s)

IIUS instrucción de información de usuario de sesión (s)

IJ identification of justification, identification de justification (f), identificación de justificación (s)

IKBS intelligent knowledge-based system

IKE IBM Kiosk for Education *tn*; Internet key exchange

IKMP Internet key management protocol

IKP Internet keyed payments (IBM)

IKS interactive knowledge system

IKZ Impulskennzeichen (pulse signalling)

IL indicateur de longueur (f), indicador de longitud (s), length indicator [LI]; indicación de liberación (s), clear indication [CLI]; integrated learning

I_L loudness impairment values

ILA image light amplifier

ILB inner lead bond

ILBM interleaved bitmap

ilc international line incoming

ILCD instrucción de lista de capacidades de documento (s)

ILD injection laser diode; interlayer dielectric

ILEC incumbent local exchange carrier (US)

ILETS International Law Enforcement Telecommunications Seminar

ILF infra-low frequency

ilg internationale Leitung gehend (g), international line outgoing

ILGP indicador de longitud de grupo de parámetros (s), parameter group length indicator [PGLI]

ILI indicador de longitud de instrucción (s), command length indicator [CLI]

ILIL input longitudinal interference loss

ilk internationale Leitung kommend (g), international line incoming

ILL identidad de la línea llamante (s), calling line identity [CLI]

ILLA identidad de la línea llamada (s), called line identity [CDI]

ILM identificación de llamadas maliciosas (s), malicious call identification [MCI]

ILMI integrated local management interface; interim local management interface

ILP indicador de longitud de parámetro (s), parameter length indicator [PLI]

ILPD instrucción de limite de página de documento (s)

ILR indicador de longitud de respuesta (s), response length indicator [RLI]; interworking location register

ILS instrument landing system; international language support; intra-cavity laser spectroscopy

ILT instructor-led training

ILTMS international leased telegraph message switching service

Im intensity modulation; inter-modulation ($2^{nd}/3^{rd}$ order)

IM immediate (prompt alarm); impulso de marcación (s), decadic pulsing [DP]; information de maintenance (f), información de mantenimiento (s), maintenance information [MI]; integrated model; interface module; interpersonal messaging, messages entre UER télex (f); inter-telex store and forward unit [SFU] messages

IM&T information management and technology

IMA individual mobilization augmented; input message acknowledgement, accusé de depôt (f), acuse de recibo de mensaje introducido (s); Interactive MIDI Association; Interactive Multimedia Association

iMac Internet Macintosh *tn*

IMACS image management and communication system

IMAO in my arrogant opinion

IMAP Internet mail application protocol (RFC 2060/2062); Internet message access protocol

IMAX maximum image

IMB Intel media benchmark

IMC identificador de método de codificación (s), coding method identifier [CMI]; (temps de) indisponibilité moyen cumulatif (f), mean accumulated down-time [MADT]; initial micro-code load; intermodulation distortion; international maintenance centre; Internet mail connector (Microsoft); Internet Mail Consortium

IMD inter-metal dielectric

IMDC intelligent multiple-disk controller

IMDS image data stream (format) (IBM)

IMDTC international multiple destination TV connection

IME input method editor

IMEI international manufacturer/mobile equipment identity

IMF interactive mainframe facility (Hewlett Packard)

IMG image; Interactive Media Group (part of Microsoft)

IMH information model handler

IMHO in my humble opinion, à mon humble avis (f)

IML incoming matching loss; initial machine load; initial micro-code/(-program) load (IBM); intermediate language

IMM idle mass management; immediate assignment message; information model manager; input message manual

IMMA ion microphobe mass analysis

IMO Information Market Observatory; in my opinion, à mon avis (f)

IMP implementation language for EMAS; improved mercury autocode; indication of microwave propagation (Ferranti); information message processor; interface message processor (ARPANET); integrated mail processor (Royal Mail letter sorters); interpretive menu processor

IMPA intelligent multi-port adaptor (DCA)

IMPATT impact (ionization) avalanche transit time (diode)

IMR integrated model repository; Internet monthly report (in ARPANET)

IMS Information Management System (IBM *tn*); image management system; information/integrated management system; intermediate maintenance standards; ion mobility spectroscopy

IMSE integrated modelling support environment

IMSI international mobile station identity

IMSP Internet message support protocol

IMSVS Information management system/virtual storage (IBM)

IMT inter-machine trunk

IMT-2000 International Mobile Telecommunications 2000

IMTA International Mobile Telecommunications Association

IMTC International Multimedia Teleconferencing Consortium

IMTI international mobile terminal identity

IMTS improved mobile telephone service

IMTV interactive multimedia TV

IM-UAPDU IP-message user agent protocol data unit

IMUI international mobile user identity

IMUL integer multiply

IMUN international mobile user number

IMUX intelligent multiplexer; inverse multiplexing

IN índice de nitidez (s), articulation index [AI]; ingress node; intelligent network (AIN); interrogating node

IN/1/1+/2 intelligent network concepts

INA Institut National de l'Audiovisuel

INA information networking architecture; integrated network access; link inhibit acknowledgement signal

INAC inactive

INAG Ionospheric Network Advisory Group

INAP intelligent network application protocol/(part)

InARP inverse address resolution protocol (IETF)

inc increment

INC international carrier; International Network of Crackers

INCAMS individual cassette manufacturing system

INCC indicateur de contrôle de continuité (f), continuity-check indicator [CCH]; International Network Controlling Centre

INCM intelligent network conceptual model

ind induction

IND indicatif national de destination (f), indicativo nacional de destino (s), national destination code [NDC]; international number dialling routing code; message de demande

d'information (f), information request [INR]

INDB intelligent network database

INDE test access code

Indeo Intel video (data compression system *tn*)

INDI-VSt Vermittlungsstelle für Information- und Ansage-Dienste (g), switching centre for information and message services

INEM identidad nacional de estación móvil (s), national mobile station identity [NMSI]

INF information file (Microsoft Mail); (domaine d')information de signalisation (f), signalling information field [SIF]; information message, message d'information (f), mensaje de información (s); ionospheric forward scatter

INFA facility information (s)

infimum greatest lowest bound

INFN Istituto Nazionale di Fisica Nucleare (i) (Italian State Research organization)

INFO information element defined at user-network interface

Infobahn Internet and other high-speed networks (information + autobahn)

Infonet multi-protocol access network service (Computer Sci Corp *tn*)

informatics computer science; electronic data processing

INFORMIX database management system *tn*

INFOSEC information (systems) security

infostrada information + autostrada (*cf* infobahn)

INFOTEX information via Telex (US)

INFSWG Information Services Working Group (IUIC, UK)

ING general forward set-up information message (f) [GSM]

INGRES relational DBMS by Computer Associates *tn*

INHD far-end inhibit

INHL local inhibit

INI inter-network interface

INIC ISDN network identification code

init old Apple Mac term for system extension

INIT initialize/initialization; initiate

INITACK initiate acknowledgement

INKA internationale Kreditkarte (international credit card)

INL indicateur de longueur (f), length indicator [LI]; interwork/-nodal link

INM intelligent network management (protocol)

INMARSAT International Maritime Satellite Organization, Organización Internacional de Telecomunicaciones Maritimas por Satélite (s)

INMC international network management centre

INMS integrated network management system

INN inter-node network

INND Internet News Demon

IN-NSM intelligent network non-switching manager

INO link inhibit signal (f) [LIN]

InP indium phosphide (used in semiconductors)

INP information request; intelligent network processor; Internet nodal processor; interfaz de nodo de red (s), network node interface [NNI]; internal network number indicator

INRIA Institut National de Recherche en Informatique et Automatique (f)

ins insulate/insulation

INS indicateur de service (f), service indicator [SI]; indirect NICS subscriber; inertial navigation system; input string; integrated network server; internal network subsystem; Internet network service

INSIS (inter-institutional) integrated services information system

IN-SL intelligent network – service logic

INSM identité nationale de la station mobile (f), national mobile station identity [NMSI]

IN-SM intelligent network – switching manager

IN-SSM intelligent network – switching state model

INT integer; interrogative prosign; internal; interrupt; mensaje de intervención (s), forward transfer message [FOT]

INTA interrupt acknowledge

INTAP interoperability technology association for information processing

INTELSAT International Telecommunications Satellite Organization

Internaute (French) Internet surfer

InterNIC Internet Network Information Centre

INTERSPUTNIK International Organization of Space Communications (Moscow)

INTIM interrupt and timing

INTO interrupt if overflow occurs

Intranet private network, intra-corporate version of Internet

INTT incoming no test trunks

INTUG International Telecommunications Users Group

INU mis-dialled trunk prefix (f)

INV inventory

INV SIG inverted signal

INWATS inward wide area telephone service/system

IO indicación de estado 'O' ('ocupado') (s), station indication B busy [SIB]; input-output; inward operator

IO.SYS hidden Microsoft Windows system files

IOB índice de cooperación (s), index of cooperation [IOC]; initial operational capability; input-output buffer; integrated optical circuit/component; inter-office channel; Inter-organization Board for Information Systems and Related Activities

IOCC input-output channel converter; input-output controller chip;

International Overseas Completion
Center

IOCCC International Obfuscated C Code
Contest

IOCS input-output control system

IOCTL input-output control

IOD identified outward-dialling

IODC international operator direct
calling

IOE indice d'opinion à l'écoute (f),
índice de opinión en la escucha (s),
listening opinion index [LOI]

IOEXT I/O extension module

IOG input-output group

IOI International Olympiad in
Informatics

IOIM input-output interface module

ION Integrated On-Demand Network
(Sprint *tn*); integrated optical network;
Inter-lending OSI Network

IONL internal organization (of the)
network layer (*cf* catenet)

IOP input-output processor, Ein-
/Ausgabe-Prozessor (g); message
d'intervention (d'une opératrice) (f),
forward-transfer message [FOT]

IOPL input-output privilege level

IOS input-output subsystem; integrated
office system

IOSA integrated optic-spectrum analyser

IOSGA input-output support gate array

IOT inter-office trunk

IOT&E initial operational test and
evaluation

IOW in other words

IOWA Inmates of the World Asylum

IP identificateur de paramètre (f),
identificador de parámetro (s), parameter
identifier [PI]; identificador de protocolo
(s), protocol identifier [PI]; indicatif de
pays (f), indicativo de país (s), country
code [CC]; information provider;
instruction pointer; intellectual property;
intelligent peripheral; inter-digital pause;
intermediate point; inter-personal;
Internet Protocol; ISDN access port;
ISDN interworking unit port

IPA intermediate power amplifier;
International Phonetic Alphabet;
interworking by port access

IPAR individual parameter settings

IPARS international passenger airline
reservation system

IPAS identificador de punto de acceso al
servicio (s), service access point
identifier [SAPI]

IP-AUTH initial program load [IPI]
authentication header (RFC 2402)

IPB inter-processor bus

IPBX international private branch
exchange

IPC indicador de prueba de continuidad
(s), continuity check indicator [CCH];
instructions per clock; inter-process
communication

IPCH initial paging channel

IPCP informe de prueba de conformidad
de protocolo (s), protocol conformance
test report [PCTR]; Internet protocol
control protocol

IPCS Informatique et Calcul Parallèle de
Strasbourg (f); informe de prueba de
conformidad de sistema (s), system
conformance test report [SCTR];
interactive problem control system

IPD indicatif de pays pour la
transmission de données (f), indicativo
de país para datos (s), data country
control [DCC]; intercambio de
parámetros dentro de banda (s), in-band
parameter exchange [IPE]

IPDN international public data network

IPDP industrial peripheral data
processing

IPDS intelligent printer data stream
(IBM)

IPDU inter-network protocol data unit

IPE in-band parameter exchange

IPEC identificador de punto extremo
conexión (s), connection endpoint
identifier [CEI]

IPEM Internet privacy enhanced mail
(standard)

IPFC information presentation facility compiler (IBM)

I-phone information phone

IPI image processing and interchange (ISO/IEC 12087 standard); intelligent peripheral interface

IPI-IIF IPI image interchange facility (part 3)

IPI-PIKS IPI programmers' imaging kernel system (part 2)

IPL information processing language; initial program load/loader; interactive services primary link; Internet Public Library (University of Michigan); ion projection lithography

IPLAN integrated planning and analysis

IPM impulses per minute; indicativo de país para el servicio móvil (s); indicativo de país móvil (s), indicativo de país de la estación (s), mobile country code [MCC]; interference prediction model; internal polarization modulation; interpersonal message, message de personne à personne (f); interruptions per minute

IPMAS interpersonal messaging abstract

IPME interpersonal messaging environment

IPMS interpersonal messaging service/system

IPMS MS interpersonal messaging system message store

IPMS UA interpersonal messaging system user agent

IPM-UA interpersonal messaging user agent

IPN indicación/indicador de plan de numeración (s), numbering plan indicator [NPI]; indicación de puntero nulo (s), null pointer indicator [NPI]; interpersonal notification; interrupción del procedimiento negativa (s), negative procedure interruption [NPI]

IPND indicador de plan de numeración y direccionamiento (s), numbering and address plan indicator [NAPI]

IPng Internet protocol next generation (RFC 1550)

IPO initial public offering; input-process-output

iPOC Internet Policy Oversight Committee

IPP international phototelegraph position; Internet print protocol; inter-processor process; interrupted Poisson process

IPPV impulse pay-per-view (on satellite/cable TV)

IPR in-pulse to register; inside plant repeater; intellectual property rights

IPR-FDM interrupción del procedimiento – fin de mensaje (s), procedure interrupt – end of message [PRI/EOM]

IPR-FDP interrupción del procedimiento – fin de procedimiento (s), procedure interrupt – end of procedure [PRI/EOP]

IPR-PMS interrupción del procedimiento – señal de multipágina (s), procedure interrupt – multi-page signal [PRI/EOM]

ips impulses per second

IPS in-place switching; in-pulse to sender; instructions per second; inter-process communication; Ionospheric Prediction Service

IPSE integrated project support environment

IPSEC IP security architecture (RFC 2401)

IPSM indicatif de pays du service mobile (f), mobile country code [MCC]

IPSS international packet switching service; inter-processor signalling system

IPT ideal process time; indicativo de país para telefonía (indicativo de país telefónico) (s), telephone country code [TCC]

IPTC International Press Telecommunications Council

IPU instruction processor; interface processing unit

IPUP intelligent peripheral user part

IPv4 Internet protocol version 4

IPX Internet packet exchange (RFC 1132)
IPX/ODI interwork packet exchange/open data-link interface
IPX/SPX Internet packet exchange/sequenced packet exchange
IQF intrinsic quality factor
IQL interactive query language
Ir iridium (used in heat-resistant alloys)
IR identificateur de réponse (f), identificador de respuesta (s), response identifier [RI]; identidad de red (s), network identity [NI]; identificador de red (s), network identifier [NI]; indicativo de red (s); índice de ruido (s), network connection [NC]; information retrieval; infrared; Immortal Riot; ingress router; instruction register; intermediate rate; internal rate; international roaming; Internet registry
IR(D)D integrated receiver (digital) decoder
IRAC Inter-department Radio Advisory Committee
IRAM intelligent RAM
IRAS infrared reflection-absorption spectroscopy; Internet routing and access service
IRC Interagency Radio Committee; international record carrier; Internet Relay Chat; inter-rate centre channel
IRCC International Radio Consultative Committee
IRD integrated receiver/decoder
IrDA Infrared Data Association
IRDATA industrial robot data
IRDS information resource dictionary system
IRE Institute of Radio Engineers (incorporated into IEEE); integrierte Reichweitenvergrösserung (g), integrated range extension
IRED infrared emitting diode
IRET interrupt return/return from interrupt
IRF inheritance rights filter (Novell); intermediate routing function

IRG International Rogues Guild
IRGB intensity red green blue
IRIDIUM satellite mobile communication system
IRIS imaging of radicals interacting with surfaces; Imaginons un Réseau Internet Solidaire (1997) (f); Institute for Research in Information and Scholarship (Brown University, US)
IRIT Institut de Recherche en Informatique de Toulouse (f)
IRIX OS on Silicon Graphics workstations
IRL industrial robot language; in real life; interactive reader language; inter-repeater link (fibre-optic link; IEEE 802.3)
IRLED infrared light-emitting diode
IRLS interrogation recording and location system
IRM indicatif de réseau mobile (f), indicativo de red móvil (s); indicativo de red para el servicio móvil (s), mobile network code [MNC]; information resource management; inherited rights mask
IrMC Infrared Mobile Communications
IRN internal recurrent neural network
IROR internal rate of return
IRP independent routing processor; internal reference point; international routing plan
IRQ interrupt request (line), requête d'interruption (f); interworking service request identifier (f)
IRS Infrared Astronomical Satellite; intermediate reference system; international repeating station
IRSD instrucción de resincronización de documento (s)
IRSG Internet Research Steering Group
IRT indicación de servicio (s), transit delay indication [TDI]; interrupted ring tone
IRTA integrated regional telematics architecture

IRTC-1 interconnect reliability test chip-1

IRTF Internet Research Task Force

IRU interworking responsible unit

IRV international reference version

IRX information retrieval experiment

IS in Betrieb (g), in service; indexed sequential (storage); indicador de servicio (s), service indicator [SI]; índice de sonoridad (s), loudness rating [LR]; information security/system; integrated systems; interface specifications/switch; interim standard (ISO); intermediate system (ISO); interrupt status; inter-system

IS&N intelligence in services and networks

IS/IT information system/information technology

IS1 (US) information separator one (unit separator)

IS2 (RS) information separator two (record separator)

IS3 (GS) information separator three (group separator)

IS4 (FS) information separator four (file separator)

ISA Industry Standard Architecture; instruction set architecture; Instrumentation Society of America; interconexión de sistemas abiertos (s) (OSI); International Smalltalk Association

ISAKMP Internet Security Association and Key Management Protocol

ISAM indexed sequential access method

ISAPI Internet services application programming interface (Microsoft *tn*)

ISB independent sideband (transmission); independencia de la secuencia de bits (s), bit sequence independence [BSI]; Internet Sharing Box (Dane-Elec *tn*)

ISBD International Standard Bibliographic Description

ISBN International Standard Book Number (bar-coded)

ISC índice de sonoridad del circuito (s),

circuit loudness rating [CLR]; Industry Steering Council; Information Systems Committee (of the UFC); instruction-set computer; International Switching Centre, internationale Vermittlung (g); inter-system communication (IBM)

ISCC international service coordination centre

ISCP ISDN signalling control part

ISD image section descriptor; instructional systems design; integridad de la secuencia de dígitos (s), digit sequence integrity [DSI]; international subscriber dialling

ISDE international data switching exchange

ISDN integrated services digital network, Dienstintegrierendes digitales Netz (g); Dienstintegriertes digitales Universalnetz (g); dienstintegriertes Digitalnetz (g); I see dollars now; I still don't know; integration subscribers don't need; It still does nothing

ISDN-BC ISDN bearer capability

ISDN-BE ISDN user forum (G Benutzerforum)

ISDN-BRI basic rate ISDN

ISDN MNP ISDN/mobile NP

ISDN PRI ISDN primary rate interface

ISDN PRM ISDN protocol reference model

ISDN-SN ISDN subscriber number

ISDN TNP ISDN/telephony NP

ISDN-UP ISDN user part

ISDS integrated switched data service

ISDX integrated services digital exchange

ISE immediate service; índice de sonoridad en emisión (s), sending loudness rating [SLR]; integrated service engineering; Interactive Software Engineering (US Corp)

ISEH índice de sonoridad del eco para el hablante (s), talker echo loudness rating [TELR]

ISEL índice de sonoridad por el efecto

local (s), sidetone loudness rating [STLR]

ISEM inspection/review specific equipment model

ISEN índice de sonoridad del enlace (s), junction loudness rating [JLR]

ISEO índice de sonoridad del eco del oyente, listener echo loudness rating [LELR]

ISET in-station echo-canceller test equipment/(tester)

ISF Information Security Foundation; information/integrated system factory

ISG índice de sonoridad global (s), overall loudness rating [OLR]

ISGC interfaz de servicio de gestión de capa (s), layer management service interface [LMSI]

ISGS interfaz de servicio de gestión de sistemas (s), systems management service interface [SMSI]

ISH information super-highway

ISI Information Sciences Institute (University of Southern California); Institute for Scientific Information; internally-specified index; inter-symbol interference

ISIS integrated systems and information services; International Satellite for Ionospheric Studies; Internationally Syndicated Information Services (US); investigative support information system

IS-IS intermediate system – intermediate system (ISO protocol)

ISKM Internet starter kit (for Apple Macintosh *tn*)

ISL interactive system language; inter-satellite link

ISLAN integrated services LAN (IEEE 802.9)

ISLisp international standard LISP

ISLM integrated services line (module)

ISLU integrated services/ISDN line unit

ISM inductor super magnetron; industrial scientific and medical (applications); Internet service manager (Microsoft *tn*);

ISDN standards management (ETSI group)

ISMC international switching maintenance centre

ISMF interactive storage management facility

ISMX integrated sub-rate data multiplexer

ISN information systems network (AT&T)

ISO International Organization for Standardization (International Standards Organization – deprecated); índice de sonoridad objetivo (s), overall loudness rating [OLR]

ISO/OSI ISO/open systems interconnect architecture

ISO-7 7-bit character code (ISO 646-1973)

ISOC Internet Society

ISOC.DE Internet Society Deutschland

ISOC-FR Internet Society France

ISODE ISO development environment

ISOE índice de sonoridad objetivo eléctrico (s), objective/overall loudness rating [OLR]; índice de sonoridad objetivo en emisión (s), transmitting objective loudness rating [TOLR]

ISO-IEC/JTC1 Joint Technical Committee of ISO (1)/IEC (1)

ISOR índice de sonoridad objetivo en recepción (s), receiving objective loudness rating [ROLR]

ISP in-service performance; instruction set processor; interactive session protocol; intermediate service part; international signalling point; Internet Services Provider; interrupt stack pointer; interrupt status port

ISPA Internet Service Providers Association (UK)

ISPBX integrated services PBX

ISPC international signalling point code; international sound-programme centre

ISPF interactive system productivity facility

ISPF/PDF interactive system

productivity facility/program development facility

ISPM in-situ particle monitor

ISPN INFO Security Product news; integrated services private network

ISPO Information Society Project Office, Oficina de Proyectos de la Sociedad de la Information (s), Büro Projekte der Informationsgesellschaft (g) (EU)

ISQL interactive SQL

ISR índice de sonoridad en recepción (s), receiving loudness rating [RLR]; information storage and retrieval; in situ rinse; International Simple Resale (UK); intermediate session routing; interrupt service routine; interrupt status register; receive loudness rating, indice sonore à la réception (f)

ISRP información suplementaria de realización de protocolo para pruebas (s) [PIXIT], información suplementaria sobre realización de protocolo para pruebas (s), protocol implementation extra information for testing [PIXIT]

ISS improved signal strength; information sending station; integrated switching system; Internet security scanner; Internet Security Systems (US Corp); ion-scattering spectroscopy; Ionosphere Sounding Satellite (Japan)

ISSA Information Systems Security Association

ISSI inter-switching system interface

ISSIP inter-switching system interface protocol

ISSLS International Symposium on Subscriber Loops and Services

ISSN integrated special services network; International Standard Serial Number (bar code)

IST Imperial Software Technology; in-service trigger; integrated switching and transmission; send loudness rating (f) [SLR]

ISTC international satellite transmission centre; international switching and testing centre

ISTN integrated switching and transmission network

ISTR I seem to recall/remember

ISU independent signal unit; información de servicio de usuario (s), user service information [USI]; initial signal unit, unité de signalisation initiale (f), unidad inicial de señalización (s); intermediate switching unit

ISUP ISDN user part

ISV independent software vendor

ISVR international simple voice relay (Internet telephony)

IT intervalle de temps (f), intervalo de tiempo (s), time slot [TS]; inactivity test; information technology, informatique (f), informática (s), Informatik (g); tietojenkäsittelyoppi (fi); information transfer/type; intelligent terminal; inter-toll; interwork termination

IT0 intervalos de tiempo 0 (s), time slot 0 [TS0]

IT CCD interline transfer CCD

ITA Independent Television Authority (UK – now the ITC); Information Technology agreement; infrastructure terminal d'abonné (f); interim type approval; International Telegraph Alphabet, alphabet international télégraphique (f)

ITAA Information Technology Association of America

ITAC-T International Telecommunications Advisory Committee – Telecommunications

ITB information technology branch; intermediate text block

ITC information transfer capability; intercept; intermediate toll centre; International Teletraffic Congress; International Television Centre; international transit centre; International Typeface Corp; intervalo de tiempo de canal (s), channel time slot [CTS]

ITCC international telephony charge card

ITD instrucción de transmisión digital (s), digital transmit command [DTC];

señal de transacción introducida aceptada para entrega (s), input transaction accepted for delivery (code)

ITE information technology equipment; integrated terminal equipment; internal terminal emulator; international telephone exchange; intervalo de tiempo estadístico (s) [STI]

ITEL program for police to process BT call data (UK)

ITEM identidad temporal de la estación móvil (s), temporary mobile station identity [TMSI]

ITF interactive test facility

ITFS instructional television fixed service

ITI Indian Telephone Industries; intermittent trouble indication

ITIL Information Technology Infrastructure Library

I-time instruction time

ITITO Information Technology Industry Training Organization (UK)

ITMC international transmission maintenance centre

ITN identification tasking and networking; Independent Telecommunication Network; Independent Television News (UK)

ITNTO IT Technology National Training Organization

ITOS improved Tiros satellite

ITP información de tarificación/tasación (s) (billing information)

ITPA Independent Telephone Pioneers Association

ITPC international television programme centre

ITPEG identificateur terminal de point d'extrémité de groupe (s), group terminal endpoint identifier [GTEI]

ITPR infrared temperature profile radiometer

ITR integrated telephone recorder; international transit exchange

ITRI Interconnection technology Research Institute

ITS incompatible time-sharing system (MIT OS); insertion test signal; Institute for Telecommunications Sciences (US); interactive terminal service; Internal Theft Syndicate; Internet telephony server/switch; inter-time switch; invitation to send

ITSC International Telecommunications Services Complex; International Telecommunications Standards Conference; International Telephone Services Center

ITSEC Information technology security evaluation criteria

ITSM identité temporaire de station mobile (f), temporary mobile station identity [TMSI]

ITSO International Telecommunications Satellite Organization

ITT information technology and telecommunications

ITT&B information technology telecommunications and broadcasting

ITTA Information Technology Training Association

ITTP intelligent terminal transfer protocol (Ericsson standard)

ITU International Telecommunication Union

ITU D ITU - Telecommunication Development Sector (formerly BDT)

ITU R ITU - Radio Communication Sector (formerly CCIR/IFRB)

ITU T ITU - Telecommunications Standardization sector

ITU TS ITU - Telecommunications Standards Section

ITU TSB ITU - Telecommunication Standardization Bureau

ITUG International Telecommunications User Group

ITV Independent Television (UK); independent transport vendors; interactive TV; virtual path identifier (s)

ITX intermediate text block

IU intervalle unitaire (f), intervalo unitario (s), unit interval [UI];

identificador de UDPS (s), SPDU
identifier [SI]; integer unit; interaction
unit; interworking unit

IUAP Internet user account provider

IUCAF Inter-Union Commission on
Frequency Allocations for Radio
Astronomy and Space Science

IUIC Inter-University Information
Committee on Computing (UK)

IUR 1.0 International Requirements for
Interception

IUR identificación del usuario de la red
(s), network-user identification [NUI];
interfaz usuario-red (s), user-network
interface [UNI]

IUS ISDN user services application
module

IUS/ITB interchange unit
separator/intermediate transmission
block

IUT implementation under test,
réalisation à tester (f)

IUU information d'usager à usager (f),
mensaje de información de usuario a
usuario (s), user-to-user information
[UUI]

IV interactive video

IV&V independent verification and
validation

IVC international video-conference
centre

IVD internal vapour deposition

IVDLAN integrated voice-data local area
network (IEEE 802.9)

IVDM integrated voice-data multiplexer

IVDT integrated voice-data terminal

IVHS intelligent vehicle highway system

IVIS interactive video interface-
information system

IVL Independent Vendor League; Intel
Verification Laboratory

IVP integrated vacuum processing

IVPC International Virus Prevention
Conference

IVR interactive and verification service;
interactive voice response

IVS interactive video-disk system;
interactive video service

IVSN initial voice-switched network

IVT interrupt vector table

IVTS international Video
Teleconferencing Service

IW Imperial Warlords; inside wire;
interworking

IWA International Webmasters'
Association

I-way information highway

IWBNI it would be nice if

IWCA inside wiring cable

IWCS integrated wideband
communications system

IWD interworking descriptions

IWF interworking function, fonction
d'interfonctionnement (f); Internet Watch
Foundation (UK)

IWM interactive workflow manager;
integrated WOZ machine

IWMSC interworking mobile switching
centre

IWP Interim Working Party of CMTT

IWU interworking unit (B-ISDN)

IWUP interworking unit user part

IWV Impulswählverfahren (g), loop-
disconnect dialling method; loop-
interrupt dialling system

IXC IntereXchange Carrier (US long-
distance telco)

IXI International X25 infrastructure

IXT interaction cross talk

IYFEG insert your favourite ethnic group

IYKWIM if you know what I mean

IYKWIMAITYD if you know what I
mean and I think you do

IYSWIM if you see what I mean

J

J justification

J/K joule per Kelvin

J/m joule per metre

J/m³ joule per cubic metre

J2 Java 2 interface

J2EE Platform Enterprise Edition *tn*

J2ME Java 2 Platform Micro Edition *tn*

J2SE Java 2 Platform Standard Edition *tn*

J64 jonction (f), 64 kbit/s interchange circuit

JA jump address; jump if above

JACAL Jaffer's canonical algebra

JACS joint area communications system

JAD joint application development/design

JADE James's DSSSL engine

JAMES Joint ATM experiment on European Services

JANAP Joint Army-Navy-Air Force Publication (US)

JANET Joint Academic Network (used by UKERNA)

Janus antenna able to receive from opposing sides

JAPE joke analysis and production engine (Edinburgh University)

japh just another Perl hacker

JAR Java archive

JASS JavaScript accessible style sheet

Java Sun Microsystems programming language *tn*; programmers' programming fluid (*ie* black coffee); just another vague acronym

JBE jump if below or equal

JBIG joint Bi-Level Image Expert Group

JBOD just a bunch of disks (*ie* not a RAID configuration)

JC jump if carry set

J-CALS Joint Computer-Aided Acquisition and Logistic Support (US DoD)

JCB justification control bit

JCDR jeu de caractères dynamiquement redéfinissables (f), juego de caracteres dinámicamente redefinibles (s), dynamically redefinable character set [DRCS]

JCENS Joint Communications-Electronics Nomenclature System

JCL job control language (IBM)

JCP(S) junction call processing (subsystem)

jct junction

JDBC Java Database Connectivity (Sun Microsystems *tn*)

JDC Japan's Digital Cellular/Communications (system)

JDISM Japanese Display Interface for Monitors Group

JDK Java development kit (Sun Microsystems *tn*)

JDP Joint Development Program

JDS synchronous digital hierarchy (s)

JE jump if equal

JECI Java external call interface (Sun Microsystems *tn*)

JEDEC Joint Electronic Devices Engineering Council

JEIDA Japanese Electronic Industry Development Association (US)

JEPI Joint Electronic Payments Initiative (W3C)

JES job entry system/subsystem

JES2/3 job entry system 2/3

JESS Java Enterprises Solutions Symposium (Paris, 1999)

JESSI Joint European Submicron Silicon

JF junction frequency; junctor frame

JFC Java foundation classes (Sun Microsystems *tn*)

JFCL jump if flag set and then clear the flag

JFET junction field-effect transistor

JFIF JPEG file interchange format

JFS journaled file system (IBM); jumbo group frequency supply

JFY justify

JG jump if greater

JGE jump if greater or equal

JGF junctor grouping frame

JIC Joint Industrial Council
JIPS JANET Internet protocol service
JIS Japan Industry Standard
JIT just in time
JJT Josephson junction transistor
JK-flip-flop flip-flop device with two inputs and two outputs
JKLAP jack key and lamp access panel (part of SMAS)
JL jump if less
JLE jump if less than or equal to
JLR junction loudness rating
JM2C just my 2 cents
JMAPI Java management application program interface
JMP jump
JMX jumbo group multiplex
JNA jump if not above
JNAE jump if not above or equal
JNATCS Joint National Air Traffic Control Services (UK)
JNB jump if not below
JNBE jump if not or equal
JNDI Java naming directory interface (Sun Microsystems *tn*)
JNET Japanese Network
JNG jump if not greater
JNGE jump if not greater or equal
JNI Java Native Interface (Sun Microsystems *tn*)
JNLE jump if not less or equal
JNO jump if no overflow
JNP jump if no parity
JNS jump if no sign
JNZ jump if not zero
JOE Java objects everywhere
JOLT Java open language toolkit (Sun Microsystems *tn*)
JOOP Journal of Object-Oriented Programming
JOSS joint overseas switchboard (US armed forces)
JOVIAL Jules' own version of IAL
JPE jump if parity even
JPEG Joint Photographic Expert Group
JPL Jet Propulsion Laboratory

JPLDIS Jet Propulsion Laboratory Display Information System
JPO jump if parity odd
JRE Java runtime environment (Sun Microsystems *tn*)
JRO Jet replication object (Microsoft)
JRST jump and restore
JS jump if sign
JSA Japanese Standards Association
JSA Java Security Alliance
JSD Jackson system development; justification service digit
JSE Java spectrum emulator (Sun Microsystems *tn*)
JSF junctor switch frame
JSN junction switch number
JSNM Japan Society of New Metals
JSP Jackson structured programming
JSSS Java script style sheet
JSTV Japanese Satellite Television
JSW junctor switch
JTA Jewish Telegraph Agency (Germany)
JTAG Joint Test Action Group
JTB jump trace buffer
JTC Joint Technical committee (ISO/IEC); junctor terminal circuit
JTC^{3}A Joint Tactical Command Control and Communications Agency
JTEC Japan Telecommunications Engineering and Consultancy
JTIDS Joint Tactical Information Distribution System (USAF jamming-proof system)
JTM job transfer and management/manipulation
JTMP Jackson Transfer and Manipulation Protocol
JTRB Joint Telecommunications Resources Board
JTS Java transaction services (Sun Microsystems *tn*)
JTSSG Joint Telecommunications Standards Steering Group
JU joint user
JUGFET junction-gate field-effect transistor

JUGHEAD Jonzy's Universal Gopher Hierarchy Excavation and Display

JUGL JANET user group for libraries (UK)

jukebox optical disk library

JUNET Japan UNIX Network; Japanese University Network (Internet)

JVM Java virtual machine

JVTOS joint viewing and tele-operation service

JWID Joint Warrior Inter-operability Demonstration

JYNRUG JANET Yorkshire and Northumbria Regional User Group (UK)

JZ jump if zero

K

k kilo, one thousand, 10^3

K black in CMYK colour scheme; constraint length of convolutional code; kelvin (thermodynamic/temperature SI(12) unit); 'over' cue in Morse code (*ie* invitation to transmit)

K&R C version of C by Kernighan and Ritchie

k/d keyboard/display

K/h kelvin hour

K6 AMD processor *tn*

KA keep-alive (signal)

KADS knowledge-acquisition domain system; knowledge analysis and design system

KAM keep alive memory

KANA Koppelnetzanschluß A (g), switching network terminal A

KAOS Kent applicative operating system

KAP kernel Andorra Prolog

KAPSE kernel Ada programming support environment

Karnaugh Karnaugh maps show states/conditions in logic circuits

KAS keep-alive sequencing

KAU keystation adaptor unit

kb kilobit

kB kilobyte

K-band microwave band between 12-40 GHz

KBD keyboard

kbit kilobit (1024 bits)

kbit/s kilobits per second

kBps kilobytes per second

KBps kilobytes per second

KBS knowledge-based system

kbyte kilobyte (1024 bytes)

KC cipher/ciphering key

KCL Kyoto common Lisp

KCMS Kodak colour management system

K-complexity Kolmogorov complexity

kcs kilocycles per second

KDC key distribution centre; Kodak digital camera

KDD knowledge discovery in databases; Kokusai Denshin Denwa (Japanese international carrier)

KDE K Desktop Environment (for Linux)

KDLOC 1000 developed (source) lines of code

KDR keyboard data recorder

KDT key definition table; keyboard display terminal

KEE knowledge engineering environment

KEFIR key findings reporter (GTE)

Kennzahl identifying code (g)

Kennzeichen identifying signal (g)

KER key-encryption resource (Wide Area Network application)

KERMIT file transfer program (Columbia University)

kernel lowest subdivision of an OS

KeV kilo-electron volt

KEY index file Microsoft Mail

KFS kommender Fangsatz (g), switched connection hold circuit incoming

KFT kilofeet

KG key generator (encryption); Koppelgruppe (g), switching network section

KGABST Koppelgruppe AB Steuerung (g), control for switching section AB

KGW Knotengruppenwähler (g), node group selector

KH11 Keyhole 11 satellites (for real-time observation, US military)

kHz kilohertz

Ki (individual) subscriber authentication key

KIBO knowledge in – bullshit out

KIF knowledge interchange format

KIGO knowledge in – garbage out

kilobaud 1000 baud

KILOSTREAM British Telecom service

KIPS thousand instructions per second

KIS Knowbot Information service

KISS keep it simple stupid

KIT key intelligence topics

KKF cross-correlation function (g)

KKI Kreditkarten-Institut (g), credit card company

KKT kleiner Konzentrator (g), small concentrator

KKTS kleiner Konzentrator-Satz (g), small concentrator unit

KKTSTG kleiner Konzentrator-Steuerung (g), controller for small concentrator

kL kilolitre

KL künstliche Leitung (g), artificial line attentuator

KL1 kernel language 1

Klg Knoten-Vermittlungsstelle-Leitung gehend (g), node switching centre line outgoing

Klk Knoten-Vermittlungsstelle-Leitung kommend (g), node switching centre line incoming

KL0 kernel language 0

KLOC 1000 lines of code

kludge inelegant but effective correction/patch (software/hardware)

KM knowledge module

kmc kilo-megacycles (*aka* gigahertz)

KMP Knuth-Morris-Pratt (algorithm)

KMS knowledge-management system

kMUP kilo mobile user part

KN KoppelNetz (g), switching network

KNI Katmai New Instructions (on Intel processor *tn*)

KnU knowledge utility

Ko kilo octet (*ie* kb)

ko/s kilo-octet par seconde (f)

kohm kilo-ohm

Konf-Fpl Konferenz-Fernplatz (g), conference trunk switchboard

KOOL knowledge-oriented object language (Bull)

KOS Knights of Shadow

KP1 terminal seizure signal in CCITT No.5 signalling

KPA key-process area; key pulse adaptors

KPI kernel programming interface

KPS transit seizure signalling in CCITT Nº 5 signalling

KPT Kai's Power Tools *tn*

KPTC Kenya Posts and Telecommunications Corporation

KQML knowledge query and manipulation language

KR keyset receiver; knowledge representation

KRC Kent recursive calculator

KRD-D keyset receiving device digital

Kremvax mythical Usenet site at the Kremlin (originally an AFJ)

KRS knowledge retrieval system

KRW Kotenrichtungswählern (g), node route selector

KS kiosk service; kommender Satz (g), incoming circuit unit

KSAM keyed sequential access method

KSDS key-sequenced data set

Ksh KornShell (version of UNIX shell)

KSL Knowledge Systems Laboratory (Stanford University)

KSL-NRC Knowledge System Lab of the National Research Council (US)

KSLOC 1000 source lines of code

KSPH 1000 keystrokes per hour

KSR keyboard send/receive (terminal)

KSU key service unit

KT Konzentrator (g), line concentrator; noise/power density

KTA key telephone adaptor (Rolm)

KTN kernel transport network

KTS key telephone system/set

KTU key telephone unit

KU Kanalumsetzer (g), channel frequency translator

Ku band 10-12 GHz range of the electromagnetic spectrum

KUIP kernel user interface package

kV kilovolt

KV Koppelvielfach (g), switching matrix

kVA kilovolt-ampere (output rating)

KVA Koppelvielfach-Stufe A (g), switching matrix A

kVAr kilo-volt-ampere reactive

KVR Koppelvielfach-Reihe (g),
switching matrix row

KVRCST Koppelvielfachreihe C
Steuerung (g), controller for switching
matrix row C

KVSt Knoten-Vermittlungsstelle (g),
node switching centre

kWh kilo-watt hour

KWIC keyword-in-context

KWRP knowledge worker resource
processor

KZU Kennzeichenumsetzer (g), signal
converter

KZUG Kennzeichenumsetzer gehend (g),
signal converter outgoing

KZUK Kennzeichenumsetzer kommend
(g), signal converter incoming

KZW Kennzahlweg (g), code route

L

L inductance; layer; límite de tiempo para una referencia y números secuenciales (s); luminance (symbol for measure of brightness of a surface); wavelength (symbol; also λ)

L network half an unbalanced T network
L&E linking and embedding
L&O logic and objects
L&T line and terminal
L1/2/3/4 level 1/level 2/level 3/level 4
L2F level 2 forwarding
L2ML layer 2 management link
L2R layer 2 relay (function)
L2R BOP L2R bit-oriented protocol
L2R COP L2R character-oriented protocol
L2TP level 2 tunnelling protocol
L6 Bell Telephone Laboratories' low-level linked list language
LA lieux d'abonné (f), renter's premises [RP]; link acknowledgement, acuse de enlace (s); location area
LAAS Laboratoire d'Analyse et d'Architecture des Systèmes (f)
Lab Lightness red-green axis yellow-blue axis
LAB line attachment base (IBM)
LabVIEW Laboratory Virtual Instrument Engineering Workbench *tn*
LAC location area code/control/controller; loop assignment centre
Lace language for assembling class in Eiffel
LACM location area configuration management
Lacrosse US radar-equipped low-orbit (spy) satellite
LADC local area data channel
LADDR layered device driver architecture (Microsoft Windows *tn*)
LADS local area data service

LADT local area data transport/transmission (Bell)
LAI location area identity/identifier
LAIS local automatic intercept system
LALL longest allowed lobe length
LALN señal de liberación por abonado llamante (s), calling party clear signal [CCL]
LALR look-ahead left recursive
LAM lobe attachment unit (token ring)
LAMA local automatic message accounting
LAMC language and mode converter
LAMMA laser micro-mass analysis
LAMMS laser micro-mass spectroscopy
LAN local area network, réseau local (f), lokales Netz (g), Ortsnetz/lokales Netzwerk (g)
LAN Manager network OS (Microsoft *tn*)
LANACS LAN asynchronous connection server
LANC LAN control bus
LANCE Local Area Network Controller for Ethernet *tn*
LANDP LAN distributed platform
LANE LAN emulation
LANL Los Alamos National Laboratory
LANlord LAN software program (Microcom Inc *tn*)
LANServer network OS (IBM *tn*)
LAP link access procedure/protocol; LISP assembly program; local access port
LAP-B link access procedure B – balanced (X.25 CCITT)
LAP-D link access procedure for D-channel (ISDN)
LAPDm link access procedures on data mobile channel
LAPF link access procedure for frame relay
LAPF-Core core aspects of LAPF
LAPM link access procedure for modems (V.42), procedimiento de acceso al enlace para módems (s)

LAPX link access procedure half-duplex (X.32)

LAR load access rights; log area ratio; lot age report

LARAM line-addressable RAM

LAS League of Arab States (General Secretariat); local area service; low-altitude (observation) satellite

LASER light amplification by stimulated emission of radiation

LaserActive interactive variant of laser disk *tn*

LASINT laser intelligence

LASS local area signalling/switching service

LAT local access terminal; local area transport (DEC)

LATA local access and transport area

LaTeX version of TeX by Lamport

LATIS loop activity tracking information system

LATS loop access test system

LAVC local area VAX cluster (DEC)

LAWN local area wireless network

LAX language example

LB log book

LBA linear-bonded automaton; logical block addressing

L-Band frequency range 0.5-1.5 GHz; 95-1450 MHz for mobile communications

LBE language-based editor; loopback enable

LBF base feeder loss

LBL Lawrence Berkeley Laboratory

LBN line balancing network

LBO line build-out (unit)

LBR laser beam recording

LBRV low bit rate voice

LBS load balance system

LBTCBS local battery talking common-battery signalling

LBX local bus accelerator

LC liberación completa (s), release complete [RLC]; limited capability; liquid chromatography (inductance-capacitance); local control; logical channel; base combiner loss

LCA life-cycle analysis; local configuration analysis; logic cell array; Lotus communications architecture *tn*

LCAMOS loop cable maintenance operation system

LCAP loop carrier analysis program

LC-BCSM link control basic call state model

LCC language for conversational computing; lead-covered cable; leadless chip carrier (Intel); loading coil case

LCD liquid crystal display; long-constrained (data) delay; loss of cell delineation; lowest common denominator

LCF logic for computable functions; low-cost fibre

LCG logical connection graph

LCGN logical channel group number

LCI La Chaine Information (f) (France)

LCID language code ID

LCK Library Construction Kit (Microsoft/FoxPro *tn*)

LCL Liga control language; longitudinal conversion loss; lower confidence limit

LCLV liquid crystal light valve projector

LCM least common multiple; line concentrator module; link control manager

LCN local communication network; logical channel number; loosely-coupled networking (Control Data Corp)

LCP link control procedure/protocol

LCR acheminement au moindre coût/optimal (f), routage optimal; least-cost routing; line control register

LCS Laboratory for Computer science (MIT); language for communicating systems; line conditioning signal, signal de conditionnement de ligne (f); Linux compatibility standard *tn*

LCSAJ linear code sequence and jump

LCSI line current status indicator

LCSU local concentrator switching unit

LCT last compliance time (used in GCRA definition); (bloque de) liberación de conexión de transporte (f), transport connection request [TCR]

LCTL longitudinal conversion transfer loss

LCU last cluster used

LD laser diode/disk; liaison de données (f), data link [DL]; link disconnect; linker directive; lista de distribución (s), distribution list [DL]; loop disconnect

LDA logical device address

LDAP light(-weight) directory access protocol

LDB location database

LDC Lotus Development Corp

LD-CELP low-delay code-excited linear prediction

LDD lightly-doped drain; logic design data; logical database design

LDDI local distributed data interface

LDE indicación de longitud máxima rebasada (s), length-exceeded indication

LDFA local data function area

LDH location data home

LDI logistics data interchange

LDL language-description language (*aka* meta-language); linear differential logic; logic-based data language; lower detection limit

LDM limited-distance modem; linear delta modulation; long-distance modem

LDMTS long-distance message telecommunications service

LDN location data (home network); national trunk network (s)

LDO local delivery operator

LDP Linux documentation project; low-density plasma

LDR line driver-receiver; load register

LDS langage de spécification et de description (f), specification and description language [SDL]; local digital switch; local (TV) distribution system (US: wideband microwave/cable)

LDS/GR specification and description language–graphic representation (f)

LDS/PR specification and description language–phase representation (f)

LDSL local delivery services licence (UK)

LDT local descriptor table; logic design translator

LDTA Long-Distance Telecommunications Administration (Taiwan)

LDV location data (visited network)

LDX long-distance xerography

L_E earphone coupling loss

LE LAN emulation; Leitungsendgerät (g), management terminal; less or equal; listener echo loss; local exchange, central local (s); logical entity; loop extender

LE/PH local exchange/packet handler (ISDN)

LEA load effective address

LEAF law enforcement access field; LISP extended algebraic facility

LEAN Language from East Anglia and Nijmegen (Universities)

LEAP language for expression of associative procedures

LE-ARP LAN emulation – address resolution protocol

LEC LAN emulation client; light energy converter; liquid-encapsulated Czochralski; local exchange carrier; location exchange carrier

LECID LAN emulation client identifier

LECS LAN emulation configuration server

LED lenguaje de especificación y descripción (funcionales) (s), specification and description language [SDL]; light-emitting diode (s), diodo emisor de luz; Lightly-Educated Dudes

LED/GR LED/representación gráfica (s), SDL/graphic representation

LED/PR LED/representación con frases textuales (s), SDL/phrase representation

LEDA library of efficient data types and algorithms

LEDG lenguaje de especificación y descripción globales (s), overall specification and description language [OSDL]

LEDS low-end disk system
LEEP large expanse extra perspective
LeFun logic equations and functions
legacy non-plug'n'play peripheral, generally 'old' (pre-Windows)
LEGSS low-end graphics subsystem
LEL link-embed and launch-to-edit (Lotus); lower explosive limit
LELR listener echo loudness rating
LEM language extension module; logical end of media
Lempel-Ziv disk compression algorithm
LEN line equipment number; low entry networking; low-end networks
LEO Lyons Electronic Office (UK 1947)
LEOS low earth-orbit satellite
LEP Laboratoire Electronique Philips; light-emitting polymer
LEQ line-equipped
LER light-emitting resistor
LERP linear interpolation
LES LAN Emulation Server
LEST low-end systems and technologies
LETB local-exchange test bed
LETI Laboratoire d'Electronique de Technologie et d'Instrumentation (f)
LEV loader-editor-verifier
Lex lexicon; Unix lexical analyser generator
LEX lexical analyser (generator for UNIX); local exchange
lexeme lexical unit of language (formed in same way as phoneme)
LF laminar flow; line feed; line filter; line finder; low frequency
LFA loss of frame alignment
LFACS loop facilities assignment and control system
LFB look-ahead for busy (information)
LFC local function capabilities
LFI last file indicator
LFL lower flammable limit
LFN long file name (*ie* not 8.3 DOS format)
LFNBK long file name back-up (utility program)
LFR less-favoured region

LFRAP long-feeder route analysis program
LFS señal de línea fuera de servicio (s), line out of service signal [LOS]
LFSID local form session identifier
LFT low function terminal (IBM)
LFU least frequently used; link forced uninhibit (signal)
LG longitud de grupo (s), group length [GL]
LGC line-group controller
LGDF large-grain data flow
LGDT load global descriptor table
LGN linear graph notation; logical group node
LGPO local group policy objects (Microsoft Windows NT)
LGQ linear Gaussian quadratic
LGRC longitudes de gamas de repeticiones codificadas (s), coded run lengths [CRL]
LGU liberación de guarda (s), release guard [RLG]
LH line hunting; link header; load high; local host; locating handler
LHCM location handler configuration management
LHFM location handler fault management
LHH locating handler home network
LHM langage homme-machine (f), lenguaje hombre-máquina (s), man-machine language [MML]
LHO location handler originating network
LHO/T location handler originating/terminating network
LHPM location handler performance management
LHS line-handling system; literate Haskell source; signal de ligne hors service (s), line out-of-service signal [LOS]
LHT locating handler terminating network
Li lithium (used in batteries for portables)

LI laser interferometry; length indicator/indication; line identity

LIA link-inhibit acknowledgement signal

LIAS library information access system

LIB line bus; line interface board; message de libération (f), mensaje de liberación (s), release [REL]

LIBD line information database

LIBER Ligue des Bibliothèques Européennes de Recherches (f)

LIBOR London inter-bank floating rate

LIC (mensaje de) liberación completa (s), release complete message [RLC]; línea de identificación de la comunicación (s), call identification line [CIL]; linear integrated circuit; line/(linear) interface coupler (IBM)

LICS Lotus International Character Set *tn*

LID leadless inverted device; link-inhibit denied (signal); message de libération differée (f), mensaje de liberación diferida (s), delayed release signal [DRS]

LIDAR light detection and ranging (technology)

LIDB line information database (ISDN)

LIDF line intermediate distribution frame

LIDO liberado (s), released [RLSD]

LIDT load interrupt descriptor table

LIEP large Internet exchange packet (Novell)

LIF line interface; low insertion force

LIFE logic of inheritance, functions and equations

LIFIA Laboratoire d'Informatique Fondamentale et d'Intelligence Artificielle (f)

LIFO last in – first out, último en entrar – primero en salir (s)

LIG libération de garde (f), release guard [RLG]

Liga control language (attribute evaluator generator)

LIGHT life-cycle global hypertext

Li-ion lithium ion (laptop re-chargeable battery)

LIJP leaf-initiated joint parameter

LIK Legends in Kaos

LIL Larch interface language; longitudinal impedance loss; loudness insertion loss

LILO Linux loader

Lily Lisp library

LIM limited distance modem; line interface module; linear (delta) modulation; Lotus-Intel-Microsoft

LIM EMS Lotus-Intel-Microsoft expanded memory specification

LIMA laser-induced mass analysis; Lotus-Intel-Microsoft-AST

LIMDOW light intensity modulation direct over-write (Fujitsu *tn*)

LIMS laser-induced mass spectrometry; library information management system; limb infrared measurements in the stratosphere

LIN link inhibit signal

LiNbO$_3$ lithium niobate (used in superconductors)

LINCOMPEX linked compressor and expander

Linda model for distributed processing

LINGOL linguistics-oriented language

LINK/1/2 data/voice network exchange (Timeplex Inc *tn*)

Linpack benchmarking routine

LINS Laboratory for Artificial Neural Systems (University of Texas)

Linux Linus Unix (implementation of Unix kernel; after Linus Torvalds)

LINX London Internet Exchange

LIP large Internet packet

LIPL linear IPL

LIPS logical/linear inferences per second

LIPX large interwork packet exchange (NetWare)

LIR location inventory report

LIS langage implementation système (f), language implementation system; logical IP subnet (RFC 1577)

LISA large installation systems administration (USENIX); local integrated software architecture

LISP list processing language (for

manipulating non-numeric data); lots of irritating superfluous parentheses
LIST list number
Listproc mailing list processor owned/developed by BITNET
Listserv automated mailing list (server) distribution system
LISU local junction switching unit
LIT message de libération terminée (f), release-complete message [RLC]
LIU line interface unit
LIX legal information exchange
LK logischer Kanal (f), logic channel
LKDM low K dielectric material
LL land line; leased line, ligne/liaison louée (f); libération (f), liberación (s), release; line leg (telegraph)
LL (parsing) left-to-right leftmost parsing
LLA llamada aceptada (s), call accepted [CA]; logical link administration; low-level action
LLAP LocalTalk Link Access Protocol (Apple Macintosh *tn*)
LLAR local loop access ring
LLB latching loopback
LLC link level control; llamada conectada (s), call connected [CC]; llamada en curso, call proceeding, call in progress [CP]; logical link control (IEEE protocol); low-layer capability/compatibility
LLCC leadless chip carrier
LLD lower limit of detection
Lldo número llamado (s), called number [Cd]
LLDT load local descriptor table
LLE large local exchange; long-line equipment
LLEN llamada entrante (s), incoming call [IC]
LLF line link frame; low-level format/formatting; low-layer functions
LLI logical link identifier; Lunatic Labs Incorporated
LLL low-level language
LLM local linear models

LLN line-link network
LLNL Lawrence Livermore National Laboratory
LLNQ least lots next queue
LLP line-link pulsing; permanent logical link, liaison logique permanente (f)
LLRA LapLink remote access *tn*
LLS localized light scatterer
LLSC link set control
LLSU low-level signalling unit
LLT link local inhibit test (signal)
LLTV low-light television
LLV llamadas virtuales (s), virtual calls [VC]
LM layer management; leg multiple; Leistungsmerkmal (g), feature; light microscope; load monitor; logical module; low mobile; traffic channel (with lower capacity than Bm)
lm lumen (SI unit of luminous flux, one candela)
LMAO laughed my ass off
LMB Le Micro Bulletin (Webzine)
LMBCS Lotus multi-byte character set *tn*
LMD line mode data; local multipoint/multichannel distribution system
LME layer management entity; LM Erisson
LMF language media format; low magnetic field; mobile feeder loss
LMH local maintenance handler
LMI local management interface; local memory interconnection
LML lazy meta-language
LMMA laser microprobe mass analysis
LMMS local message metering service
LMN location mark number
LMNP land mobile numbering plan
LMOS loop maintenance operations system
LMR land mobile radio
LMS least mean square; local measured service; Lotus messaging switch *tn*
LMSI layer management service interface; local mobile station identity; local mobile subscriber identity

LMSS land mobile satellite service

LMSW load machine status word

LMT (very narrow band) linear modulation technology; local magnetic time; local maintenance terminal; location management in terminal

LMU LAN management utilities (IBM); line monitoring unit

LMX L-type multiplex

LN link attention; load number

LN:DI Lotus Notes: document imaging *tn*

LNA launch numerical aperture; link attention acknowledgement; local numbering area; low-noise amplifier

LNB low-noise blocking (down-) converter

LNC layer network coordinator; Leszynski naming convention

LND last number dialled; local number dialled

LNFR liaison numérique fictive de référence (f), hypothetical reference digital link [HRDL]

LNNI (standardized) LAN emulation network-to-network interface

LNO line-number – traffic circuit number

LNP local number portability

LNR low-noise receiver

LNS linked numbering system

LO line occupancy; linear objects; local oscillator; lock-out

LOC level of concern; lines of code; linked-object code; loop on-line control; loss of cell delineation

LOCANA loosely orthogonal class-activity notation for analysis

LOCAP low capacitance (cable)

LOCIS library of Congress Information System

LOCOS local oxidation of silicon

LOD Legion of Doom; level of detail

LOD/H Legion of Doom/Hackers

LODSB load string byte

LOF lock off-line; loss of frame (UNI fault management); lowest-operating-/-observable frequency

LOGO Lisp-like programming language for children

LOH line overhead (SONET)

LOI listening opinion index

LOL laughing out loud; Legion of Lucifer; longitudinal output level

LOLITA Language for On-Line Investigation and Transformation of Abstractions

LOMS log mobile station

LONAL local off-net access line

LOOPE loop while equal

LOOPNE loop while not equal

LOOPNZ loop while not zero

LOOPS Lisp object-oriented programming system

LOP line of position; loop prevention; loss of pointer

LOPZ loop while zero

LORAN long-range radio navigation (using time differences of pulsed transmissions)

LOS line of sight; line-oriented editor; line out of service; loss of incoming signal; loss of selectivity

LOTIS logic timing (and) sequencing

LOTOS Language of temporal ordering specification (ISO 8807)

LP linear programming; linearly-polarized (mode); link printer; linking protection; Lodge-pole pine (telephone pole); log-periodic (array/antenna); longitud de parámetro (f), parameter length [PL]; longitudinal parity; low pass (filter); propagation loss

LP$_{01}$ fundamental mode of an optical fibre

LPA linear power amplifier; loopback acknowledgement message

LPAC London Parallel Applications Centre

LPB local packet bus

LPC laser particle counter; linear predictive coding/encoding; liquid-borne particle counter; local procedure call; low particle concentration

LPCDF low-profile combined distributing frame

LPCVD low-pressure chemical vapour deposition

LPD light point defect; line printer daemon (Unix); link protocol discriminator; low probability of detection

LPDA link problem determination aid (IBM)

LPDN local public data network

LPE liquid-phase epitaxy

LPF League for Programming Freedom; low-pass filter

LPG langage de programmation générique (f), linguaggio procedure grafiche (i) (Bologna); Linux Programmers' Guide

LPGB llamada potenciada a grupo de barcos (s), fleet group call

lpi lines per inch; longitudinally-applied paper insulation

LPI low-pressure isolation; low probability of interception

LPL list programming language; Lotus programming language *tn*

LPLMN local PLMN

lpm lines per minute

LPM linearly-polarized mode

LPN logical page number

LPP lightweight presentation protocol; link peripheral processor

LPR line printer remote

LPRA Low-Power Radio Association

LPS licensable programme services; low-power Schottky

LPSM Levenson phase-shift mask

LPT line printer (terminal)

LPT port port for printer (parallel port)

LPT1/2/3 first/second/third parallel printer port

LPTV low-power television

LPV low-pressure vent

LPVS link-packetized voice subsystem

LQ letter quality

LQA link quality analysis

LQM link quality monitoring; local queue manager

LR link register; link request; location register/registration; loudness rating

LR (parsing) left-to-right, rightmost

LRAP long route analysis programme

LRC local register cache; local routing centre; longitudinal redundancy check; low-rate coding

LRCC longitudinal redundancy check character

LRE low-rate encoding

L$_{RET}$ returned echo level

LRF long-range facility

LRFAX low-resolution facsimile

LRGP loudness rating guard-ring position

LRIR limb radiance inversion radiometer

LRL least redundancy loaded

LRLTRAN Lawrence Radiation Laboratory Translator

LRM language reference manual; least-recently-used master

LRN location reference number

LRP long range plan

LRS laser Raman spectroscopy; line repeater station

LRT link remote inhibit test signal; Logic Replacement Technology (UK company)

LRU least-recently used; least-replaceable unit; line replacement/replaceable unit; lowest replaceable unit

LS language system; Leitungssatz (g), line unit; letter shift; ligne/liaison spécialisée (f) (permanent, leased Web link); line switch; line sync; link set; link state; loading splice

LS byte least-significant byte

LS0/LS1 locking shift 0/1

LS1R locking shift 1 right

LS2 locking shift 2

LS2R locking shift 2 right

LS3 locking shift 3

LS3R locking shift 3 right

LSA LAN and SCSI adaptor (IBM); limited-space charge accumulation; line sharing adaptor; link state advertisement; Linux Standards Association; local security authority (Microsoft); local serving area

LSAC signalling link activity control

LSAPI license services application program interface

L-sat European Space Agency communications satellite

LSB least significant bit, niedrigstwertiges bit (g); Linux standard base; lower sideband

LSC least significant character; line signalling channel; link state control; loopback select code

LSCP low-speed card punch

LSCU local service control unit

LSD data signalling link (f); least-significant digit; line-sharing device; line signal detector

LSDA signalling data link allocation

LSDS standby data link selection

LSE latex sphere equivalent; Language Sensitive Editor (DEC *tn*)

LSF line switch frame

LSHI large-scale hybrid integration

LSI Großintegration (g), large-scale integrated (circuit); line status indication

LSK line skip

LSL Larch shared language; large-scale integration; link support layer; load segment limit

LSLA signalling link activation

LSLD signalling link de-activation

LSLR signalling link restoration

LSM laser scanning microscope; line selection/switch module; line service marking

LSML lazy standard metalanguage

LSP library specification process; link state packet (NetWare); Liskov substitution principle; semi-permanent link (f)

LSPK loudspeaker

LSPSD low-speed packet switched data

LSPTR low-speed paper tape reader

LSR leaf set-up request

LSRP link state routing protocol; local switching replacement planning

LSS local synchronous subsystem; loop switching system

LSSD level-sensitive scan design

LSSU link status signal unit

LST line start code; loud-speaking telephone

LSTA signalling terminal allocation

LSTR listener sidetone rating

LSU leading signal unit; local synchronization utility; lone signal unit, unité de signalisation solitaire (f), unidad aislada de señalización (s)

LSV line status verifier

LSY Leitungssysyem (g), bus line system

LSYD language for systems development

LSYD Leitungssystem direkt (g), direct line system

LSYE Ersatzschalteleitungssystem (g), line system with standby change-over

LSYP peripheres Leitungssystem (g), system of lines to peripherals

L-system Lindenmeyer system (g), way of generating infinite sets of strings

LSYZ zentrales Leitungssystem (g), central bus system

LT letter telegram; line termination, link terminal; link trailer; link transfer; lower tester; low tension

LTA load-transfer-acknowledgement signal

LTAB line test access bus

LTB last trunk busy

LTC last-trunk capacity; line-terminating circuit; line-traffic coordinator; linear timecode (frame detector on audio tapes); lithographic test chip; local telephone circuit

LTCS long-term capability set

LTCVD low-temperature chemical vapour deposition

LTE line terminating equipment; line transmission equipment; local terminal emulator

LTF line trunk frame

Ltg Leitung (g), line/circuit

LTG Leitungsende (g), line trunk group
LTL longitudinal transfer rate/ratio; longueur totale de ligne (d'exploration) (f), longitud total de la línea de exploración (s), total scanning line-length [TLL]; lot-to-lot
LTM live traffic model
LTO low-temperature oxidation/oxide
LTP local test port; logical terminal profile; long-term prediction
LTPCU local test port control unit
LTPD lot tolerance percent defective
LTR langage temps-réel (f), real-time language [RTL]; load task register; load-transfer signal
LTRS letters shift
LTS local telephone system; loop-test system
LTU line test unit
LTV local thickness variation
LU line unit; local unit; logical unit; location update
LUA link uninhibit acknowledgement signal; location updating accepted; logical unit application
LUB least upper bound
Lucid query language (*also* dataflow language) (Lucid Inc *tn*)
Lucid Emacs text editor for X-Window system (FSF)
LUCM location update configuration management
L-UDM location related – user data management
LUF lowest-usable (high) frequency
LUFM location update fault management
LUH location update handler
LUI local user input
LUL longueur utilisable de la ligne d'exploration (f), usable scanning line length [ULL]
LULT line-unit-line termination
luma luminance
LUN link uninhibit signal; logical unit number
LUNI LANE UNI
LUNT line-unit-network termination

LUPM location update performance management
LUS large ultimate size > 6 000 lines
LUSTRE Lucid synchrone temps-réel (f), synchronous real-time Lucid
LUT look-up table
LUW logical unit of work
LV latent variable; length and value; logical volume (IBM); low voltage
LVA limit value availability
LVD low-voltage directive
LVDS low-voltage differential signalling
LVDT linear variable/voltage differential transformer
LVI low-voltage inverter
LVM line verification module; locating verification module; logical volume management (IBM)
LVR low voltage relay
LV-ROM optical disk for analogue and digital (Philips *tn*)
LVS layout verification of schematic
LW lazy write; leave word; Leitungswähler (g), line selector
LWA light-wire armoured
LWE lower window edge
LWG logistics working group
LWP lightweight process
LWR line width reduction
LWS Large Wafer Study
lx lux (Si unit of illuminance)
LYCOS Internet search engine
LYRIC language for your remote instruction by computer
LZH Lempel-Ziv and Haruyasu (compression algorithm)
LZW Lempel-Ziv-Welch (compression algorithm)

M

M mandatory, obligatoire (f), obligatorio (s); mega; metre; messagerie (f), message handling [MH]; million - 10^6; modem; modifier function bit; múltiple (s), multiple; parámetro obligatorio (s), mandatory parameter; mutual inductance

M&E music and effects

M/MU monitor/matrix unit

M/N/T main/satellite/tributary network

M13 DS1-to-DS3 multiplexer

M2 Mondo 2000

M2FM modified modified frequency modulation

M2M manager-to-manager

M2toM3 Modula 2 to Modula 3

M6 Metropolis 6 (France)

mA milli-ampere, 10^{-3}

MA mass announcement; medium adaptor; MJU alert; mobile allocation; monitor de tasa de errores en la alineación (s), alignment error rate monitoring [AERM]

Ma Bell nickname for AT&T

MAA major synchronization point

MAB mensaje de abonado libre (s), calling party-free message [CPM]

MABLR modulation d'amplitude à bande latérale réduite (f)

Mac Macintosh: Apple Mac (Apple)

MAC message authentication code, code d'authentification de message (f); mandatory access control; market approved correction; medium access component; medium/media access control (IEEE 802.4 token bus); medium access control sublayer; multiple access computer (MIT); multiple analogue component; multiplexed analogue component

MAC layer media access control layer

Mac OS X Macintosh OS X (*ie* ten) (Apple *tn*)

MAC1 machine-aided cognition (project at MIT)

MACBS multi-access cable billing system

MACC multiplier-accumulator

MACF multiple association control function

MACH multi-layer actuator head (Epson)

Mach variant of UNIX OS by Carnegie-Mellon University

MACL Macintosh Allegro Common Lisp

MACLisp dialect of LISP

MacMinix Apple Mac version of Minix

MACN mobile allocation channel number

MACOM major command

macro macro-instruction (Windows intra-application code)

MACSYMA MAC's symbolic manipulator

MACT maximum achievable control technology

MAD demande de modification d'appel (f), call modification request signal [CMR]; magnetic anomaly detector (Russian 'grusha' = pear-shaped); Michigan algorithm decoder (University of Michigan); mixed analogue and digital; modem access device; modulation adaptative delta (f), adaptive delta modulation [ADM]; retardo medio administrativo (s), mean administrative delay

MADAP Maastricht automatic data processing and display system

MADE manufacturing and automated design engineering

MADGE microwave aircraft digital guidance equipment

MADN multiple access directory numbers

MADT mean accumulated downtime; micro-alloy diffused transistor

MADTRAN FORTRAN-to-MAD translator

MADYMO mathematical dynamic modelling

MAE marcación automática de extensiones (s), direct dialling in [DDI]; mean absolute error; message de modification d'appel effectuée (f), call modification complete [CMC]; metro area Ethernet; metropolitan area exchange

MAF management application function; mobile additional function; mode addition flag

mag magnetic/magneto

MAGE mechanical aerospace ground equipment

MAGIC multi-service applications governing integrated communications

mah milli-ampere-hour

MAH mobile access hunting (supplementary service)

MAHO mobile assisted handover

MAI mail message (encrypted) Microsoft Mail; mobile allocation index; multiple applications interface

MAID Market Analysis and Information Database

MAIDT mean accumulated intrinsic downtime

MAIHF mobile assisted inter-frequency handoff

Mailbot e-mail server that automatically responds to requests

MAINBOL macro implementation of SNOBOL 4

MAINSAIL machine independent SAIL

MAIO mobile allocation index offset

MAJ mise à jour/majoration (f)

MAJC Microprocessor Architecture for Java Computing (Sun *tn*)

Majordomo freeware mailing list processor

MAL micro-assembly language

MALDI matrix-assisted laser desorption and ionization

malloc C library routine: storage allocation

MALT mobile telephony A interface line terminal

malware malicious/malevolent/malign software (*ie* viruses)

MAMI modified alternate mark inversion

man manual (Unix command)

MAN metropolitan area network

MAP maintenance administration panel; maintenance analysis procedures; major synchronization point; management application protocol; Manufacturing Automation Protocol (ISO-OSI standard); mathematical analysis without programming; memory allocation map; minimum acceptable performance; mobile application part

MAPI messaging applications programming interface (Microsoft)

MAPICS manufacturing accounting and production information control system

MAPS Mail Abuse Protection System

MAPTC mobile telephony application part incoming transaction coordinator

MAR memory address register; mercury-arc rectifier; message de refus de modification d'appel (f), call modification reject message [CMRJ]; micro-program address register; mobile autonomous registration

MARCO map rasterization and conversion system

MARISAT maritime satellite system

Mark 1 automatic sequence-controlled calculator Mark 1

MAROTS maritime orbital test satellite

marquee web page banner with scrolling text

MARS Maritime Mobile Access and Retrieval System (ITU); Military Affiliated Radio System; multicast address resolution server

MARSYAS Marshall System for Aerospace Simulation

MART tiempo medio de reparación activa (f), mean active repair time

MARTS metallic access remote test system

MARVEL machine-assisted realization

of virtual electronic library (US Library of Congress)

MAS maintenance subsystem; Modula 2 algebra system

MASCOT modular approach to software construction, operation and test (UK MoD)

MASE message administration service element

MASER microwave amplification by stimulated emission of radiation

MASM macro-assembler for MS-DOS (Microsoft)

MASS maximum availability and support subsystem (Parallan)

MAT machine-aided translation; maintenance access terminal; metropolitan area trunks

MATD maximum acceptable transit delay

MATE multifunction asynchronous terminal emulator

MATFAP metropolitan area transmission facility analysis program

MATV master antenna television

MAU maintenance unit; medium/media attachment unit (Ethernet); multi-station access unit (token ring)

MAUC master authentication centre

MAUTH mobile telephony authentication handler

MAVDM multiple application virtual Dos machine [VDM]

MAWP maximum allowable working pressure

max maximum

max CR maximum cell rate

max CTD maximum cell transfer delay

MAXE maintenance of AXE

MAXIT maximum interference threshold

maxwell unit of magnetic flux

MB machine batch; Magnetband (g) (magnetic tape); megabit; megabyte; message buffer

MB SAP MAC broadcast SAP

MBA maintenance-oriented group blocking-acknowledgement message

MBA multiple beam antenna

MBASIC Microsoft BASIC *tn*

MBC machine bath collection; Middle East Broadcasting Centre; mobile bearer control

MBCF mobile bearer control function

MBCS multi-byte character set (IBM)

MBE molecular beam epitaxy

MBF mobility bearer function

MBG Microsoft mail header/mail file

M-bit more-data bit

Mbit/s megabits per second

MBMS molecular beam mass spectrometry

MBONE Multicast Backbone (virtual internet backbone for Multicast IP)

MBPC model-based process control

Mbps megabits per second

MBps megabytes per second

MBR master boot record; memory buffer register

MBS master boot sector; maximum burst size; M-bit sequence; mobile base station; mobile broadband services/system; multi-block synchronization signal unit, unité de signalisation de synchronisation des multiblocs (f), unidad de señalización de sincronización de multiblocs (s)

MBSSD mobile telephony base station system data

MBT maximum burst tolerance; mobile bothway trunk

MBTC model-based temperature control

MBTC mobile bothway trunk circuit

MBUR maintenance-oriented circuit group blocking and unblocking receipt

MBUS maintenance-oriented circuit group blocking and unblocking sending

MBX mailbox; message bus exchange

MByte megabyte

MByte/s megabytes per second

mc megacycle, milli-coulomb

MC main cross-connect (DEC); maintenance centre; maintenance connector; maritime centre; megacycle; message category; message centre; Mikrocomputer-Steuerung (g), microcontroller; mini-cartridge; multi-carrier; C with modules; multi-copier

MCA manual-changeover-acknowledgement signal, signal d'accusé de réception de commutation manuelle sur liaison de réserve (f); measurement capability analysis; MicroChannel (bus) Architecture (IBM *tn*); multicast address

MCAD mechanical CAD

MCAE mechanical CAE

MCAV modified constant angular velocity

MCB memory control block; miniature circuit breaker; multicasting box

MCBA mean cycles between assists

MCBF mean cycles between failures

MCBI mean cycles between interrupts

MCBS micro-cell base station

MCBSE mean cycles between scrap event

MCC maintenance connector controller; maintenance control circuit; measurement computing and communications; Microelectronics and Computer Technology Corporation (US); mobile country code; mobile cross-talk control; mobile telephone control channel; multi-cell cabinet configuration; Mosaic Communications Corporation (*now* Netscape *tn*)

MCCF mobile call control function

MCCN micro-cell-/(cellular) control node

MCCU multi-system channel communications subsystem

MCD maintenance cell description; mini-client driver; mobile call delivery; module de coopération (f), index of cooperation [IOC]

M-CDR mobility management CDR

MCDV maximum cell delay variance

MCE Music Choice Europe

MCEB Military Communications-Electronics Board

MCF mensajes de control de flujo del tráfico de señalización (s), signalling traffic flow control [TSFC]; message confirmation; message control

functionality; meta-content file/format; meta-content framework; mobile control function; monitorización de la calidad de funcionamiento (s), performance monitoring [PM]

MCG multi-colour graphics

MCGA multi-colour graphics adaptor/array

MCH mobile call charge; mobile call handler; mobile control handler

MCH(O/T) mobile call handler (originating/terminating network)

MCHB maintenance channel buffer

MCHCM mobile call handler configuration management

MCHFM mobile call handler fault management

MCHO mobile call handler originating

MCHO/T mobile call handler originating/terminating network

MCHPM mobile call handler performance management

MCHT mobile call handler terminating

MCI machine check interruption; maintenance command interpreter; making communications improbable; malicious call identification (supplementary service); media control interface (Microsoft Windows); message type indicator; Microwave Communications Inc

MCIAS multi-channel intelligent announcement system

MCIC modulación casi instantánea compansorizada (s), nearly-instantaneous compandored modulation [NIC]

MCIS Microsoft Commercial Internet System

MCIXS Maritime Cellular Information Exchange Service

MCK multi-coupler kit

MCL Macintosh Common Lisp; Microsoft Compatibility Labs

MCLR maximum cell loss ratio

MCLV modified constant linear velocity

MCM multi-carrier modulation; Monte Carlo Music; maintenance control module; manufacturing cycle management; multi-chip module (Intel)

MCN maintenance connector network; master customer number; micro-cell network; mobile control node

MCNE Master Certified Novell Engineer

MCNS multimedia cable networks systems

MCO signal de commutation manuelle sur liaison de réserve (f), manual change-over signal

MCO SAP MAC connection-oriented SAP

MCP main call process; mega-chips per second; motion-compensated prediction; multi-chip package

MCPA multicarrier power amplifier

MC-PGA metallized ceramic – pin grid array

MCPN mobile customer premises network

MCPR multi-media communication processing and representation

MCPS mini-core processing subsystem

MCR modem control register

MCS maintenance control subsystem; master/material control system; meta-class system; minimum cell rate; multinational character set; multipoint conference service; multiple congestion states

MCSD Microsoft Certified Solution Developer

MCSE Microsoft Certified Product Specialist

MCT malicious call trace/(tracing); mercury-cadmium-telluride detector; motion-compensated transform (compression algorithm); múltiplex por compresión en el tiempo (s), time-compression multiplex [TCM]

MCTD maximum cell transfer delay

MCU master control unit; measuring coupling unit; microprocessor control unit; mobile calibration unit; multi-chip unit (DEC); multi-coupler unit (antenna);

multipoint conference unit; multipoint control unit

MC-UDM mobile call – usage data metering

mcvax Mathematisch Centrum (was EUnet backbone node, Amsterdam)

MCVD metal chemical vapour deposition; modified chemical vapour deposition

MCVFT multi-channel voice-frequency telegraphy

MCW modulated continuous wave

MCXO microcomputer-compensated crystal oscillator

MD macro diversity; major delivery; make directory (DOS command); management domain; mediation device; message digest function; message discriminator/distributor; Mini Disk (Sony *tn*); minutes dégradées (f), minutos degradados (s), degraded minutes [DM]; molecular dynamics; multiple dissemination; multiplexor digital (s), digital multiplexer [DM]; clase manipulación de documento (s), document manipulation class [DM]

MDA mail delivery agent; manufacturing defects analyser; mensaje de dirección hacia adelante (s), forward address message [FAM]; message digest algorithm 5 (encryption algorithm – RFC 1321); modulación delta adaptativa (s), adaptive delta modulation [ADM]; modular digital architecture; monochrome display (and printer) adaptor (IBM); multi-dimensional analysis; multiple digit absorbing

MDAC Microsoft data access components

MDB management/multimedia database

MDBS mobile database station

MDC mail distribution centre; mensaje de dirección completa (s), address complete message [ACM]; Meridian Digital Centrex; multiplicación digital de circuitos (s), digital circuit multiplication [DCM]

MDCU message distributor control unit
MDCUB MDCU bus
MDD Marshals of Dynamic Discord
MDDBMS multidimensional database management system
MDE marcación directa de extensiones (s), direct dialling-in [DDI]; mobile data equipment
MDEQ mobile telephony digital equipment
MDF main distribution frame; modulation par déplacement de fréquence (f), modulación por desplazamiento de frecuencia (s), frequency shift keying [FSK]; multiplexación por división de frecuencia (s), múltiplex con división en frecuencia (S), frequency division multiplexing [FDM]
MDHMS McDonnell Douglas human modelling system *tn*
MDI medium-dependent interface; memory display interface; more-data indicator; multiple document interface (Microsoft Windows API)
MDIC Manchester decoder and interface chip (AT&T)
MDIO message distributor input/output
MDIOR message distributor input/output repeater
MDK Multimedia Developers' Kit (Microsoft *tn*)
MDL minimum detection limit; communication between management entity and DLL
MD-MOS multi-drain metal-oxide semiconductor
MDN message disposition notification
MDNS managed data network service
MDP mensaje de paso de largo (s), pass-along message [PAM]; message discrimination process; modulación por desplazamiento de fase binaria codificada diferencialmente y filtrada (s), binary-phase shift keying [BPSK]
MDPSK modified differential PSK
MDQ market-driven quality

MDR memory data register; minimum design requirement; Mittel Deutscher Rundfunk (g)
MDRAM multi-bank dynamic RAM
MD-ROM playback-only data Mini-Disk
MDS microprocessor development system; mobile data subscriber/subsystem; modify device status; multipoint distribution services; Microsoft Download Service *tn*; minimum discernible signal
MDSE message delivery service element
MDSL moderate speed digital subscriber line
MDT mean downtime; multiplexación por división en el tiempo (s), time-division multiplexing [TDM]
MDT/AMDT múltiplex por división en el tiempo/acceso múltiple por división en el tiempo (s), time-division multiplexing/time-division multiple access [TDM/TDMA]
MDTRS mobile digital trunked radio system
MDU modular dispensing unit
MDUS medium data utilization station (Australia)
MDX modular digital exchange
ME maintenance entity; mean error; message element; mobile equipment
MEA maintenance entity assembly
MEB memory expansion board
MEBS medium energy back-scattering spectrometry
MEC mensaje hacia atrás para información de establecimiento fructuoso de la llamada (s), successful backward set-up information message [SBM]; mensajes de establecimiento de la comunicación (s), call set-up message [CSM]; mobile equipment console
MECU millions of ECU
MED mensaje de orden de conexión del enlace de datos de señalización (s), signalling data link connection order message [DLM]; modelling for

equipment design; modem equivalent device

Medienrat.de (German) Internet Media Council

MEECN minimum essential communications network

MEF maintenance entity function; máquina de estado finito (s), finite state machine [FSM]; measurement function

MEFA extended finite-state machine (s)

MEG mensaje hacia adelante para información de establecimiento general (s), forward set-up message [FSM]

Meganet LAN service (General DataCom Inc *tn*)

Megastream BT's digital leased-circuit service

MEGO my eyes glaze over

MEHO mobile evaluated handover

MEI maintenance event information; unsuccessful backward-set-up acknowledgement message (UBM)

MEIT máximo error en el intervalo de tiempo (s), maximum time-interval error [MTIE]

MEL mensaje hacia adelante para establecimiento de la llamada (s), forward set-up message [FSM]

mem memory

MEM manufacturing enterprise model; manufacturing equipment monitor; member (of workgroup post office – Microsoft Mail *tn*); micro-electro-mechanical (system)

MENL maximum external noise level

MEOS medium earth-orbit satellite

MEP mensaje de paso de emergencia a enlace de reserva (s), emergency change-over message [ECM]; multiple exchange paging

MER nominal error rates

MERCAST merchant ship broadcast system

MERIE magnetically-enhanced reactive ion etching

MERISE Methode d'étude et de réalisation informatique pour les systèmes d'enterprise (f)

MERIT intelligent network operating module; máximo error relativo en el intervalo de tiempo (s), maximum relative time interval error [MRTIE]

Merlin Sharp PDA *tn*

MERS most economic route selection

MES manufacturing execution systems

M-ES mobile end system

MESFET metal semiconductor field-effect transistor

MESI modified exclusive shared invalid (protocol)

MET memory enhancement technology (Hewlett Packard *tn*); multi-button electronic telephone; multi-emitter transistor

METAFONT system complementing TeX for designing digital fonts

METAL mega-extensive telecommunications application language

méta-moteur meta-search

METL multi-element two-layer

METS Materials and Equipment Trading Service

MEU mensajes entre UAR télex (s), inter-Telex SFU messages [IM]

MeV mega-electrovolt

mf micro-farad (μf)

MF marca de fin (s), end of TSDU mark [EOT]; media filter; mediation function; medium frequency (between 300-3 000 kHz); Mehrfrequenzverfahren (g) (multi-frequency dialling); mobility functions; more fragments (flag TCP header); multi-frame; multifrequency

MFB marcador de fin de bloque (s), end of block mark [EOB]

MFBT multifrecuencia bitono (s), dual-tone multifrequency [DTMF]

MFC mass flow controller; Microsoft Foundation Class *tn*; multifrequency code, code multifréquence (f); multifrequency (signalling) compelled

MFCE mode field concentricity error

MFC-LME Ericsson version of MFC signalling

MFC-R2 channel-associated signalling standard

mfd micro-farad

MFD mode field diameter; multi-function device

MFDAP mobile telephony fetch subscriber data mobile application part

MFE maximal free expression

MFENET Magnetic Fusion Energy Network

MFFS Microsoft flash file system

MFJ Modified Final Judgement (of Bell divestiture decree, US)

Mflops mega-flops: million floating-point operations per second

MFM mass-flow meter; maximally-flat magnitude; modified/multiple frequency modulation (encoding)

MFPB multifrequency pushbutton (signalling)

MFPI multifunction peripheral interface

MFPT (señalización) multifrecuencia por teclado (s), multifrequency pushbutton [MFPB]

MFR multifrequency receiver

MFS Macintosh File System *tn*; magnetic (tape) field search; memory file system; metropolitan fibre systems; modified filing system (Revelation Technologies); bearer capability (f) [BC]

MFT mainframe termination; master file table (Microsoft Windows NT); metallic facility terminal

MFTL my favourite toy language

MFTM multifrecuencia de tonos de marcación (s), dual-tone multifrequency [DTMF]

MFV Mehrfrequenzverfahren (g) (multifrequency dialling)

Mg magnesium (used in light-metal alloys)

MG manufactured goods; MicroGnuEmacs

MGA monochrome graphics adaptor

MGB maintenance-oriented group blocking message

MGC manufactured goods collection; metafichero de gráficos por computador (s), computer graphic metafile [CGM]

MGCP media gateway control protocol

MGE mensaje general de información para el establecimiento (s) [GSM]; modular GIS environment

MGMT management system

MGN multi-grounded neutral

MGP mensaje general de petición para el establecimiento (s), general forward set-up information message [GRQ]

MGT mobile global title

MGU maintenance-oriented group unblocking message; modem gateway unit

MH message handling

mh milli-henry

MHD moving head disk

MHDL microwave hardware description language

MHE message-handling environment

MHEG Multimedia and Hypermedia Information Coding Expert Group

MHF medium-high frequency

MHI material hazard index

MHO mobile telephony handover

MHP message handling processor

MHPCC Maui High Performance Computing Center

MHPHC media de la hora diaria definida por horas completas (s), average of daily peak full hour [ADPFH]

MHS message handling system/service (ITU-T X.400); mise hors service (f), isolation/shutdown

MHS-SE message-handling system service element

MHTML MIME (encapsulated) HTML

MHz megahertz (megacycles per second)

MI machine interface; maintenance information; management interface; marketing identifier; multiple inheritance

MI/MIC mode indicate/mode indicate common

MIA mensaje inicial de dirección con información (s), initial address message with additional information [IAI]; message initial d'adresse (f), initial address message [IAM]; minor synchronization acknowledgement

MIB management information base (RFC 1156)

MIB 2 management information base 2 (RFC 1213)

mic microphone

MIC medium interface connector (FDDI); message identification code, code d'identification de message (f); message integrity check; Microsoft International Corp; microwave integrated circuit; minimum ignition current; mobile interface controller; modulation par impulsions et codage (f), modulación por impulsos codificados (s), pulse code modulation [PCM]; mutual interface chart; monolithic integrated circuit

Mic 1/2 micro-programming language 1/2

MICA modem ISDN channel aggregation

MICD modulation différentielle par impulsions et codage (f), modulation par impulsions et codage différentiel (f), modulación por impulsos codificados diferencial (s), differential pulse code modulation [DPCM]

MICDA modulation par impulsions et codage différentiel adaptatif (f), modulación por impulsos codificados diferencial adaptativa (s), adaptive differential pulse code modulation [ADPCM]

MICDA-SB diferencial adaptativa de subbanda (s), sub-band ADPCM [ADPCM-SB]

MICE modular integrated communications environment; Multimedia Integrated Conferencing for Europe

MICR magnetic ink character recognition

micro one millionth, 10^{-6}

Micro EMACS EMACS as used on IBM compatible PCs

Micro VAX operating system (DEC *tn*)

MicroDrone a Microsoft employee

MicroSloth derogatory reference to Microsoft

MICS macro-interpretive commands; MVS integrated control system

MID material ID; maritime identification digit, cifras de identificación marítima (s); mean interruption delay; message identifier; mensaje inicial de dirección (s), initial address message [IAM]; multiplexing identifier

MIDA message interchange distributed application

MIDAS multiple input data acquisition system; multi-tier distributed application service suites

MIDI musical instrument digital interface

MIDR mosaicked image data record

MIDS matrix information directory services; multiple information distribution system

MIE magnetron ion etching; mensaje de información por el establecimiento (s), general forward set-up information message [GSM]

MIF Maker Interchange Format (FrameMaker Corp *tn*); management information format; minimum interworking facility/functionality; modo de predicción intratrama forzada (s), intraframe prediction code [IFM]

MIFR master international frequency register

MIG mensaje de inhibición de gestión (s), management inhibit message [MIM]; metal-in-gap

MII major industry identifier; mensaje inicial de dirección con información adicional (s), initial address message with additional information [IAI]; Microsoft-IBM-Intel

MIKE micro interpreter for knowledge engineering

MIL machine interface layer (Go Corp); mensaje de identidad de la línea que llama (s), calling-line identity message [CLI]

MILIA International Content Market for Interactive Media

milli one thousandth, 10^{-3}

MILNET (US) Military Network – part of DDN

MIL-STD military standard

MIL-STD-188-131 H320 compatible videoconferencing (US military)

MIM management inhibit message; message input module; metal-insulator-metal; Microsoft Instant Messenger (MSN *tn*); modem interface module

MIMAC measurement and improvement of manufacturing capacity

MIMD management information of metrology data; multiple instruction (stream) multiple data (stream) processor

MIME multipurpose Internet mail extensions (RFC 1521)

mimencode MIME encode (binary data)

MIMO multi-input/multi-output

min minute

MIN mensaje de identidad de la línea llamante no disponible (s), calling-line identity unavailable signal [CLU]; mobile identification number; mobile intelligent network; multi-stage interconnection network

MIND modular interactive network designer (CONTEL)

MINIT minimum interference threshold

MINS mobile intelligent network services

MINT MinT is not TRAC

Minuet Minnesota Internet users essential tool

MINX Multimedia Information Network Exchange

MIOS modular input-output system

MIP medium interface point; minor synchronization point; (servicio de) mensajería interpersonal (s), interpersonal message (service) [IPM]

MIPL multi-mission image processing laboratory

mips million instructions per second

MIPS microprocessor without interlocked pipeline stages

MIR magnification/reduction; maximum information rate; micro-instruction register

MIRANDA functional programming language (similar to ML)

MIRFAC mathematics in recognizable form automatically compiled

MIS maintenance information system; management information system; message initial d'adresse avec informations supplémentaires (f), initial address message with additional information [IAI]; metal insulator silicon; mobile telephone interworking subsystem

MISCF miscellaneous frame

MISD multiple instruction (stream) single data (stream) processor

MISFET metal insulator semiconductor FET

MISG Market Information Services Group

MIT Massachusetts Institute of Technology

MITER modular installation of telecommunications equipment racks

MITI Ministry of International Trade and Industry (Japan)

MITS micro-computer interactive test system; Micro Instrumentation and Telemetry Systems (US Corp)

miu metallic interface unit

MIU multi-station interface unit; multi-system interconnect unit

MIX member information exchange

MJ multi-junctor

M-JPEG motion JPEG

MJTC multi-junction thermal converter

MJU multipoint junction unit

MK mittlerer Konzentrator (g), medium-size line concentrator

Mkdir make directory

MKR marker
MKSA metre-kilogram-second-ampere system
MKT mittlerer Konzentrator (g), medium-size line concentrator
MkTypLib make type library (Unix command)
ML machine language; mean life; message length; metalanguage; Manipulator Language (IBM robot-handling language *tn*); motivo de la liberación (s), release cause
MLAB modelling laboratory
MLABT mobile telephony line terminal label translator
MLAPI multilingual application programming interface
MLC major local company; mensaje modificación de llamada completada (s), call modification complete [CMC]; multilayer ceramic; multilink control field (X.25), campo de control multienlace (s)
MLCAP mobile telephony locating cancellation mobile application part
MLCC multilayer ceramic capacitors
MLD modelos lineales dinámicos (s), dynamic linear model [DLM]; retardo medio logístico (s), mean logistic delay
MLDS medium large disk systems
MLE multiline editor
MLHG Multiline Hunt Group
MLID multilink interface driver; multiple link interface driver
MLM mailing list manager; medium-level language; modify lot location; multi-level metal; multi-longitudinal mode
MLOC mobile telephone location coordinator
M-loop tape path in VHS
MLP multi-layer perceptron; multi-layer protocol; procedimiento multienlace (s), multilink procedure
MLPC model-level planning committee
MLPP multilevel precedence and pre-emption

ML-PPP multilink point-to-point protocol
MLR message de libération retardée (f), delayed release message [DRS]; message log report; multi-channel linear recording
MLS multiple listing service
MLSE maximum likelihood sequence estimation/estimator
MLSO mode-locked surface-acoustic-wave oscillator
MLT mechanized line/loop testing; multi-line terminal; multiple logical terminal
MLUAP mobile telephony location updating mobile application part
MLV modify logging versions
Mm módem (s), modem
MM manufacturing methods; man-machine; mémoire des messages (f), memoria de mensajes (s), message store [MS]; Mensch-Maschine-Sprache (man-machine language); mixed mode, modo mixto de funciónamiento (s); mobility management
MM/AM memoria de mensajes/almacenador de mensajes (s), memory/message store [MS]
MMA Microcomputer Managers Association (US); MIDI Manufacturers' Association
MMBS multimedia base station
MMC man-machine communications; Manufacturing Methods Council; matched memory cycle; Mickey Mouse Club; Micro-computer Marketing Council; Microsoft management console (Microsoft Windows NT); MML command; mobile-mobile card
MMCD multimedia compact disk
MMCS mass memory control subsystem
MMCX Multimedia Communication Exchange (Lucent Technologies *tn*)
MMD micro-lithographic mask development program
MMDF multi-channel memorandum distribution facility
MMDS microwave multipoint distributed

180

systems; multi-channel multipoint
distribution system

MME man-machine entity; mobile
management entity

mmf magnetomotive force;
micromicrofarad (*also* μμf)

MMF mail message file (Microsoft
Mail); mobility management function;
multi-mode fibre interface/fibre-optic
cable

MMFS manufacturing message format
standard (EIA)

MMGM mensaje múltiple de gestión de
red y de mantenimiento (s), multi-unit
network, management and maintenance
message [MMM]

MMH maintenance man-hours

MMI man-machine interface; multimedia
interface

MMIC monolithic microwave integrated
circuit

MMIN multi-path multi-stage
interconnection network

MMIS maintenance management
information system; materials manager
information system

MML man-machine language, Mensch-
Maschine-Sprache (g)

MMM material movement management;
message multiple de gestion et de
maintenance du réseau (f), multi-unit
network, management and maintenance
message; mobile-telephony mobility
management

MMMC machine material movement
component

MMMS Material Movement
Management Standard; multimedia mail
service

MMO multi-model optimization, mobile
module (Intel)

MMOS modified MOS

MMPM multimedia presentation manger

MMR méthode de mesure de référence
(f), reference test method [RTM]

MMS main memory status;
manufacturing message specification;

memory management system; mobile
mobility and radio subsystem

MMSI maritime mobile services
identities

MMSQ Microsoft message queue server

MMST Microelectronics Manufacturing
Science and Technology

MMU mass memory unit; memory
management unit; mensaje múltiple (s),
multi-unit message [MUM]

MMW millimetre wave

Mmx Matrix math extensions *tn*

MMX mastergroup multiplex; matrix
manipulation extensions; multimedia
extensions (Intel processor instructions
tn)

Mn manganese (used in batteries and
ceramics)

MN minimum network; multiplexeur
numérique (f), digital multiplexer [DM]

MNBF mobility non-bearer function

MNC mobile network code;
multiplicative noise compressor

MNCC mobile network configuration
check

MNCS multipoint network control
system

MND mobile network design

MNN main network node; mensaje del
nodo-a-nodo (s), node-to-node message
[NNM]; mini-link network manager

MNOS metal-nitride-oxide
semiconductor

MNP micro-computer networking
protocol; Microcom Networking (error-
correction) Protocol *tn*

MNPVC multi-network permanent
virtual circuit

MNR non-delivery notification (f)
[NDN]

MNRU modulated noise reference unit

MNS metal-nitride semiconductor;
mobile network signalling

MNSAN mobile telephony number series
analysis

MNT maintenance, mantenimiento (s)

Mo méga-octet (f), megabyte (MB)

MO magneto-optical; managed object; market operations; metal-organic; mobile originated; Messobjekt (g), item under test

MO/CD-R magneto-optics compact disk – re-writable

MO/PP mobile-originated point-to-point message

MOA memorandum of agreement

MOAD mother of all demos

MOAT measurement and operations analysis team (at NLANR)

MOB moveable object block (*aka* sprite)

MOBILIPREV reverse tunnelling for mobile IP (RFC 2344)

MOBITEX mobile packet data system

MOBO motherboard

MOBSSL Merritt and Miller's own block-structured simulation language

MOC mobile-originated call

MOCA Merisel Open Computing Alliance

MOCN machine-outil à commandes numériques (f), numerical-control machine tool

MOCVD metallo-organic chemical vapour deposition

mod modify/modification

MOD magnetic-optical disk; movies on-demand (on Internet); storage format for Amiga module music files

modem modulator-demodulator

Modula modular high-level language developed from Pascal (Modula 2,3)

MODULE master deliverables list

MOF maximum-observed frequency

MOHLL machine-oriented high-level language

Mohm mega ohm, megohm

MOI managed-object instance

Mol mole (measure of resistance)

MOL multiplexeur optique local (f), local optical multiplexer

MOLENE multiplexeur ligne économique en noeud étoile (f), star-node economical line multiplexer

MOLP Microsoft Open Licence Pack *tn*

MOM message output module; Microsoft Office manager; Microsoft Office Manager *tn*

MOMS microwave radio operation and maintenance system

MOO multi-user object oriented

MOP maintenance operations protocol (DEC); modem part; modify operating procedures; multiple on-line programming; multiple original print-outs (HP)

Mops méga-octets par seconde (f), megabytes per second [MBps]

MOPT mean one-way propagation time

MoRE Masters of Reverse Engineering (Norwegian hacker group)

MOS metal-oxide-silicon; metal-oxide semiconductor; mean opinion score; multiplexer out-of-sync

Mosaic Internet browser software (NCSA *tn*)

MOS-C metal-oxide semiconductor capacitor

MOSFET metal-oxide semiconductor field-effect transistor

MOSS maintenance and operator subsystem

MOST metal oxide semiconductor transistor; metal-oxide-semiconductor FET

MOTAS member of the appropriate sex

motd message of the day

moteur de recherche search engine

MOTIS message-oriented text interchange service

MOTOS member of the opposite sex

MOTSS member of the same sex

MoU memorandum of understanding (European protocol)

MoU-3GIG MoU-third generation interest group

MoU-BARG MoU-billing and accounting group

MoU-CONIG MoU-conformance of network interfaces group

MoU-EREG MoU-European Roaming Group

MoU-P MoU-procurement

MoU-RIC MoU-radio interface coordination

MoU-SERG MoU-service group

MoU-SG MoU-security group

MoU-TADIG MoU-transfer account data interchange group

MoU-TAP MoU-type approval administrative procedures

MOV metal-oxide varistor

MOVPE metal organic vapour pressure epitaxy

MOVS Microsoft Office and VBA Solutions (conference); move string

MOX mixed oxide (nuclear reactor fuel)

Mozilla original name of Netscape's Internet browser and mascot (cousin of Godzilla)

mp melting point

MP manipulador de paquetes (manejador de paquetes) (s), packet handler/handling [PH]; massively parallel; Messplatz (g), measuring equipment; multilink point-to-point protocol; multiple processors

MP tape metal-particle tape

MP/M multi-programming monitor; multi-tasking program for microcomputers

MP1 modo procesable número 1 (s), processable mode Nº 1 [PM1]

MP3 MPEG-1 audio layer 3 (developed by Fraunhöfer Institute)

MPA mensajes de paso a enlace de reserva y retorno al enlace de servicio (s), change-over and change-back service [CHM]; método de prueba alternativo (s), alternative test message [ATM]

MPAG mobile telephony paging

MPC marker pulse conversion; memory protection check; Mercury Personal Communications (UK); message-passing co-processor; miniature protector connector; Multimedia PC (MPC Marketing Council standard); multimedia personal computer; multi-path channel; multi-process communications

MPCA máquina de protocolo de control

de asociación (s), association control protocol machine [ACPM]

MPCC multi-protocol communications controller

MPCH main parallel channel

MPCS multi-party connection subsystem

MPDM modular processor data module

MPDN mobile public data network

MPDU message protocol data unit

MPE mean percent error; mensaje hacia atrás para petición de establecimiento de la llamada (s), backward set-up message [BSM]; Multiple Programming Executive (Hewlett Packard)

MPEG Moving/Motion Picture Experts Group (CCITT-ISO/IEC group)

MPEG2 Motion Picture Experts Group (TV to DVD standard)

MPES mensaje de prueba de enlaces de señalización (s) [SLTM]

MPF multi-channel peak factor; multilink packet fragmentation

MPG mensaje de petición general (s), general request message [GRQ]

MPG/MIE mensaje de petición general y mensaje hacia adelante de información general para establecimiento (s), general forward set-up information message [GRQ/GSM]

MPGA mask programmable gate array

MPH (mobile) management entity-physical (layer)

MPI mensaje de petición de información (s), information request message [INR]; message passing interface; multi-precision integer

MPK multi-processing kernel (NetWare 5)

MPl Messplatz (g), measuring equipment

MPL mensaje de paso lo largo (s), pass-along message [PAM]; message-passing library (IBM); micro-programming language; Motorola Programming Language *tn*; multimedia personal (computer); multi-schedule private line

MPLA mobile project line assembly

MPMD multiple processor-multiple data

MPOA multi-protocol over ATM (ATM Forum)

MPOD máquina de protocolo de operaciones a distancia (s), remote operations protocol machine [ROPM]

MP-OES multipoint optical emission spectroscopy

MPOW multiple-purpose operator workstation

MPP máquina de protocolo de presentación (s), presentation protocol machine [PPM]; massively parallel processor/processing (Intel); message posting protocol; multiple parallel processing; multi-phase printing

MPPD multi-purpose peripheral device

MPQP multi-protocol quad port (IBM)

MPR mensaje de prueba de conjunto de rutas de señalización (s), signalling route set test message [RSM]; método de prueba de referencia (s), reference test method [RTM]; mis-dialled trunk prefix; multiple provider router (Microsoft Windows 95); multi-protocol router (Novell NetWare); Swedish National Board for Measurement and Testing

MPRES modular plasma reactor simulator

M-Prolog Marseilles Prolog

MPS máquina de protocole de sesión (s), session protocol machine [SPM]; master power supply; mobile priority subscriber; multi-page signal; multiprocessor specification

MPSC multi-protocol serial controller

MPSX mathematical programming system extended

MPTF máquina de protocolo de transferencia fiable (s), reliable-transfer protocol machine [RTPM]

MP-TMD máquinas de protocolo de TMD (s), document transfer and manipulation – performance monitoring [DTAM-PM]

MPTN multi-protocol transport network

MPTS multi-protocol transport services/stream

MPTY multi-party supplementary service

MPU microprocessor unit

MPV closed-user group selection and validation check request (s) [CVS]

MPWD machine-prepared wiring data

MPX multiplex; multiplexer, multiplexor (s); multiprocessor extension

MQG multi-threaded query gate

MQI message queue/(queuing) interface (IBM)

MQIS message queue information store

MQS loss of incoming signal (f) [LOS]

MQW multi-quantum well

Mr message reference

MR magneto-resistive; memory (address) register; message rate; modem ready

MRBS micro-cell radio base station

MRCC maritime rescue coordination centre

MRCF Microsoft Realtime compression format *tn*

MRCI Microsoft Realtime compression interface *tn*

MRCT mobile telephone radio channel tester

MRD manual ring-down; modo de restauración por demanda (s), demand refresh mode [DRM]

MRDS Multics Relational Data Store *tn*

MRF maintenance reset function; message refusal (signal); multiplexage par répartition en fréquence (f), frequency-division multiplexing [FDM]; signal de refus de message (f), message refuse

MRG mensaje de reinicialización de grupo de circuitos (s), circuit group reset message [GRS]

MRI magnetic-resonance imaging; measurement requirements and interface

MRIR medium resolution infrared radiometer

MR-ISA modelo de referencia de interconexión de sistemas abiertos (s), open system interface – reference model [OSI-RM]

MRJE multileaving remote job entry

MRN mobile roaming/routing number; mobile terminal roaming number

MRNAP mobile telephony roaming number provision mobile application part

MRNR mobile telephony roaming number routing

MROM mask read – only memory

MROUTER Multicast Router

MRP manufacturer resource planning; materials requirements planning; message routing process; mouth reference point; multiple re-use pattern

MRPL main ring path length

MRP-RDSI modelo de referencia de protocolo RDSI (s), ISDN protocol reference model [ISDN-PRM]

MRR measurement result recording

MRRC mobile radio resources control

MRRM mobile telephony radio resource management

MRS Materials Research Society; media recognition system; modifiable representation system

MRSE message retrieval service element

MRT mean repair time, tiempo medio de reparación (s); memoria tampón de retransmisión (s), retransmission memory buffer [RTB]; multiplexage par répartition dans le temps (f), time-division multiplexing [TDM]

MRTIE maximum relative time interval error

MRTR mobile radio transmission and reception

MRU maximum receive unit; most recently used (files)

MRVA MTP routing verification acknowledgement

MRVR MTP routing verification result

MRVT MTP routing verification test

MS mass spectrometer/spectrometry/spectroscopy; measurement signal; memory system; message store; micro-second; Microsoft; mobile service; mobile station; mobile subscriber

MSA message subséquent d'adresse (f),

subsequent address message [SAM]; mobile subscription/subscriber administration

MSACM Microsoft audio-compression manager

MSAS Japanese military satellite constellations

MSAU multi-station/(status) access unit

MSAV Microsoft anti-virus

MSB mid-range systems business; most significant bit, bit más significativo (s); multi-block synchronization signal unit

MSBF mean swaps between failures

MSBVW magneto-static backward volume wave

MSByte most significant byte

MSC Malaysian (Multimedia) Super-Corridor (new hi-tech city); Maritime Switching Centre; mensaje de sobrecarga (s), overload message [OLM]; mensaje de supervisión del circuito (s), circuit supervision message [CCM]; message sequence chart; message switching computer/centre; mobile (services) switching centre; multi-strip coupler

MSC-A mobile service switching centre-A; MSC with call control at handover

MSC-B mobile service switching centre-B

MSC-B' MSC to which a subsequent handover is done

MSCDEX Microsoft CD extensions (CD driver)

MSCF mobility service control function

MSCM mobile station class mark

MSCP mass storage control protocol (DEC); mobile service/station control point

MSCU mobile station control unit

MSD mass storage device; Micro Systems Development; mensaje subsiguiente de dirección (s), subsequent address message [SAM]; Microsoft system diagnostics (ie MSD.EXE tn); most significant digit

MSD1-7 mensaje subsiguiente de dirección Nº 1-Nº 7 (s)

MSDAORA Microsoft Data Access Oracle *tn*

MSDASQL Microsoft Data Access SQL *tn*

MSDF mobility service data function

MSDN Microsoft Developer Network *tn*; mobile station directory number

MS-DOG irreverent name for MS-DOS

MS-DOS Microsoft disk operating system (Microsoft *tn*)

MSDP mobility service data point

MSDR multiplexed streaming data request

MSDS Microsoft developer support

MSDSE mobile satellite data switching exchange

MSE maintenance sub-entity/entities

MSED minimum signal element duration

MSEM Metrology Specific Equipment Model

MSF Microsoft solution framework; mobile storage function

MSFR minimum security function requirements (IBM)

MSFVW magneto-static forward volume wave

msg message

MSG management steering group; mensaje de supervisión de grupo de circuitos (s), circuit group supervision message [GRM]

msgGUI GUI for Gnu Smalltalk

MSGSM GSM mobile station

MSI manufacturing support item; marine/maritime safety/security information; MBus-to-SBus interface (Sun Microsystems *tn*); medium-scale integration; Microsoft system information; Mobile Systems International (US corporation)

MSID mass spectrometer lead detector

MS-ID mobile station random identification pattern

MSIDN mobile station international data number; mobile subscriber identity number

MSIDXS Microsoft Index Server *tn*

MSIN mobile station identification number; mobile station identity

MSINOF32 system information for Microsoft Windows 98

MSISDN mobile station international ISDN number

MSK minimum shift keying

MSL main signalling link; marcaje de servicios de la línea (s), line service marking [LSM]; map specification library; maximum segment lifetime; mirrored server link; mensaje de supervisión de la llamada (s), call supervision message [CSM]; modify system logging

MSLD mass spectrometer leak detector

MSM mass storage magazine; Micronetics Standard MUMPS

MSMQ Microsoft message queue

MSN Microsoft Network (ISP *tn*); mobile services node; mobile station number; multiple subscriber number; multi-subscriber numbering (ISDN)

MSNAN mobile telephony subscriber number analysis

MSNB mobile subscriber number

MSNF multi-system networking facility (of SNA *qv*)

MSO mobile switching office; multiple systems operator (*eg* cable operators)

MSP maintenance service provider; marketing and sales productivity; multi-thread simultaneous processing; Microsoft Paint *tn*

MSR magnetic stripe reader

MSRN mobile station roaming number

MSS maintenance support system; mass storage subsystem (voice mail); maritime satellite service; maximum segment size (TCP); message subséquent d'adresse avec un seul signal (f), subsequent address message with one signal [SAO]; mobile satellite system; mobile services switching; mobile switching subsystem/subscriber; modify system state; multi-protocol switched services (IBM); multi-spectral scanner

MSSC maritime satellite switching centre; mobile satellite switching centre

MSSCP mobile (service) switching and control point

MSSE message submission service element

MSSFU maritime-satellite store-and-forward unit

MSSP mobile (service) switching point

MSSW magnetostatic surface wave

MST multi-service terminal; multi-slotted token

MSTAB Manufacturing Systems Technical Advisory Board

MSTP multi-mission software transmission project

MSTS mobile station test system

MSU main switching unit; message signal/switching unit; microwave sounding unit; multi-block synchronization signal unit; mobile signalling variant

MSVC meta-signaling virtual channel

MSW machine status word; magnetic surface wave; magneto-static wave; microwave spectrometer

MSX Microsoft Extended Basic *tn*

MSYNC master synchronization

MT memoria tampón de transmisión (s), memoria tampón (s), transmission buffer [TB]; maritime terminal; measured time; message transfer; message type; mobile terminal/terminated; modified tape armour

MT/PP mobile-terminated point-to-point message

MTx modo de trama x (s), frame mode Nº x [FMx]

MTA mail/message transfer agent; maintenance task analysis; major trading area; mensaje de tasación (s), charging message [CHG]; Message Transfer Architecture (AT&T *tn*); mobile telephony traffic coordinator A-subscriber; mobile terminal address; mobile trading area; mode de transfert (temporel) asynchrone (f), modo de transferencia asíncrono (s), asynchronous transfer mode [ATM]; multi-functional adaptor; multiple terminal access; multi-protocol terminal adaptor

MTAE message transfer agent entity, entité agent de transfert de messages (f)

MTAS message transfer abstract service

MTAU metallic test access unit

MTB magnetic tape billing; mobile telephony traffic coordinator B-subscriber

MTBA mean time between assists

MTBAp mean (productive) time between assists

MTBD mean time between degradations

MTBF mean time between failure(s), moyenne des temps entre défaillances (f); moyenne des temps de bon fonctionnement (f); temps moyen entre pannes (f), tiempo medio entre fallos (s); fiabilidad (s)

MTBFp mean (productive) time between failures

MTBI mean time between interrupts/incidents

MTBO mean time between outages

MTBPM mean time between preventive maintenance

MTBSF mean time between system failures

mtc Modula 2 to C translator

MTC MIDI time code; mobile terminated call

MTCE maintenance

MTCN minimum throughput class negotiation

MTD magnetic tape drive

mtd mounted

MTDP Minority Telecommunications Development Program (US)

MTE mastergroup translating equipment, équipement de traduction de groupe tertiaire (f); message transfer event

MTF message transfer facility; modulation transfer function; Microsoft tape format *tn*

mtg mounting

MTG magnetic tape group
MTI message type indicator; Micro Technology Inc; moving target indication
MTIE maximum time interval error
MTL message transfer layer, couche de transfert de messages (f)
MTM magnetic tape magazine; micro-cell transceiver module; mobile-to-mobile (call)
MTOL mean time off-line/mean time on-line
MTP message transfer part/protocol; message transmission/transport part
MTP-SAP message transfer part service access point
MTR memoria tampón de retransmisión (s), retransmission buffer [RTB]; mensajes de prohibición, de autorización y de restricción de transferencia (s), transfer-prohibited and transfer-allowed messages [TFM]; mobile traffic recording
MTRN mobile terminal roaming number
MTRS mean time to restore service, tiempo medio hasta el restablecimiento de servicio (s)
MTS main trunk system; material tracking standard; message telecommunications service (AT&T); message toll service; message transmission system/subsystem; message transfer system; Michigan terminal system; Microsoft transaction server *tn*; mobile telephone subscriber/subsystem; mode de transfert (temporel) synchrone (f), modo (de) transferencia síncrono (s), synchronous transfer mode [STM]; multi-channel TV sound
MTSD module de transmission en mode semi-duplex (f), half-duplex transmission mode [HDTM]
MTSE message transfer service element
MTS-N módulo de transporte síncrono de nivel N (s), synchronous transport module level N [STM-N]
MTSO mobile telephone switching office
MTSR mean time to service

restoral/restoration, temps moyen de rétablissement du service (f)
MTT magnetic tape terminal; maritime test terminal; multi-transaction timer
MTTA mean time to assist
MTTC mean time to correct
MTTD mean time to diagnose
MTTF mean time to failure, durée moyenne de fonctionnement avant défaillance (f), tiempo medio hasta el fallo (s)
MTTFF mean time to first failure, durée moyenne de fonctionnement avant la première défaillance (f), tiempo medio hasta el primer fallo (s)
MTTR mean time to recovery/repair, mean time to restoration, moyenne des temps pour la tâche de réparation (f), durée moyenne de reprise, tiempo medio hasta el restablecimiento (s)
MTTRS master telephone transmission reference system
MTTS multi-tone test signal
MTTSR mean time to service restoral, temps moyen de rétablissement du service (f)
MTU magnetic tape unit; maximum transmission unit; mobile test unit; multi-terminal unit
MTUP mobile telephone user part
MTV mobile telephony visiting subscriber; Music Television *tn*
MTVP mobile telephony visiting subscriber pointer
MTWX mechanized teletypewriter exchange
MTX mobile telephone exchange
MTX/HLR mobile telephone exchange/home location register
MTXG gateway MTX
MTXH home MTX
MTXV visited MTX
MU mark-up, matrix unit; mobile user; multiple destination unidirectional, destinos múltiples (s); Multiplexeinrichtung (g), multiplex equipment

MUA mail user agent; maintenance-oriented group unblocking-acknowledgement message

MUD multi-user dialogue/dimension/domain/dungeon

MUD SAP MAC U-plane data SAP

MUF maximum usable frequency

MUI multimedia user interface

mul multiply

MUL digital multiplexer plus demultiplexer; modify user login

mu-law voice encoding standard (ISDN)

MULDEM multiplexer-demultiplexer

MULTICS Multiplexed Information and Computing System (GE *tn*)

MultiMon multi-monitors (Microsoft Windows 98)

Multiplan spreadsheet program (Microsoft *tn*)

MUM multi-unit message, message multiple (f), mensaje múltiple (s)

MUMPS Massachusetts General Hospital Utility Multi-Programming Service

MUMS multi-user mobile station

Münz-Fw Münz-Wählfernsprecher (g), coinbox – automatic telephone

MUP mobile user part; modify user password; multiple universal naming convention provider

MUR radio relay message unit

MUS message signal unit; monitor de tasa de errores en las unidades de señalización (s), signal unit error rate monitor [SUERM]

MUS SAP MAC U-plane speech SAP

MUSA multiple-unit steerable antenna

MUSE monitor of ultra-violet solar energy; multiple sub-Nyquist sampling encoding; multi-user shared/simulated environment

MUSH mail user's shell; multi-user shared hallucination

MUSICAM masking universal sub-band integrated coding and multiplexing

MUSL Manchester University systems language

MUT mean up-time; monitor under test; multi-terminal

MUX multiplex/multiplexing equipment, Multiplexeinrichtung (g); multiplex/multiplexer

MUXing multiplexing

MV megavolt

mv millivolt; move/rename file (Unix)

MVB MicroVax Business

MVC mobile telephone voice channel; model view controller

MVD motion vector data

MVDM multiple virtual DOS machine

MVDS/MMDS microwave/multipoint video distribution service

MVGA monochrome VGA

MVI multi-vendor interaction

MVIP multi-vendor integration protocol

MVP mobile voice privacy; modular voice processor; multimedia video processor

MVS multiple virtual storage/(store) (IBM)

MVS/ESA multiple virtual storage/enterprise systems architecture

MVS/SP MVS (1)/system product

MVS/TSO MVS (1)/time-sharing option

MVS/XA MVS (1)/extended architecture

MVTR moisture vapour transmission rate

mw milliwatt

MW medium wave; megaword; message waiting; molecular weight; multi-wink

MWBC mean wafers between cleans

MWD message waiting data

MWI message waiting indicator/indication

MWN message waiting notification

MWS microwave scatterometer

MWT monitor wafer turner

MWTDPS microwave team data processing system, système de traitement de données du groupe micro-onde (f), sistema de proceso de datos del grupo micro-ondas (s), Datenverarbeitungssystem der Mikrowellen-Arbeitsgruppe (g)

MWTN multi-wavelength transport network

mx matrix

MXE message system

MXS Microsoft Exchange server

MXU message transfer unit; multiplexer unit

MZ1 two-byte executable header (Mark Zbikowski DOS programmer)

N

n frequency symbol; number density as particles per m^3 (electrons in semiconductor); refractive index

n nano, one thousand millionth, 10^{-9}

N entities, entités (f), entidades (s); network; número máximo de retransmisiones (s)

N(R) receive sequence number/transmitter receive sequence number, número secuencial en recepción (s)

N(S) send sequence number; transmitter send sequence number, número secuencial en emisión (s)

N(S)N national (significant) number

N/A not applicable, no aplicable [N/A]

N/I noise/interference ratio; non-interlaced

N/W network

N$_0$ spectral noise density; sea level refractivity

N x niveau x (f), nivel x (s), level x [Lx]

NA Nebenanschluß (g), extension; network adaptor/address/architecture; no access; not applicable; numerical aperture; numéro d'abonné (f), número de abonado (s), subscriber's number [SN]; ouverture numérique (f)

N-A numérique-analogique (f), digital analogue [D-A]

NAA neutron activation analysis

NAB National Association of Broadcasters (US)

NABER National Association of Business and Educational Radio (US)

NAC network adaptor card; null attachment concentrator

NACD National Association of Computer Dealers (US); network automatic call distribution

NACK negative acknowledgement

NACOM National Communications System (US civil defense)

NACS National Advisory Committee on Semiconductors (US); NetWare Asynchronous Communication Services (Novell *tn*)

NACSEM National Communications Security Emanation Memorandum (US)

NACSIM National Communications Security Information Memorandum (US)

NACSIS National Centre for Science Information Systems (Japan)

NACSOM NASA Communications Network

NAD network access device; network administrator; noeud d'accès en distribution (f), distribution access node [DAN]

NADF North American Directory Forum

NADO noeud d'accès de distribution optique (f), optical distribution access node

NAE network address/addressing extension

NAEC Novell Authorized Education Centre

NAFF not allocated

Nafis National Automated Fingerprint Identification System (UK)

NAG Network Administrator's Guide (to Linux)

NAG Numerical Algorithm Group (UK)

NAHM network access handler mobile terminal; network authentication handler mobile terminal

NAHN network access handler network side; network authentication handler network

NAIG North American Interest Group

NAK negative acknowledgement, acuse de recibo negativo (s)

NAl Nebenanschlußleitung (g), PBX subscriber line

NAL Novell application launcher

NAM number assignment module; numéro d'abonné multiple (f), multiple subscriber line [MSN]

NAMPS narrowband

analogue/(advanced) mobile phone service (Motorola)

NaN not a number

NANBA North American National Broadcasters' Association

NAND gate logical operator

nano nanosecond; one billionth, 10^{-9}

NANP North American Numbering Plan

NAP network access point

NAPE National Pirate Exchange

NAPI numbering and addressing plan indicator

NAPLPS North American Presentation-Level Protocol Syntax (AT&T, ANSI)

NAPPA Northern American Phreakers/Piraters association

NAPSS numerical analysis problem-solving system

NARC non-automatic relay centre; Nuclear Phreakers/Hackers/Carders

NA-RDSI número de abonado de RDSI (s), ISDN subscriber number [ISDN-SN]

NARUC US National Association of Regulatory Utility Commissioners

NAs Nebenanschluß (g), extension

NAS network access server, Network Applications Support (DEC *tn*); networked attached storage; Numerical and Atmospheric Sciences Network

NASA-IS NASA Information Services

NASCOM NASA Communications Network

NASDAQ US National Association of Securities Dealers Automated Quotations

NASHD high bit-rate service access node (f)

NASI NetWare asynchronous services interface (Novell *tn*); Nebenanschlußleitung (g), PBX subscriber line

NASTRAN NASA stress analysis program

NAT network address translation

NATA North American Telecommunications Association

NAU network access/addressable unit

NAVASTAR/GPS navigational satellite global positioning system

NAVSAT Satellite Navigation Corp

Navspasur Navy Space Surveillance system (US)

NAVSTAR-GPS navigational satellite timing and ranging – GPS

NAWK new AWK

Nb Nachbildung (g), load; niobium (superconductive when alloyed with tin or aluminium)

NB narrowband (< 2 Mbit/s); narrow beam; no blancos (g), no break here [NBH]; normal burst

NB3 National Band Three (UK)

NBACR network announcement request

nbb number of bytes of binary

NBC National Broadcasting Company (US)

NBCD natural binary-coded decimal

NBE not below or equal

NBF NetBEUI frame (Microsoft)

NBFCP NetBIOS frame control protocol

NBFM narrowband FM

NBG von Neumann-Bernays-Goedel (set theory)

NBH network busy hour; no break here, corte prohibido aquí (s)

NBI nothing but initials

NBIP nine-bit inter-processor protocol

NBISDN narrowband ISDN

NBMA non-broadcast multiple access

NBMA/NHRP non-broadcast media ARP/next hop resolution protocol

NBOC network building-out capacitor

NBOR network building-out resistor

NBP name binding protocol (AppleTalk *tn*)

NBR number, numéro (f), número (s)

NBRVF narrow-band radio voice frequency

NBS US National Bureau of Standards (old name for NIST)

NbSk Nachbildung einer Sprechkapsel (g), load-simulating microphone insert

NBSP no-break/non-breaking space

NBSV narrowband secure voice

NBT NetBIOS over TCP/IP
NBTC non-basic terminal capabilities
NBTR narrowband tape recorder
NC network computer/congestion/connect/connection/controller; no circuit; noise criterion; number of unallocated channels at node; numerical control; ordinateur de réseau (f), network computer
NC1 nivel de congestión 1 (f), congestion level 1 [CL1]
NCA National Command Authorities (US); network communications adaptor; network computing architecture (Internet)
NCAP non-linear circuit analysis program
NCC national (telephone) code change – to 10-digit numbers (UK); National Coordinating Center for Telecommunications (US); national colour code; National Computing Centre (UK); negociación de la clase de caudal (facilidad) (s) [TCN]; network-centric computing; network (PLMN) colour code; network control centre; Norwegian Cracking Company
NCCD network-dependent call connection delay
NCCE native computing and communications environment (TINA)
NCCF network communications control facility (IBM)
NCCH Norwegian Computing Centre for the Humanities
NCCID network-dependent call clear indication delay
NCCM negociación de la clase de caudal mínimo (facilidad) (s) [MTCN]
NCCSAP network layer call control SAP
NCD Network Computing Devices Inc
NCELL neighbouring cell
NCET National Council for Educational Technology (US)
NCF NetWare command file; network connection failure

NCGA National Computer Graphics Association (US)
NCI Network Computer Inc (Oracle)
NCID network clear indication delay, temps d'indication de libération dans le réseau (f)
NCIS National Criminal Intelligence Service (UK)
NCL node compatibility list
NCM network configuration management; numerical control machine-tool
NCMS National Center for Manufacturing and Science
NCOS network computer OS
NCP NetWare Core Protocol (Novell *tn*); NetWare cross-platform services (Novell *tn*); network configuration process (NORTEL); network control protocol/point; not copy protected
NCR National Cash Registers (US); network call references; no carbon required/non-carbon ribbon
NCR-DNA NCR Corporation Distributed Network Architecture
NCS National Communications System (US DoD); National Environment Research Council Computer Service (UK); neighbouring cell selection and handling/support; net control station; network communication standard; network computing system (DEC/HP); network coordination station; no-checking signal
NCSA National Center for Supercomputing Applications (US); National Computer Security Association (US); network coordination station assignment
NCSC National Communications Security Committee; National Computer Security Centre (US); network coordination station common
NCSI Network Communications Services interface (Network Products Corp); network coordination station interstation

NCSS network coordination station spot-beam

NCT night-closing trunks

NCTA National Cable Television Association; notación combinada tabular y de árbol (s), tree and tabular combined notation [TTCN]

NCTA-GR NCTA en forma gráfica (s), TTCN graphic representation [TTCN-GR]

NCTE network circuit/channel terminating equipment

Nd neodymium (used to make artificial rubies for lasers)

ND natural deduction

NDA non-disclosure agreement

NDB non-directional (radio) beacon

NDC national destination code; normalized device coordinates, cordonnées normalisées/normées (f)

NDCS network data control system

NDDK Network Device Development Kit (Microsoft *tn*)

NDE non-destructive evaluation

NDER National Defense Executive Reserve (US)

NDES normal digital echo suppressor

N$_{DEV}$ nivel del eco devuelto (s), returned echo level [L$_{RET}$]

NDF new data flag

NDI installation directory number (f)

NDIAG Norton Diagnostics *tn*

NDIR non-dispersive infrared spectroscopy

NDIS network driver interface specifications (Microsoft/3Com)

NDL National Database Language (US); network definition language (Burroughs)

NDM National Distribution Network (US); negative delivery notification; non-delivery (status) notification; normal disconnected mode, modo desconectado normal (s)

NDM requester Network DataMover Requester

NDM server Network DataMover Server

NDP neutron-depth profiling; numeric data processor

NDPS Novell distributed print services *tn*

NDR network data representation; nombre distinguido relativo (s), relative distinguished number [RDN]; Nord Deutscher Rundfunk (g)

NDRO non-destructive read-out

NDS additional directory number (other than main DDI number); NetWare Directory services (Novell *tn*)

NDSE national data switching exchange

NDT net data throughput; non-destructive testing

NDUB network-determined user busy

NDUV non-dispersive ultraviolet spectroscopy

NDV número de datos del vector (s) [VDN]

NE network element/entity; new executable (two-byte executable header); notificación de entrega (s), delivery (status) notification [DN]

NEACP National emergency Airborne Command Post (US)

NEARnet New England Academic and Research Network

NEAT Novell Easy Administration Tool *tn*

NEBS network equipment building system

NEC National Electric Code (US); Nippon Electric Corp

NECA Network Exchange Carrier Association

NECCSD NEC Computer System Division

NECOS network coordinating station

NED NetView-DECnet Interface Option *tn*

NEE near-end error

NEF network element function

NEF1 Normes d'exploitation et de fonctionnement (f), (French) operating and performance standards

neg negative/negate

negater NOT gate

NEHO network-evaluated handover

NEI Normes d'exploitation et d'ingénierie (f), French operating and engineering standards

NEL network element layer

NELIAC Navy Electronics Laboratory International ALGOL

NEM network element management

NEMA National Electrical Manufacturers Association (USA)

NEMP nuclear electromagnetic pulse

NEP network entry point; noise equivalent power; notificación de entrega positiva (s), positive delivery notification [PDN]

NEQ non-equivalence

NEQ gate non-equivalence gate; XOR gate

NERECO network remote communications

NES noise equivalent signal

NESAC National Electronic Switching Assistance Center (US)

NESC National Electric Safety Code (US)

NESHAP National Emissions Standards For Hazardous Air Pollutants (US)

NESP near-end signalling point

NESS National Environmental Satellite Service (US)

NEST Novell embedded systems technology *tn*

NESTA National Endowment for Science Technology and the Arts (UK)

Net the Internet

net neural network

NET Norme Européenne de Télécommunications (Eu)

NET 1000 unsuccessful AT&T network

Net BEUI network BIOS extended user interface

Net BIOS network BIOS

net nanny self-appointed politically-correct Internet arbiters

net police *v* net nanny

NetBEUI NetBIOS extended user interface (IBM)

NetBIOS network basic input/output system (token ring interface)

netCDF network common data form

NETID network identifier

netiquette Internet etiquette

Netlink high-speed multiplexer (DCA)

netmon network monitor (Microsoft Windows)

Netnews Usenet

NETNORTH Canadian segment of BITNET

NetRPC Network Remote Procedure Call (Sun Microsystems *tn*)

NetView network management/control architecture (IBM *tn*)

NetWare LAN operating system (Novell *tn*)

NETZC existing mobile network in Germany

NEU network expansion unit

NEUA National Elite Underground Alliance

NEWP NEW Programming Language (Burroughs *tn*)

NeWS Network Extensible Window System (Sun Microsystems *tn*); Novell Electronic Webcasting Service

NEXT near-end cross-talk

NeXTSTEP OS by NeXT *tn*

NF network function; neue Funktionauffeilung (g), new distibution of functions; Niederfrequenz (g), sub-audio and audio frequencies: 0,3–3,4 kHz; node feature; noise factor/figure; normal form; not finished

NFA non-deterministic finite-state automaton

NFAS non-associated facility signalling

NFB negative feedback

NFF no fault found

NFFVk NF-Fernverbindungskabel (g), audiofrequency trunk cable

NFlUe NF-Leitungsübertrager (g), audiofrequency line transformer

NFM network fault management

NFNT new font (Apple Mac font resource)

NFOM near-field optical microscopy

NFPA National Fire Protection Association

NFS Network File System (Sun Microsystems *tn*)

NFSP National Federation of Software Pirates

NFSWI SAP network layer frame switching SAP(2)

NG número de grupo (s), group number [GN]

NGDLC next-generation digital loop carrier

NGE not greater or equal

NGFL National Grid For Learning (UK schools on the Internet)

NGI next generation Internet; numéro de groupe fermé d'usagers international (f), número de GCU internacional (s), international CUG number [ICN]

NGI/O next-generation I/O

NH Nazi Hackers

NHK Nippon Hoso Kyokai (Japanese Broadcasting Corporation)

NHM network-management and maintenance signal

NHOH never heard of him/her

NHR non-hierarchical routing

NHRP next-hop routing/resolution protocol; non-hierarchical routing protocol

Ni nickel

NI network identifier/identity; network identification code; network information/interface

NI-1 National ISDN 1 (US standard)

NIA next instruction address

NIAL nested interactive array language

NIAM natural language information analysis method

NIB negative impedance booster

nibble half a byte (four bits *ie* US 'quarter'; *also* nybble)

NIC nearly-instantaneous companding (code); nearly-instantaneous compandored modulation, modulation quasi instantanée avec compresseur-extenseur (f); negative impedance converter; network identification code; network-independent clock, reloj independiente de la red (s); network information centre; network interface card; numeric intensive computing

NIC.DDN.MIL Network Information Centre for DDN

NiCad nickel-cadmium (used in batteries)

Nicam near-instantaneous compandored audio-multiplex

NICC network intelligence call centre

NICE Network Information and Control Exchange (DECnet *tn*); Non-profit International Consortium for Eiffel

NICS NATO Integrated Communications System

NICU network interface control unit

NID network information database; network interface device; network inward dialling; New Interactive Display (NEC monitor *tn*)

NIDL network interface definition language

NIE Newton Internet Enabler (Apple *tn*)

NIEM número de identificación de estación móvil (s), mobile station identification number [MSIN]

NIFTP network independent file transfer protocol (UK Blue Book)

NIHCL National Institutes of Health class library for C++ (US)

NII National Information Infrastructure (US)

NIIAC National Information Infrastructure Advisory Board (US)

NIIG National ISDN Interface Group (US)

NIL new implementation of Lisp

NIM network installation management (IBM)

NIMH nickel-metal hydride (batteries)

NIMS near-infrared mapping spectrometer

NIMT National Institute for Management Technology (Ireland)

NING Novell interworking group
NIO native input/output
NIOD network inward/outward dialling
NIP notificación interpersonal [IPN]; número de identificación personal (PIN)
NIPG National ISDN Parameter Group (US)
NIPS network I/O per second
NIR network information retrieval
NIRA near-infrared reflection analysis; negotiation of intermediate allowance
NIRR negotiation of intermediate rate required/requested
NIS Network Information Services (Sun Microsystems *tn*)
N-ISDN narrowband ISDN
NISM numéro d'identification de la station mobile (f), mobile station identification number [MSIN]
NISO National Information Standards Organization (US)
NISS National Information Services and Systems (UK)
NISSBB NISS bulletin board
NISSPAC NISS public access collections (UK)
NIST National Institute of Standards and Technology (US)
N-ISUP narrowband ISDN User Part
NIT network information table; network integration test (service); non-repetitive information type
NITC National Information Technology Center (US)
NITEL Nigerian Telecommunications Authority
NIU network interface unit (T-carrier; *v* smartjack)
NIUF North American ISDN Users' Forum
NJCL network job control language
NJE network job entry
NJS noise jammer simulator
NL natural language; network layer; new-line character; nueva línea (s), new line; number of unallocated channels on link
NLB number of lines of binary

NLDM network logical data manager
NLE non-linear editing; not less or equal
NLM NetWare Loadable Module (Novell *tn*)
NLP National Level Program; natural language processing; non-linear processor
NLPID Network Layer Protocol Identifier
NLQ near letter quality (of printer output)
NLR noise load ratio
NLS native language system; on-line system
NLSP NetWare Link Services Protocol (Novell *tn*)
NLST non-listed name
NLT not less/lower than
NLT-Vr negative Leitung Transistor-Verstärker (g), negative line transistor amplifier
NLV national language version (IBM)
Nm nanometer
NM combined delivery/non-delivery notification (f) [CN]; network management (entity); network module
NMA network monitoring and analysis
NMA números múltiples de abonado (s), multiple subscriber number [MSN]
NMAS network management application area in TMOS
NMBC Nihon Mobile Broadcasting Co.
NMC network management centre
NMCS National Military Command System (US)
NMEA National Maritime Electronics Association
NMG network management gateway
NMI Nautilus memory interconnect *tn*; non-maskable interrupt
NMIP network management interface processor (mini-computer)
NML National Media Lab (US); network management layer
NMM NetWare management map (Novell *tn*); network-management and

maintenance signal, signal de gestion et de maintenance du réseau (f)

NMMW near-millimetre wave system

NMN national mobile number

NMO nota media de opinión [MOS]

NMOD notas medias de opinión sobre las degradaciones (s) [DMOS]

NMOS negative channel metal-oxide semiconductor

N-MOS n-channel MOS

NMP network management processor (mainframe); network management protocol (AT&T); nivel máximo permitido (s) [PML]

NMR normal mode rejection, nuclear magnetic resonance

NMRE nivel máximo del ruido exterior (s) [MENL]

NMRM network management reference model

NMS network management signal; network management station; network management system; noise measuring set

NMSI national mobile station/subscriber identification number

NMT Nordic Mobile Telephone system *tn*; not more than

NMU network management unit

NMVT network management vector transport protocol

NN national network; national (significant) number, numéro national (significatif) (f); network node; neural network; node-to-node message

N-N numérique-numérique (f), digital-to-digital [D-D]

NN(S) número nacional (significativo) (s), national number (significant) [NN]

NNC national-network-congestion signal

NNE notificación de (estado de) no entrega (s), non-delivery status notification [NDN]

NNG national number group (telephone number prefix)

NNI network node interface (ATM); network-to-network implementation; network-to-network interface

NNM network node manager; node-to-node message

NNR national number routed; notification de non-remise (f), non-delivery notification [NDN]; notificación de no recepción (s), non-receipt notification [NRN]

NNRC Neural Networks Research Centre (Helsinki University of Technology)

NNTP network news transfer protocol (Usenet)

NNU signal de numéro non utilisé (f), unallocated-number signal [UNN]

NO network operator

Nº de YR-TU sequence number response (field) [YR-TU-NR]

NOA Network Out-Sourcing Association (UK)

NOC network operations centre

NOCC network operations control centre

NOCELL neighbouring outer cell

NOCN-IT The National Open College Network – Information Technology (UK)

Noctovision TV system using infrared light *tn*

NOD network outward dialling, news on demand (on Internet)

NODAL Norsk Data's NORD 10 (language)

NODAN noise-operated device for anti-noise

NOF not on file

NOI Notice of Inquiry (FCC)

NOK not OK; incorrect

NOM note d'opinion moyenne (f), mean opinion score [MOS]

nom nominal

nomen nomenclature

NON notificación nacional (s), national notification

NO-OP no-operation instruction

NOp network operator

NOPAC network OPAC

NOR not OR

NORC network operators RACE committee

Nortel Northern Telecom (UK)

NOS network OS

NOSC (JANET) network operations service centre

NOSFER new fundamental system for (the) determination of reference equivalents, nouveau système (fondamental) pour la détermination des équivalents de référence (f)

NOT! Nation of Thieves

NOTAL not sent to all addresses

Np neper, 8.6 dB (unit of attenuation of A_2/A_1); number of packets per burst

NP negotiation procedure; network operator; network performance; non-deterministic polynomial time; northern pine (*ie* telephone pole); numbering plan; number portability

NPA network performance analyser; network printer alliance; nivel de presión acústica (s), sound pressure level [SPL]; Northern Phreakers Alliance; numbering plan area

NPAC number of paging and access channels

NPAI network protocol addressing information, information d'addresse protocole de réseau (f)

NPAN numbering plan area

NPC network parameter control; non-deterministic polynomial complete; normalized projection coordinates; Northern Phun Company

NPCI network protocol control information

NPCID network portion clear indication delay

NPD network protection device

NPDA network problem determination application/analysis

NPDB number portability database

NPDSA National Public Domain software archive

NPDU network layer protocol data unit; network protocol data unit

NPI network printer interface; numbering plan identifier/indicator; null pointer indication

NPIP new product integration process

NPL National Physical Laboratory (UK); network planning; New Programming Language (IBM *tn*); non-procedural language

NPM network performance management/monitor

NPN negative-positive-negative; Notes Public Network; NPN transistor

NPO network performance objectives

NPPL network picture processing language

NPR no preparado para recibir (s), receive not ready [RNR]; noise-power ratio; non-photo-realistic rendering (3D graphics); non-polarized radial

NPRM Notice of Proposed Rulemaking (FCC)

NPS Novell Productivity specialist (Novell)

NPSI NCP Packet Switching Interface; network packet-switching interface

NPSS NASA packet switching system

NPTD network-dependent data-packet transfer delay

NPTN National Public Telecommuting Network (US)

NPU natural processing unit

NPV net present value

NPX numeric processor extension

NQd non-quadded

NQR nuclear quadrupole resonance

NQS network queuing system (Cray)

NR notification de remise (f), delivery notification [DN]; notificación de estado de recepción (s), receipt status notification [RN]; noise rating; non-reactive

NRA nuclear reaction analysis

NRAT SAP network layer rate adaption transfer SAP

NRC National Research Council (US); non-recurring charges

NRCS national replacement character set

NRD network resource management; número de referencia de documento (s), document reference number [DRN]

NRE non-recurring engineering

NREN National Research And Education Network (US)

N_{RES} niveau d'écho résiduel (f), nivel de eco residual (s), residual echo level $[L_{RES}]$

N_{RET} niveau de retour d'écho (f), returned echo level $[L_{RET}]$

NRF network routing facility

NRI net radio interface

NRM network reference model; network resource manager; normal response mode, modo de respuesta normal (s)

NRN non-receipt notification

nroff new roff (Unix)

NRP network terminating point; (signal de) nouvelle réponse (f), re-answer signal [RAN]

NRPC número de referencia de punto de comprobación (s)

NRRC Nuclear Risk reduction Centre (US)

NRT network routing table; non-real-time

NRTEE notificación del retardo de tránsito de extremo a extremo (s), end-to-end transit delay notification [EETDN]

NR-UDPT número de UDPT DT (s), TPDU data [TPDU NR]

NR-UDPT-DA número de UDPT DA (s), TPDU expedited acknowledge [ED TPDU]

NR-VT-UT respuesta de número secuencial (s), sequence number response field [YR-TU-NR]

NRZ non-return-to-zero, sin retorno a cero (s)

NRZ1 non-return-to-zero change at logic 1

NRZI non-return-to-zero inverted

NRZL non-return to zero level

NRZM non-return-to-zero mark

NRZS non-return-to-zero space

ns nanosecond, one thousand millionth, 10^{-9}

NS Nassi-Scheiderman program-execution chart; National Semiconductor Corporation; network service; network supervisor/synchronization

NS/DOS networking services/DOS

NS/EP (US) National Security/Emergency Preparedness

NSA National Security Agency (US); National Security Anarchists; numéro de séquence vers l'avant (f), forward sequence number [FSN]

NSA.1 notación de sintaxis abstracta uno (s), abstract syntax notation N° 1 [ASN.1]

NSA-A NSA attendu (f), forward sequence number expected [FSNX]

NSA-C numéro de séquence vers l'avant de la dernière trame sémaphore de message acceptée par le terminal distant (f), FSN of last MSU accepted by remote level 2 [FSNC]

NSA-D FSN of the last MSU in the RTB [FSNL]

NSA-E FSN of the last MSU transmitted [FSNT]

NSAnl Nebenstellenanlage (g), private branch exchange

NSA-P FSN of the oldest MSU in the RTB [FSNF]

NSAP network layer service access point; network service access point (ISO)

NSAPA network service access point address (ISO)

NSAPI Netscape server API (Netscape *tn*)

NSA-R NSA reçu (f), FSN received [FSNR]

NSBA número secuencial del bloque de que se acusa recibo (s), block-acknowledged sequence number [BASN]

NSBC número secuencial de bloques completos (s), block-completed sequence number [BCSN]

NSC National Security Council (US); National Semiconductor Corp; National Switching Centre; network specialized centre; network switching centre; non-standard facilities command

NSD número secuencial directo (hacia

adelante) (s), forward sequence number [FSN]

NSDA NSD (hacia adelante) de la última unidad de señalización de mensaje aceptada por el nivel 2 distante (s), FSN of last MSU accepted by remote level 2 [FSNC]

NSDB network system/services database

NSDD National Security Decision Directive (US)

NSDE FSN expected [FSNX]

NSDI National Spatial Data Infrastructure (US)

NSDKNSF National Science foundation (US)

NSDP FSN of the oldest MSU in the RTB [FSNF]

NSDR FSN received [FSNR]

NSDT FSN of the last MSU transmitted [FSNT]

NSDU FSN of the last MSU in the RTB [FSNL]; network service data unit

NSE neutral stream etch

nsec nanosecond

NSEC número secuencial del estado del circuito (s), common-channel signalling number [CSSN]

NSF National Science Foundation (US); non-standard facilities

NSFnet National Science Foundation Network (US)

NSI Network Solutions Inc (US equiv of UK's Nominet); non-SNA interconnect; numéro secuencial inverso (hacia atrás) (s), backward sequence number [BSN]

NSID NICS/STANAG interface device

NSIR número secuencial inverso (hacia atrás) recibido (s), backward sequence number received [BSNR]

NSIT número secuencial inverso (hacia atrás) de la próxima unidad de señalización que ha de transmitirse (s), backward sequence number of next SU to be transmitted [BSNT]

NSL native structured language; net switching loss; niveau de sortie longitudinale (f), tensión (nivel) de salida

longitudinal (s), longitudinal output level [LOL]

NSM Netscape server manager (Netscape *tn*)

NSMN national significant mobile

NSO network service order

NSP national signalling point; native signal processing (Intel); network service part; network status presentation; Network Services Protocol (DEC *tn*)

NSP SAP network layer speech SAP

NSPC National Sound-Program Center (US)

NSR non-source routed; numéro se séquence vers l'arrière (f), backward sequence number [BSN]

NSRD National Software Re-use Directory (US)

NSR-E numéro de séquence vers l'arrière à émettre (f), backward sequence number of next SU to be transmitted [BSNT]

NSR-R numéro de séquence vers l'arrière reçu (f) [BSNR]

NSS native storage services (Microsoft); network subsystem; network support/synchronization subsystem; network switching subsystem; new storage system (MULTICS); nodal switching system (NSFnet); non-standard facilities set-up; part number; NetWare storage services (Novell *tn*); numéro de sous-système (f), número de subsistema (s), subsystem number [SSN]

NSSDU normal data session service data unit

NST niveau de sortie transversale (f), tensión (nivel) de salida transversal (s), transverse output level [TOL]

NstA Nebenstellenanlage (g), private branch exchange

NSTAC National Security Telecommunications Advisory Committee (US)

NSTL National Software Testing Laboratory (US)

NSTN non-standard telephone number

NStnl Nebenstellenanlage (g), private branch exchange

NT New Technology (Microsoft *tn*); network operator; network terminal option; network terminating equipment; network termination (ISDN), Netzabschluß (g); network termination S2M-UK2, Netzabschluß S2M-UK2 (g); Nippon Telephone and Telegraph Public Corp; node type (SNA); non-transparent; Northern Telecom (UK); Teilnehmer-Endeinrichtung S2M (g), subscriber network termination

NT-1 network termination type 1

NT1 Netzabschluß S2M-UK2 (g) (network termination S2M-UK2)

NT2 Teilnehmer-Endeinrichtung S2M (subscriber network termination)

NTA near-term architecture; Nokturnal Trading Alliance

NTAAB new type approval advisory board

NTAS New Technology Advanced Server (Microsoft Windows *tn*)

NTC National Television Center (US); National TV Standards Commission (525-line picture resolution)

NTCA National Telephone Cooperative Association

NTCN National Telecommunications Coordinating Network (US)

NTDS Naval Tactical Data System (US)

NTE network (channel) terminating equipment

NTF network transaction function

NTFA network transaction function area

NTFS NT file system (Microsoft *tn*)

NTI network terminating interface, noise transmission impairment

NTIA National Telecommunications and Information Administration (US)

NTIC Nouvelles Technologies de l'Information et de la Communication (f)

NTISSC National Telecommunications and Information Systems Steering Committee (US)

NTLDR Microsoft Windows NT loader

NTM network traffic management; NT LAN manager (Microsoft)

NTMBS null-terminated multi-byte string

NTMS National Telecommunications Management Structure (US)

NTN network terminal number/option

NTP network time protocol (RFC 1119)

NTPF number of terminals per failure

NTR national transit exchange; numéro terminal du réseau (f), número terminal de red (s), network terminal number [NTN]

NTRAS NT remote access services (Microsoft)

NTS network test system; Microsoft Windows NT advanced server (Microsoft Windows *tn*); Windows NT file system (Microsoft Windows *tn*)

NTSA NetWare telephony services architecture (Novell *tn*)

NUI Notebook User Interface (Go Corp *tn*)

NTSC National Television System Committee (US)

NTT Nippon Telegraph and Telephone; Nippon Telephone and Telecommunications; numbered test trunk

NTU nephelometric turbidity units; network terminating unit

NTUP national telephone/telephony user part

n-tuple data object containing two or more elements

NTV Nachrichten Television

nty notify

NU network user identification; number unobtainable

NU tone number unobtainable tone

NUA network user address (X.121)

NuBus expansion bus used by Apple Macs (Texas Instruments *tn*)

NUC nailed-up-connection

NUDETS National Detonation Detection And Reporting System (US)

NUI network user identification (X.25),

identificación de usuario de red (s); network user interface

nuit night delivery (f)

NUL null, nul (f), nulo (s)

NUMA non-uniform memory access

Numéris French Télécom ISDN network

NUP national user part (SS7); network user part

Nuprl nearly ultimate PRL

NURBS non-uniform rational B-spline

NUT number-unobtainable tone

NV no overflow

NVE network-visible entities

NVIS near-vertical-incidence skywave

NVN National Videotex Network; non-volatile memory

NVOD near video-on-demand

NVON non-Von Neumann (architecture)

NVP nominal velocity of propagation

NVR niveau vocal de référence (f), nivel vocal de referencia (s), reference vocal level [RVL]; non-volatile residue

NVRAM non-volatile RAM

NVSIMM non-volatile SIMM

NVT network virtual terminal; Novell virtual terminal (Novell *tn*)

NW network; not-white

NWC net weekly circulation

NWNET NorthWestNet

NWP New World Payphones – company (UK)

NWS NetWare web server (NetWare *tn*)

NXT new technology loudspeakers

NYC New York Crackers

NYM anonymous

NYNEX New York/New England (Stock) Exchange

NYPS National Yellow Pages Service (US)

NYPSC New York Public Service Commission

NYSERNET New York State Education Research Network

NYSHII New York Safehouse II

NZPO New Zealand Post Office

O

O obligatorio (s), mandatory [M]; operational immediate (precedence); optional; oxygen

O&M operations and maintenance

O/D originador/destinatario [O/R]

O/E optical-to-electrical, opto-electronics

O/G outgoing, gehend (g)

O/P output

O/R originator/recipient

O2 object-oriented

OA office automation; operational amplifier; originator address; outgoing access

OA&M operations administration and maintenance

OAC acheminement des messages (f), message routing [HMRT]

OACSU off-air-call-set-up

OAD open architecture driver (Bernoulli)

OAGEE Official Airlines Guide – electronic edition

OAI open applications interface

OAl Ortsanschlußleitung (g), local subscriber line

OAM operations and management; operations administration and maintenance; oscillator activity monitor

OAM&P operations administration maintenance and provisioning

OAMC operations administration and maintenance centre

OAMP former designation for OMAP

OAO Orbiting Astronomical Observatory

OAP outside awareness port

OAS one-to-all scatter

OASIS On-line Article Status Information System [Elsevier]

OAsl Ortsanschlußleitung (g), local subscriber line

OATH object-oriented abstract type hierarchy

ob- obligatory (Usenet netiquette)

OB output buffer; outside broadcast

OBA object behaviour analysis; Overall Business Architecture (BAe)

OBAI object-based architecture for integration

O-BCSM originating basic call state model (incoming half-call)

OBE Office by Example (IBM *tn*)

OBEM Object-Based Equipment Model

OBERON programming language, successor to Modula 2

OBEX object exchange (Borland)

OBIC optical-beam-induced current

OBJ object; executable specification language

OBL object-based language

OBM Objektiver Bezugsdämpfungs-MPl (g), objective reference equivalent meter

OBN out-of-band noise

OBP Open Bridging Protocol

OBS organization breakdown structures

OC operations centre; optical carrier (SONET)

OC1 optical carrier level 1: 51.84 MBps

OC3 optical carrier level 3: 155.52 MBps

OCAP orange cats are pretty

OCB outgoing calls barred

OCC operations control centre; other common carriers (*ie* non-Bell); signal (électrique) d'abonné occupé (f), subscriber-busy signal [SSB]

OCCA open cooperative computing architecture

occam high-level language for parallel processing

OCD out-of-cell delineation (UNI 3.0 Section 2.1.2.2.2)

OCE Open Collaborative Environment (Apple); other common equipment

OCF object components framework (Borland)

OCHC operator call-handling centre

OCI operator controlled input; protocol indicator

OCIS Organized Crime Information Systems

OCL on-line computing system;

operator/operation control language; overall connection loss

OCLC On-Line Computer Library Center (international organization)

OCMS optional calling measured service

OCN operational carrier number

OC-n optical carrier-n (SONET transmission rate)

OC$_N$ optical channel specifications for SONET

OCP order code processor; overload control process

OCR optical character recognition

OCS object compatibility standard; Old Chipset (Amiga); on-card sequencer; overload control subsystem

octal base-8 number system

octet eight contiguous bits (eight-bit byte)

OCTOPUS Control Data Corp data network

OCU office channel unit; order-wire control unit

OCVCXO over controlled-voltage controlled crystal oscillator

OCWR optical continuous wave reflectometer

OCx optical carrier level x (SONET)

OCX OLE custom control

OD oceanographic data station; operaciones a distancia (s), remote operation [RO]; optical density; out of order; out-dial; outside diameter; overload detection

ODA on-line delivery acknowledgement, acuse de recibo de entrega en línea (s); open document architecture (ISO 8613)

ODA/ODIF open document architecture/open document interchange format

ODAI origin/destination address assignor indicator

ODAPI open database application programming interface (Borland)

ODAS Ocean Data Acquisition Systems (UNESCO)

ODB operator-determined barring

ODBC open database connectivity (Microsoft)

ODBMS object-oriented database management system

ODC discrimination des messages (f), message discrimination [HMDC]

ODCP one-digit code point

ODD operator distance dialling

ODDD operator direct distance dialling

ODER unidad de datos de protocolo de aplicación OD-ERROR (s), REMOTE OPERATION error application protocol data unit [ROER]

ODgVSt Ortsdurchgangs-VSt (g) (local transit switching centre)

ODI open data-link interface (Novell); open device interconnect (Novell); optical digital image

ODIF open document interchange format

ODIV unidad de datos de protocolo de aplicación OD-INVOCACION (s), REMOTE OPERATION invoke application protocol data unit [ROIV]

ODM object data manager (IBM); original design manufacturer

ODMA open document management API

ODMG object database management group

ODN out-dial notification

ODP open data processing (OSI); open distributed processing; originator detection pattern

ODPR Office of the Data Protection Registrar

ODR operator data register

ODRCH unidad de datos de protocolo de aplicación OD-RECHAZO (s), REMOTE OPERATION reject application protocol data unit [RORJ]

ODRS unidad de datos de protocolo de aplicación OD-RESULTADO (s), REMOTE OPERATION result application protocol data unit [RORS]

ODS open data services (Microsoft); overhead data stream; ozone-depleting substances

ODSA Open Distributed System Architecture (UK DTI)

ODSI Open Directory Services Interface (Microsoft)

ODSS Open Data Stream Structure

ODT distribution des messages (f), message distribution [HMDT]; open desktop

ODVSt Ortsdurchgangs-VermittlungsStelle (g) (local transit switching centre)

OE Operating Environment (Amiga *tn*)

OEBR optical edge bead removal

OEC signalling link congestion (f) [HMCG]

OEE overall equipment effectiveness

OEIC opto-electronic integrated circuit

öEl örtliche Endvermittlungsleitung (g) (line on local terminal switching centre)

OEM original equipment manufacturer

OEP operand execution pipeline

OES optical emission spectroscopy; outgoing echo suppressor

OF overflow flag

OFB output feedback

OFC optical fibre conductive

OFCP optical fibre conductive plenum

OFCR optical fibre conductive riser

OFG Originally Funny Guys

OFHC oxygen-free high-conductivity copper

OFM optical frequency multiplexing

OFMT output format for numbers

OFN optical fibre non-conductive

OFNP optical fibre non-conductive plenum

OFNR optical fibre non-conductive riser

OFS object file system (Microsoft); output field separator

Ofta Office of the Telecommunications Authority (Hong Kong)

OFTEL Office of Telecommunications (UK telecommunications registrar), Office britannique des télécommunications (f), Agencia de telecomunicaciones (s)

OFX Open Financial Exchange (Intuit *tn*)

OG Ortsgruppenwähler (g) (local trunk selector)

OGC outgoing trunk circuit

OGJ outgoing junction

OGP outgoing message process

OGR outgoing repeater

OGT outgoing trunk/toll

OGVT outgoing verification trunk

OGW Ortsgruppenwähler (g) (local trunk selector)

OGWS outgoing/wink start

OH off-hook

OHDH old habits die hard

OHG Operators' Harmonization Group (mobile networks)

öHl örtliche Hauptvermittlungsleitung (g) (line on local main switching centre)

ohm electrical resistance unit of measurement

OHP overhead projector

OHS off-hook service

OIC oh, I see!

OID object interaction diagram; object identification/identifier; identifiant (f)

OIDL object interface definition language

OIEA Organismo Internacional de Energía Atómica [IAEA]

OIG OSPF Interoperability Group

OIL operator identification language

OIRT International Radio and Television Organization

OIS bureautique communicante OIS (f); octeto de información de servicio (s), service information octet [SIO]; office information system (network computing)

OITT out-pulse identifier trunk test frame

OJ originating junctor

OJT on-the-job training

öKart öffentliches Kartentelefon (g) (public card-phone)

öKl örtliche Knotenvermittlungsleitung (g) (line on local node switching centre)

OL other line; overlay; oscillateur local (f)

OLAP on-line analytical processing

OLB outer lead bond

OLC on-line computing system; optical loop carrier

OLDAS on-line digital–analogue simulator (IBM)

OLe Ortsleitungsübertrager (g) (local line transformer)

OLE object linking and embedding, on-line environment

OLEC other local exchange carrier

OLECI OLE automation code inserter (Corel)

OLI on-line interface; optical line inlet; optical line interface (AT&T)

OLIT OPEN LOOK Intrinsics Toolkit (X Window System)

OLL open-loop loss

OLM overload message

OLMC output logic macro-cell

OLO optical line outlet

OLP on-line processor; Open Licence Program (Microsoft)

OLR objective/overall loudness rating; off-line recovery, overall loudness rating

OLS ordinary least squares

OLSP on-line service provider

OLSS on-line support software

OLTM optical line terminating multiplexer

OLTP on-line transaction processing

OLü Ortsleitungsübertrager (g) (local line transformer)

OLUC OCLC Online Union Catalog

OLUD on-line update

OLUM on-line update control module

OLWM OpenLook Window Manager *tn*

OM object manager; operational modelling; optical microscopy

OMA object management architecture (Microsoft)

OMAP operation maintenance and administration part

OMB Office of Management and Budget (US govt)

OMC operation and maintenance centre

OMCR operation and maintenance centre-radio part

OMC-R/S OMC für Radio/Switching SS (g)

OMCS operation and maintenance centre-switch part

OME open messaging environment

OMF object module format (Microsoft); open media framework; open message format

OMFS office master frequency supply

OMG Object Management Group (OOP consortium setting standards)

OMI open messaging interface (Lotus *tn*)

OML operations and maintenance link; outgoing matching loss

OMM output message manual

OMNS open network management system

OMPF operation and maintenance processor frame

OMR optical mark reading

OMS optical mass spectroscopy; object management system; operations and maintenance subsystem; orientation des messages (f), signalling message handling [SMH]

OMT object modelling technique; operation and maintenance terminal; orthogonal mode transducer (satellite)

ON Ortsnetz (g) (local network)

ONA open network architecture

ONAL off-net access line

ONC Open Network Computing (Environment) (Sun Microsystems *tn*); Ordinateur à jeu d'instructions complexes (f), complex instruction set computer [CISC]

ONC+ Open Network Computing plus (Sun Microsystems *tn*)

ONDS open network distribution services (IBM)

ONE Open Network Environment (Netscape *tn*); optimized network evolution (Vision)

ONI operator number identification

ONKZ Ortskennzahl (g) (local code)

ONMS open network management system

ONP open network provision (EU standard), fourniture d'un réseau (f), offener Netzzugang (g)

ONP/IN open(-systems) network product for the IN

ONS on-premises stations

ONSD optional network specific digit

ONU optical network/networking unit

OO object-oriented (*v* O2)

OOA object-oriented analysis

OOB out of band

OOBE object-oriented business engineering

OOD object-oriented design

OODB object-oriented database

OODMS object-oriented database management system

OOF object-oriented FORTRAN; out of frame (SONET)

OOFNet our own fucking net

OOGL object-oriented graphics language

OOL object-oriented language

OOO out of order

OOOS object-oriented OS

OOP object-oriented programming

OOPL object-oriented programming language

OOPS object-oriented programming system

OOPSLA (Conference on) Object-oriented programming systems language and applications

OOR operator override

OOS object-oriented systems; off-line operating simulator; operational operating system; out of service; out of sync

OOSD object-oriented structured designs

OOT object-oriented technology; object-oriented Turing

OOUI object-oriented user interface

OOZE object-oriented extension of Z

OP operation; operational process; optical; option; output; output processor

op amp operational amplifier

op code operational code

OPAC On-line Public Access Catalogue (British Library)

OPAL On-Line Presentation Access Library *tn*

OPC optical particle counter; optical photo-conductor; optical proximity correction; organic photo-conducting cartridge (laser printer drum); originating/origination point codes (SS7)

OPCODE operational code

OPD operand

OPDU operation protocol data unit, unité de données de protocole de l'opération (f)

Open GL computer-graphics system *tn*

OPEN open protocol enhanced networks

OPERATORS optimization program for economical remote trunk arrangements and TSPS operator arrangements

OPGW overhead optical ground-wave

OPI open pre-press interface; overall performance index

OPINE overall performance index of network evaluation, modelo de índice de calidad global para evaluación de la calidad de una red (s)

OPM operations per minute; outage performance monitoring

OPMODEL operations model

OPP Office of Plans and Policy (US FCC); opposite

OPR outside plant repeater; señal de orden de paso a enlace reserva (s), change-over order signal [COO]

OPS off-premises station; on-line process synthesizer; open-profiling standard (Web browsers); operations; operator's subsystem

OPSEC operations security

OPSP originating participating supplier parameter

OPT Open Protocol Technology (Novell *tn*)

OPUS octal program updating system

OPX off-premises extension

O-QAM orthogonally-multiplexed quadrature amplitude modulation

OQPSK offset quad-phone shift keying
O-QPSK offset quadrature phase-shift keying
OR off-route (aeronautical mobile) service; operations research; OR gate/operation; orange; owner's risk
OR25E objective R.25 equivalent
ORA Opportunities for applications of information and communication technologies in Rural Areas (EU)
Oracle IBA version of teletext (UK *tn*)
ORACLE relational DBMS from Oracle Corp *tn*
ORAVAC automatic switchboard routiner
ORB object request broker (Microsoft)
ORD ordinary (subscriber)
ORF Österreichischer Rundfunk (g)
ORI on-line retrieval interface
ORK Office Resource Kit (Microsoft *tn*)
ORKID open real-time kernel interface definition
ORM other regulated material; optical remote module; optically-remote switching module
ORMS operating resource management system
ORNL Oak Ridge National Laboratory
O-ROM optical ROM
ORP optical reference point
ORRCH unidad de datos de protocolo de aplicación OD-RECHAZO (s), REMOTE REJECT application protocol data unit [RORJ]
ORS output record separator; señal de orden de retorno al enlace de servicio (s), change-back declaration signal [CBD]
OS operating system
OS/2 Operating System/2 (IBM 1980s *tn*)
OS/2 Warp IBM operating system (1994 *tn*)
OS/E operating system/environment
OSA open scripting architecture; open service architecture
OSAC operator services assistance centre
OSB output signal balance (loss)
OSBL object super-base language
OSC oscillator

OSCAR Off-Line Script And Character Recognition (University of Essex); Optical Submarine Communications by Aerospace Relay (US DoD); Orbiting Satellite Carrying Amateur Radio; Oregon State Conversational Aid to Research
osc-mult oscillator-multiplier
OSD on-screen display; open software description
OSDL overall specification and description language
OSE open-systems environment
OSEH OS Emulation HomePage
OSF Object Support Framework; Open Software Foundation (US); operating system functions
OSF/Motif GUI from OSF *tn*
OSF-DCE Open Software Foundation – distributed computing environment
OSGI Open Service Gateway Initiative
OSHA Occupational Safety and Health Administration (US)
OSI open switching interval; open-systems interconnection, interconnexion des systèmes ouverts (f)
OSI net OSI network
OSI NS OSI network service
OSI RM Open Systems Interconnection Reference Model (ISO, ITU-T)
OSI/CS OSI/Communications Subsystem
OSI/FS OSI/file services
OSI/NMF OSI/Network Management Forum
OSIE open systems interconnection environment
OSM off-screen model
OSO originating screening office
OSP on-line service provider; on-screen programming; optical storage processor
OSPE Open Service Provisioning Environment
OSPF open shortest-path first (interior gateway protocol)
OSQL Object-Oriented Structured Query Language (HP *tn*)

OSRI originating stations routing indicator

OSRM Office of Standard Reference Materials

OSS object services standard; one-stop shopping; operations support system

OSSL operating systems simulation language

OSSN originating stations serial number

OSSU operator services switching unit

OST off-line store file; operation station task

OSTEST operating system test

OSTL operating system table loader

OSV offset scan voting; voice-synthesizing unit (f)

OSWS operating system workstation

OT object technology

OT&E open test and evaluation

OTA operation-triggered architecture

OTAM over-the-air management of automated HF network nodes

OTAR over-the-air re-keying

OTC operating telephone company; originating toll centre; Overseas Telecommunications Commission (Australia)

OTDM optical time-division multiplex

OTDR optical time-domain reflectometer (SONET)

OTE Hellenic Telecommunications Organization (Greece)

OTF Open Token Foundation; optical transfer function

OTH over-the-horizon (scatter of signal)

OTI Organización de la Televisión Ibero-Americana (s)

OTID Originating TID

OTLP zero dBm transmission level point

OTOH on the other hand

OTP Office of Telecommunications Policy (US); one-time programmable

OTP EPROM one-time programmable EPROM

OTP ROM one-time programmable ROM

OTR Organization and Technology Research

OTR1-200 work plans for RACE programmes

OTS off-the-shelf; operator telephone system; own time switch

OTSS off-the-shelf system

OTT over the top

OTTS outgoing trunk testing system

OU organizational unit

OUI organizationally-unique identifier

OURS open user-recommended solutions

OUTS output string

OUTWATS outgoing wide-area telecommunications/telephone system

OV overflow

OVD outside vapour deposition

OVE overall R.25 equivalent

OVE$_{LR}$ overall loudness rating

OVIDE EU-wide public videotex network

OVk Ortsverbindungskabel (g) (local distribution cable)

öVl örtliche Verbindungsleitung (g) (local distribution line)

OVl Ortsverbindungsleitung (g) (local distribution line – circuit)

OVL overlay

OVl-F Ortsverbindungsleitung für Fernverkehr (g) (local distribution line for trunk traffic)

OVL-file overlay file

OVSt Ortsvermittlungsstelle (g) (local switching centre)

OVW OpenView Windows

OW order-wire (circuit); over-write

OWC one-way communication

OWF optimum working frequency

OWL object Windows language; Object Windows library (Borland); Office Workstations Ltd (UK company)

Ox oxide

P

p peta; one thousand million million, 10^{15}

P page, página (s); Pentium (Intel *tn*); person-to-person; poll bit, bit de petición (s); predicted pictures (MPEG); priority; protocol; tiempo de persistencia (s), persistence time [R]

P&D plug and display

P(R) número secuencial de paquete recibido (s), receive sequence number of packet

P(S) número secuencial de paquete transmitido (s), send sequence number of packet

P* French kiss (emoticon)

p/ar peak-to-average ratio

P/F poll/final (bit), bit de petición/final (s)

P/H/A/ Phreakers/Hackers/Anarchists

P/I pair/impair (f), even/odd

P/ISDN primary rate ISDN

P/L-CHS physical/logical disk addressing: cylinder/head/sector

p/s pulses per second

P1 message transfer protocol; temporary storage for external file Microsoft Mail

P2 inter-personal messaging protocol

p2c Pascal to C translator

P2P person to person

P3 platform for privacy preferences; submission and delivery protocol

P3P platform for personal privacy preference (encryption)

P5 Intel Pentium 586 processor *tn*; teletext access protocol

P55C Intel's code name for its MMX processor

P6 Intel Pentium Pro processor 686

PA paging area; paso alto (s), high pass [HP]; physical architecture; power amplifier; preamble (LAN); presentation architecture; proceso de aplicación (s), application process [AP]; process allocator; public-address (system)

PAB personal address book

PABX private automatic branch exchange

PAC paging area control/controller; personal access communication; photo-active compound

PACE personal audio computer editing; Perspectives for Advanced Communications in Europe (RACE); priority access control enabled (3Com)

packet information grouped for transmission

PACM paging area configuration management

PACS personal access communications system; picture archiving and communication systems

PACT parte aplicación de capacidades de transacción (s), transaction capabilities application part [TCAP]; Partnership in Advanced Computing Technologies (UK); PCTE added common tools; prefix access code translator

PACUIT packet-switching system (Computer Transmission Corp)

PACVD plasma-assisted chemical vapour deposition

Pacxnet packet switching network (Gandalf)

PAD packet assembly/disassembly (facility); portable access device

PAD talk programming language used with HyperPAD

PADS pen application development system (Slate Corp); position and azimuth determining system

PADSX partially-automated digital signal cross-connect

PADT post-alloy diffused transistor

PAE prêt à émettre (f), ready for sending [RFS]

PAF personal ancestral file

PA-FTIR photo-acoustic Fourier transform infrared spectroscopy

PAG Phreaks Against Geeks; protocolo de acceso a la guía (s), directory access protocol [DAP]

PAGIS posición del anillo de guarda para

la determinación de índices de sonoridad (s), loudness rating guard-ring position [LRGP]

PAGS processus d'application de gestion de signalisation (f), proceso de aplicación de gestión de señalización (s), signalling management application process [SMAP]; proceso de aplicación de gestión de sistemas (s), systems management application process [SMAP]

PAI paging area identifier

PAIN Pirating Artistically Intrepid Newsmongers

PAL page alternating line; Paradox Application Language (Borland *tn*); pedagogic algorithmic language; phase alternating line (TV standard); process asset library; process automation language; programmable array logic (Monolithic Memories); programming assembly language

PALCD plasma-addressed liquid crystal display (Sony/Sharp *tn*)

PAL-M phase alternation by line – modified

PAM parte (de) aplicación móvil (s), mobile application part [MAP]; pass-along message; Peachtree Accounting Macintosh *tn*; phone access to mail; pluggable authentication module (Sun Microsystems); process application module; pulse amplitude modulation

PAMA pulse address multiple access

PAMR public access mobile radio (SMR)

PAMS pre-selected alternate master-slave

PAN personal/personalized area network

Panamsat Pan-American Satellite Corp

PANDA processed narrow deviation audio *tn*

PANS 'pretty amazing new stuff' (*ie* digital; *cf* POTS)

PAO publication assistée par ordinateur (f), desktop publishing [DTP]

PAOM parte aplicación de operaciones y mantenimiento (s), operation maintenance and application part [OMAP]

PAP packet-level procedure; password authentication protocol; printer access protocol; process application protocol

PAPAG Phreaks Against Phreaks Against Geeks

PAPER portable all-purpose electronic rendering

PAPS premier arrivé – premier sorti (f), first in – first out [FIFO]

PAQ plan d'assurance de qualité (f), quality-assurance plan

PAR parallel; peak-to-average ratio; performance analysis and review; petición de actualización rápida (s), fast update request [FUR]; positive acknowledgement and/with retransmission

PARADE parallel applicative database engine

PARADISE piloting a researcher's directory services in Europe

Paradox DBMS from Borland *tn*

Paralog parallel version of Prolog

PARAM X parameter X, parámetro X (s)

paramp parametric amplifier

Parasol parallel systems object language

PARC Palo Alto Research Center (Xerox)

pare-feu firewall

PARI protocolo de aplicación de red inteligente (s), intelligent network application protocol

Paris parallel instruction set

PA-RISC Precision Architecture RISC (HP workstations *tn*)

PARMACS package of macros for parallel programming

PARS programmed airlines reservation system

Parsley Pascal extension for constructing parse trees

PARULEL parallel rule language

PAS Panamsat Corp; photo-acoustic spectroscopy; publicly-available submitter; punto de acceso al servicio (s), service access point [SAP]

Pascal programming language

Pascal S simplified Pascal

PASED punto de acceso al servicio de enlace de datos [DLSAP]

PASF punto de acceso al servicio físico (s), physical service access point [PhSAP]

PASI production availability shipments inventory

PASP punto de acceso al servicio (de la capa) de presentación (s), presentation layer service access point [PSAP]

PASR punto de acceso al servicio de red (s), network service access point [NSAP]

PASRO Pascal for robots

PASS punto de acceso al servicio de sesión (s), session service access point [SSAP]

Passim simulation language based on Pascal

PAST point d'accés au service de transport (f), punto de acceso al servicio de transporte (s), transport service access point [TSAP]

PASU punto de asociación de subred (s), sub-network point of attachment [SNPA]

PAT palabra de alineación de trama (s), frame alignment word [FAW]; pérdida de la alineación de trama (s), loss of frame alignment [LFA]; personalized array translator; port address translation; power alarm test; process action team

PATO partial acceptance and take-over date

PATU Pan-African Telecommunications Union, Union panafricaine des télécommunications (f)

PAU pause

P-AVI videotex gateway

PAW Peachtree accounting for Windows *tn*; physics analysis workbench; Pirates Analyze Warez

PAWS portable acoustic wave sensor

PAX portable archive exchange (Unix)

pb pointer to a string of bytes

PB paso bajo (s), low pass [LP]; push button

PBC peripheral bus computer; Steuerung für Peripherieplatinen (g), peripheral board controller; programme booking centre

PBD Pacific Bell Directory; programmer brain damage

PBE prompt by example

PBEM play by electronic mail

PBER pseudo-bit-error-ratio

PBET performance-based equipment training

PBGA plastic ball grid array

PBIS Peachtree Business Internet Suite *tn*

PBL poly-buffered LOCOS

PBM play by mail; portable bitmap Unix file extension

PBMS Pacific Bell Mobile Services

PBP packet-burst protocol (NetWare); personal business service; pole broken; photon back-scattering; portable base station

PBSRAM pipeline-burst static RAM

PBX private branch exchange centralita privada (s)

PBXI PBX internacional (s), international PBX

P$_C$ family of content protocols

PC carrier power (of a radio transmitter); parallel C; path control; peg count; peripheral controller; personal computer; petición de conexión (s), connection request [CR]; port control; primary centre; printed circuit (board); process control/controller; production control; program counter; programmable controller

PC card *v* PCMCIA

PC/AT IBM personal computer – advanced technology

PC/XT IBM personal computer – extended technology

PCA Peachtree Complete Accounting *tn*; performance and coverage analyser; Personal Computer Association; printed circuit assembly; protective connecting arrangement; signal d'accusé de réception de passage sur canal sémaphore de secours (f), change-over acknowledgement signal [COA]

PCACIAS personal computer automated calibration interval analysis system

PCAD packaging computer-aided design

PCAP protocolo de capa paquete (s), packet layer protocol [PLP]

PCAV partial constant angular velocity

PCB port check bit; power circuit breaker; power control box; printed circuit board, Schaltungsplatine (g); process control block; product configuration baseline; protocol/program control block; public coinbox

PCBC plain cipher-block chaining

PCC processor control console

PCCA Portable Computer and Communications Association

PCCH physical control channel

PCCM private circuit control module

PCCS parte control de la conexión de señalización (s), signalling connection control part [SCCP]; rearranque (s), reset [SCCP]; reinicialización (s), restart [SCCP]; usuario de extremo (s), end-user [SCCP]

PCCTS Purdue compiler-construction tool set

PCD photo CD; port control diagnostic; presentation capabilities descriptor; punto de coordinación para datos (s), data coordinating point [DCP]

PCDA program-controlled data acquisition

PCDDS printed circuit digital data service

PCDMA Personal Computer Direct Manufacturers' Association (UK)

PC-DOS personal computer disk operating system (IBM)

PCE picture control entity

PCEB PCI to EISA bridge (Intel); presentation connection end-point

PCER position de conversation pour/de l'équivalent de référence (f), posición de

conversación para el equivalente de referencia (s), reference equivalent speaking position [RESP]

PCFS PC file system

PCG physical connection graph

pch punch/punched

PCH paging channel; parallel channel; prises par circuit et par heure (f), seizures per circuit hour [SCH]

PCHK parity check

PCI panel call indicator; pattern correspondence index; peripheral component interconnect (local bus) (64-bit path); personal computer interface; protection contre les interruptions (f), interrupt protection; protocol control information, indicateur de commande de protocole (f), indicador de control de protocolo (s)

PC-I/O program-controlled/controller I/O

PCIC PC-card interrupt controller

PCIO peripheral component interconnect I/O controller (Sun *tn*)

PCIUG PC Independent User Group

PCL page/printer control language (HP); peripheral conversion language; plug-compatible module; portable common loops; prueba de continuidad-llegada (s), continuity check incoming [CCI]

PCM personal computer manufacturer; physical configuration management; port command area; printer cartridge metric (Hewlett Packard); process control module; pulse code modulation, Pulscodemodulation (g)

PCM 30 30-channel pulse code modulation, 30-Kanal Pulscodemodulation (g)

PCMC PCI cache memory controller (Intel)

PCMCIA people can't memorize computer industry acronyms; Personal Computer Memory Card International Association

PCMIM PC media interface module

PCMP post chemical-mechanical polishing

PCMS plasma chemistry Monte-Carlo simulation

PCN Personal Communications Network (over 1.5 GHz, EU); PointCast Network (US corp); program composition notation; prueba de continuidad (s), continuity check [CC]

PCNE protocol converter for non-SNA equipment

PCNFS PC network file system

PCO photo-catalytic oxidation; plant control office; point of control and observation, point de contrôle et d'observation (f), punto de control y observación (s); signal d'ordre de passage sur canal sémaphore de secours (f), change-over order signal [COO]

p-code pseudo-code

PCP packet control process; point de contrôle protégé (f), protected monitoring point [PMP]; port call processing; protocolo de capa de paquete (s), packet layer protocol [PLP]; pruebas de compatibilidad [CPT]; Psycho Corporate Productions

PCP/X.25 protocolo de capa de paquetes X.25 (s) [X.25/PLP]

PCR pass card reader; peak cell rate (UNI 3.0); Personal Computer Rats; preventive cyclic re-transmission; principle component regression; program clock reference; punto de control del restablecimiento (s), restoration control point [RCP]

PCS patchable control store; Personal Communications Services (US digital mobiles; GSM competitor); personal communications system; personal conferencing specification; personalization centre; plastic-clad silica (fibre); point de commande de service (f), service control point [SCP]; print contrast signal; process control systems; production cost savings; proxy cache server; prueba de continuidad-salida (s), continuity check outgoing [CCO]

PCSA personal computing system architecture

PCSR parallel channels signalling rate

pct pulse count

PCT petición de congelación de trama (s), freeze frame request [FFR]; (bloque de) petición de conexión de transporte (s), transport connection request [TCR]; private communication technology (network security); Process Change Teams

PCTE portable common tool/table environment

PCTR protocol conformance test report, rapport de test de conformité au protocole (f)

PCU packet communications unit; paging control unit; peripheral control unit; power control unit; premises/protocol control unit

PCWG Personal Conferencing Work Group (Intel)

PC-XT PC extended

pd potential difference

PD paging data; Panasonic Drive *tn*; peripheral decoder; petición de desconexión (s), disconnect request [DR]; phase-change dual; photo-detector; physical density; por destinatario (por cada destinatario) (s), per recipient [PR]; propagation delay; physical delivery; protocol discriminator; public data/domain

PDA personal digital assistant; pre-failure detection and analysis; push-down automaton

PDAU physical delivery access unit

PDB process descriptor base

PDC Participatory Design Conference; passive data collection; personal digital cellular; Personal/Pacific Digital Cellular Standard (Japan); primary domain controller; (Microsoft) Professional Developers' Conference; Prolog Development Centre

PDCH physical data channel

PDD physical device driver; post-dialling delay

PDEL partial differential equation language

PDES product data exchange using STEP

PDF package definition file; Portable Document Format (Adobe *tn*); post-detection filter; probability density function; processor-defined function; program development facility

PDFA praseodymium-doped fibre amplifier

PDH plesiochronous digital hierarchy (34 and 140 Mbit/s)

PDI parte del dominio inicial (s), initial domain part [IDP]; picture descriptive information; power and data interface

PDIAL public dial-up Internet access list

PDI-S Projektbüro für digital-Dienste (g) (project office for ISDN)

PDL page description language; program design language; propositional dynamic language; push-down list; Process Design Language 2 (Texas Instruments *tn*)

PDM physical medium dependent; processor data module; product data management; pseudo-degraded minute; pulse delta modulation; pulse-duration modulation

PDN positive delivery notification, avis de remise positive (f), notificación de entrega positiva (s); public data network

PDO packet data optimized; portable distributed objects (Next)

PDP peripheral data processing; plasma display panel; Programmable Digital Processor (series; DEC OS *tn*)

PDP n Programme Data Processor Model n (DEC *tn*)

PDQ Peachtree data query

PDRS public digital radio service (packet switching)

PDS packet driver specification; partitioned data set; penultimate digit storage; personal document scanner; physical delivery system; planetary data system; portable document software; Premise Distribution System (AT&T); processor direct slot (Apple); program data source; protected distribution system; public domain software

PDS/MaGen problem descriptor set/matrix generator

PDSA plan do see approve (production mantra)

PDSI peripheral device serial interface (Bull DPS7)

PDSL prescribed diffusion service licence; Public Domain and Shareware Library

PDSP peripheral data storage processor

PDSS post-development and software support

PDT performance diagnostic tool (IBM); programmable data terminal; programmable drive table; petición de transmitir (s), request to send [RTS]

PDU plug distribution unit; protocol data unit (name given to SNMP messages), unidad de datos de protocolo (s) [UDP]

PDV presentation data value

PDVC phase-dependent voltage contrast

PE paging entity; parity even; Peripherieeinrichtung (g) (peripheral equipment); petición de estado (s), status request [SRQ]; phase encoded/encoding (recording); physical entity; pictorial element; portable executable (two-byte header); pre-emption; processing element; proceso ejecutivo (s), executive process [EP]; protect enable; protocol elements; protocol entity; public enemy

PEA pocket Ethernet adaptor

PEAA unidad de datos del protocolo de aplicación de PETICIÓN A-ASOCIACION (s), A-ASSOCIATE REQUEST application-protocol-data-unit [AARQ]

PEAN Pan-European ATM Network

PEARL performance evaluation of amplifiers from remote location; process and experiment automation real-time language

PEB paridad de entrelazado de bits (s), bit interleaved parity [BIP]; post-exposure bake

PEB-N paridad N de entrelazado de bits (s), bit interleaved parity N [BIP-N]

PEC packed encoding rules; photo-electric cell; printed electronic circuit

PECC partially error-controlled connections

PECP punto extremo de la conexión de presentación (s), presentation connection endpoint [PCEP]

PECVD plasma-enhanced chemical vapour deposition

PED parte específica del dominio (parte especificación de dominio) (s), domain specific part [DSP]; prediction error data

PEDS plasma-enhanced deposition system

peek BASIC instruction to access memory

PEEL programmable electrically-erasable logic (ICTI)

PEELS parallel electron energy loss spectrometry

PEG programmable event generator

PEGASYS programming environment for graphical analysis of systems

PEIPA Pilot European Image Processing Archive UK

PEL permissible exposure level

pel pixel (picture element)

PELI unidad de datos del protocolo de aplicación de PETICIÓN A-LIBERACION (s) [RLRE], A-RELEASE application protocol data unit

PEP packetized ensemble protocol (Telebit); Pan-European paging; partitioned emulation program; peak envelope power (of radio transmitter); performance enhancement package; Privacy Enhanced Mail *tn*; product error message

PEPS premier entré – premier sorti (f), primero entrado – primero salido (s), first in – first out [FIFO]

PER packed encoding rules; partial equivalence relation; señal de orden de paso de emergencia a enlace de reserva (s), emergency change-over order signal [ECO]

PERL pathologically eclectic rubbish lister; Practical Extraction and Report (programming) Language

Perm pre-embossed rigid magnetic (tape)

PERT program/performance evaluation (and) review technique

PES packetized elementary stream; photo-electron spectroscopy; positioning error signal; processor enhancement socket; programmable electronic system; prueba (de) estado (del) subsistema (s), subsystem status test [SST]

PESS prueba de estado de subsistema (s), subsystem status test [SST]

PET Planning Exercise in Telecommunications (technologies – pre RACE); positron emission tomography; post-etch treatment (of chip); Print Enhancement Technology (Compaq *tn*); series of Commodore Business Machines personal computers *tn*

PEV peak envelope voltage

PEX PHIGS extension to X Window system

pf power factor/frame

pF pico-Farad

PF power factor

PFA mensaje de petición de facilidad en adelante (s), facility request message [FAR]; Peachtree First Accounting *tn*

PFB PostScript font binary *tn*; Provisional Frequency Board

PFCC paper-feed control character

PFD power flux density

PFE portable Forth environment

PFL persistent functional language

Pfl Pflichtenheft (g) (specifications document)

PFM PostScript font metrics *tn*; pulse-frequency modulation/measuring

PFMD período fijo de medidas diarias (s), fixed daily measurement period [FDMP]

PFPU processor frame power unit

PFR plug-flow reactor; power-fail restart; programmer's file editor

PFS page format selection, selección de formato de página (s); primary frequency supply

PG parameter group; permanent glow; port group; program generic

PGA pin-grid array; professional graphics adaptor; programmable gain amplifier; programmable gate array

PGC port group control; program group control (Microsoft)

PgDn page down (key)

PGES position de l'anneau de garde de l'équivalent pour la sonie (f), loudness rating guard-ring position [LRGP]

PGH port group highway

PGHTS port group highway timeslot

PGI parameter group identifier; port group interface; purge gas inlet

P-GILD projection gas immersion laser doping

PGL peer group leader

PGLI parameter group length identifier

PGM program; portable Graymap Unix

PGML Precision Graphic Mark-up language (Adobe *tn*)

PGN portable game notation

PGP Pretty Good Privacy (Zimmerman – RSA encryption program *tn*); Programme Garant du Privé (f)

PGS general supervision panel

PGTC pair gain test controller

PgUp page-up (key)

pH parallel Haskell

PH packet handler/handling, traitement des paquets (f), manejo de paquetes (s), tratamiento de paquetes (s); paging handler; parity high; phantom circuit; phase hit; physical; physical layer

PHA Post-Hauptanschluß (g), main exchange line

PHACT Phreakers Hackers Anarchists Cyberpunk Technologists

PHANTSY Phantasy Magazine

PHARE Programme of Harmonized Air Traffic Management Research in Eurocontrol, programme de recherche sur la gestion harmonisée du trafic aérien d'Eurocontrol (f), Programa de investigación sobre la gestión armonizada del tráfico aéreo de eurocontrol (s), EUROCONTROL-Programm einer abgestimmten Forschung auf dem Gebiet des Flugverkehrsmanagements in Europa (g)

PHCM paging handler configuration management

PhD Phreak/Hacker Destroyers

PHFM paging handler fault management

PHI packet handler interface

PHIGS Programmers' Hierarchical Interactive Graphics System

PHM paging handler mobile terminal

Pho photonic

phon unit of sound

PHOTAC photo-typesetting and composing

PhotoCD standard for digitally storing 35mm photo (Kodak/Philips *tn*)

photonics fibre-optics

PHP physical plane

PHPM paging handler performance management

PHS Personal Handy Phone System (Japan)

PHT paging handler terminating network

PHTh phase hit threshold

PHUCK Phone Hackers United Crash Kill

PHUN Phreaker Hackers Underground Network

PHW public highway environment

PHY OSI physical layer

PI identificador de petición de servicio de interfuncionamiento (s), interworking service request identifier [IRQ]; parameter identifier; performance index; physical interface; Pirates Incorporated; presentation indicator; program interruption; programming infrastructure; proportional integral; protection interval;

protocol identifier; prueba de inactividad (s), inactivity test [IT]

PI2G périphérique intelligent de 2ième génération (f), second generation intelligent peripheral

PIA peripheral interface adaptor; personal imaging assistant (Nikon); programmable interface adaptor

PIB processor interface buffer

PIC PCM interface controller, Steuerung der PCM-Schnittsstelle (g); peripheral interface controller; plastic insulated cable; point in call; polyethylene insulated cable/conductor; position independent code; preferred inter-exchange carrier; primary independent carrier; priority interrupt controller; problem isolation code; program/programmable interrupt controller

PICA Project geIntegeerde Catalogus Automatisering (Holland)

PICH pilot channel

PICK OS used on mainframes and PCs *tn*

pico one millionth of a millionth; 10^{-12}

picocell small area within a communications system

PICS platform for Internet content selection; plug-in (inventory) control system; protocol implementation conformance statement, déclaration de conformité d'une mise en oeuvre de protocole (f)

PID personal ID device; port identification; private/personal identifier; process identification/identifier; process-induced defect; proprietary information disclosure; proportional integral derivative/differential; protocol identification/identifier

PID controller proportional – integral – derivative controller

PIDX Petroleum Industry Data Exchange

PIE Public Internet Exchange (Thailand)

PIER Procedure for Internet/enterprise re-numbering

PIG packet Internet groper; protocolo de información de gestión (s), management information protocol [MIP]

PIGC protocolo de información de gestión común (s) [CMIP]

PIGUI platform-independent graphical user interface

PII program-integrated information

PIII plasma immersion ion implantation

PIKS programmer's imaging kernel system

PIL Pirates in Legion; precision in-line; procedure implementation language; señal de petición de identidad de la línea que llama (s), calling-line identity request signal [CIR]

PILA présentation d'identification de la ligne appelante (f), presentación de (la) identificación de la línea llamante (s), calling line identification presentation [CLIP]; présentation d'identification de la ligne connectée (f), presentación de la identificación de la línea conectada (s), connected line identification presentation [COLP]

PILOT programmed inquiry learning or teaching

pilote (software) driver (for peripheral)

PIM personal information manager (organizer); primary interface module; protocol independent multicast; pulse interval modulation

PIMM prefijo interurbano mal marcado (s), mis-dialled trunk prefix [MPR]

PIN personal identification number; Pond Information Network; positive-doped insulating/negative-doped; positive-intrinsic-negative (fibre-optics); procedure interrupt negative; process identification number (UNIX); P-type + I-type + N-type

PIND particle impact noise detection

PINE program for Internet news and e-mail

PING packet Internet grouper (RFC-792)

ping-pong time-compression multiplexing (TCM)

Pink abandoned Apple project for object-oriented OS

Pink Book defines network service operating over a CSMA/CD bearer network

PINT PSTN/Internet Interworking

PIO parallel/private input-output; processor/programmed input-output

PIP packet interface port; peripheral interchange program; picture-in-picture display; plug-in-and-play; problem isolation procedure; process-induced particles; procedural interrupt positive; programmable interconnect point

PIPO parallel input/parallel output

PIRE puissance isotropique rayonnée équivalente (f) [EIRP]

PIRL pattern information retrieval language

PIS procedure interrupt signal

PISO parallel input/serial output

PIT programmable interval timer

PITA pain in the ass

PITO Police Information Technology Organization

PITS pie in the sky

PIU path information unit (SNA); plug-in unit

PIV peak inverse voltage; post-indicator valve

pixel picture element (one dot on a monitor screen; *v* pel)

pixelblt pixel block transfer

PIXIT protocol implementation extra information for testing, informations complémentaires sur la mise en oeuvre du protocole destinées au test (f)

PIXT protocol implementation extra information for testing

PJ/NF projection-join/normal form

PJPEG progressive JPEG

PKC public key cryptosystem

PKCS public key cryptography system; public key cryptosystem

PKE public key encryption

PKI public key infrastructure (encryption)

PKUNZIP file un-compression utility (PKWARE Inc *tn*)

PKZip file compression utility (PKWARE Inc *tn*)

pl pico-litre, 10^{-12}

PL parameter length; parity low; petición de liberación (s), clear request [CLR]; physical layer; private line; program logic; progresión de llamada (s), call progress [CP]; project line

PL OU physical layer overhead unit

PL/1 Programming Language 1 (IBM *tn*)

PL/360 (first) machine-oriented high-level language

PL/C; PL/CT Cornell University versions of PL/I

PL/M programming language/microcomputers (Digital/Intel)

PL/S programming language/systems (IBM)

PL/Z programming language Zilog (a family of systems)

PLA plain language address; primitiva local abstracta (s), abstract local primitive [ALP]; programmed/programmable logic array

PLACE programming language for automatic check-out equipment

PLAIN programming language for interaction

Plait Pupils' Learning and Access to Information Technology (UK)

PLAN Programming Language for ICL 1900 series *tn*

PLAS private line assured service

PLATO Programmed Logic for Automatic Teaching Operations *tn* (*ie* CAI)

PLB picture level benchmark

PLBC pipeline burst cache

PLC programmable logic controller

PLCC plastic leadless/leaderless/lead carrier

PLCP physical layer convergence procedure/protocol (IEEE 802.6)

PLD partial line down, descenso parcial de renglón (s); Petersburg Long Distance

(US corporation); programmable logic device

PLE prohibición de llamadas entrantes (s), incoming calls barred [ICB]; public local exchange

PLEDM phase-state low-electron-number drive memory (Hitachi *tn*)

PLI parameter length indicator

Plisp pattern LISP

PLITS programming language in the sky

PLK primary link

PLL petición de llamada (s), call request [CR]; phase-locked loop; plasma lockload; progresión de la llamada (s), call progress [CPG]

PLMN public land mobile network

PLMNGSM GSM public land mobile network

PLMNO PLMN operator

PLMS private line message switch (MCI)

PLN private line network

PL-OAM physical layer-operation and maintenance (cell)

PL-OU physical layer overhead unit (UNI physical layer definition)

PLP packet layer protocol/procedure; pulse link relay/repeater

PLS partial least squares; physical signalling sub-layer; primary link station; private line service; programmable logic sequencer; prohibición de llamadas salientes (s), outgoing calls barred [OCB]; projection of latent structures

PLSME programmable logic state machine entry

PLT public low traffic (rural)

PLTL propositional linear temporal logic

PLTTY private line teletypewriter service

PLU partial line-up; primary logical unit; procédure (de) liaison unique (f), single link procedure [SLP]

PLUS private line universal switch

PLV production level video

PLY photo-limited yield

Pm phase modulation

PM mean power; performance-management; performance monitoring;

peripheral module; per message, por mensaje (s); physical medium (sub-layer); preventive maintenance; Presentation Manager (IBM *tn*); process manager/module

PM 1 processable mode Nº 1, mode retraitable Nº 1 (f)

PMA physical medium attachment; prompt maintenance alarm

PMAC peripheral module access controller

p-machine pseudo-machine

Pmail postal mail (*v* snail mail)

PMB pilot-make-busy circuit

PMBX private manual branch exchange, handbediente Nebenstellenanlage (g)

PMC PCI mezzanine card; personal mobile communication; petición de modificación en el curso de la llamada (s), connection mode rejection [CMR]; phantom maintenance connector; private meter check; process module controller

PMCC Pensky-Martens closed cup

PMD packet mode data; physical medium-dependent; polarization mode dispersion

PME petites et moyennes entreprises (f), small and medium-size enterprises [SME]

PMFJI pardon me for jumping in

PMI performance monitoring integrator; phase-measuring interferometer; private memory interconnect (DEC); protected mode interface

PMJI pardon my jumping in

PML mensaje de petición de modificación de llamada (s), call modification request message [CMR]; permitted maximum level; procédure multiliaison (f), multi-line procedure [MLP]

PMM pool maintenance module

PMMU paged memory management unit

PMO program management office

PMOS positive channel metal-oxide semiconductor (type of MOSFET)

PMP point-to-multipoint (UNI 3.0);

project management plan; protected monitoring point, punto de monitorización protegido (s)

PMPO peak music power output

PMR pressure-modulated radiometer; private mobile radio; problem management report (IBM); señal de paso manual a enlace de reserva (s), manual change-over signal [MCO]

PMS Pantone (colour) Matching System *tn*; Paranoid Media Scrutinization; particle measuring system; permitted maximum signal; policy management system; processor-memory-switch (notation); Public Message Service (Western Union *tn*)

PMT photomechanical transfer; photo-multiplier tube; program management tool

PMTS programmierbares Modul-Testsytem (g) (programmble module test system)

PMU precision measurement unit

PMUX programmable multiplex

PMX packet multiplex

PMXA Primärmultiplexanschluß (g) (primary multiplex subscriber)

PN permanent nucleus (by CEPT for RACE); private network; processing node; pseudo-noise

P$_n$ X.400 protocol

p-n junction area of contact between two semiconductor materials

PNA Parallel Netzwerk Architektur (g) (programming network access)

PNB Pacific Northwest Bell

PNC Police National Computer (UK)

PNC 2 Police National Computer version 2 (UK)

PND program network diagram

PNE Présentation des Normes Européennes (f)

PNG portable network graphics (format)

PNI packet network interface; permit next increase

PNIC private data network identification code

P-NID precedence network in-dialling

PNL processeur non linéaire (f), non-linear processor [NLP]; procesador no lineal (s), non-linear processor [NLP]

PNNI private network node interface/private network-to-network (ATM Forum)

PNO public network operator

PnP plug-and-play; prêt à l'emploi (f)

PNP private numbering plan, plan de numeración privado (s); p-type material sandwiching n-type material

Po polonium (used in equipment to prevent static electricity)

PO parity odd; post-pay coin telephone

PoA point of attachment; Power Open Association; punto de origen de asignación (s), assignment source point [ASP]

POC point of contact; processor outage control

POCSAG Post Office Code Standards Advisory Group (US)

POD piece of data; power-on diagnostics; programmable option devices

POE PowerOpen Environment

POEMS mode S secondary radar pre-operational program

POETRY Processing of electronic requests (EC translation service)

POETS push-off early tomorrow's Saturday

POH path overhead (SONET); power-on hours

POI path overhead indicator; point of initiation; point of interconnection (with PSTN); point of interface; probability of intercept

POINTEL French Telepoint network

POIX point-of-interest exchange language specification

POL polarized; problem-oriented language

poly-si polycrystalline silicon (LCD display)

POM pool operational module

POMA parte operaciones mantenimiento

y administración (s), operation maintenance and administration part [OMAP]

POMS printed output management system

PON passive optical network

PONG PC World On-line Glossary

POOL parallel object-oriented language (Philips)

POP post office protocol (RFC 1125); point of presence, point de présence (f)

POP-11 programming language for AI, combining LISP and POP2

POP-2 high-level programming language (University of Edinburgh)

POP3 post office protocol 3

POPA pop all registers

POPF pop flags

POPL Principles of Programming Languages (annual conference/ACM)

POPLOG programming environment for POP-11 and Prolog

POPS program for operator scheduling

POPUS Post Office processing utility subsystem

POR process-of-record; point of return; power-on reset

PORTAL process-oriented real-time algorithmic language

pos position/positive

POS point of sale; product of sums – expression; programmable option select

Pose a language for posing problems

poset partially-ordered set

POSI Promotion (conference) for OSI

POSICE portable operating system interface for computer environments (*now* Posix)

POSIT Profiles for Open Systems Interworking Technologies

POSIX portable operating system interface for computer environments (IEEE standard – based on Unix)

POST power-on self-test

POSTNET postal numeric encoding technique

PostScript system/device-independent PDL (Adobe *tn*)

pot potential; potentiometer

POT point of termination

POTS plain old telephone service

POU point-of-use

POUCG point-of-use chemical generation

Power PC 64-bit RISC micro-processor chip (IBM, Apple, Motorola)

POWER performance optimization with enhanced RISC (IBM Power PC)

p-p peak-to-peak

pp push-pull

PP partial page; payload pointer; person to person; polarization-preserving (optical fibre); portable part; power patroller (Trinitech *tn*); procesador de paquetes (s), packet handler [PH]

PPC peak power control; primary point code; señal de petición de prueba de continuidad (s), continuity check request signal [CCR]

PPCI presentation protocol control information

PPCM predictive PCM

PPD parallel presence detect; peripheral pulse distributor

PPDS personal printer data stream (IBM)

PPDU presentation protocol data unit

PPE personal protective equipment; primitive procedure entity

PPGA plastic pin grid array

PPI (European) data-processing project proposal, proposition de projet informatique (f), vorgeschlagenes Informatikprojekt (g); método del proceso de Poisson interrumpido (s), interrupted Poisson process [IPP]; plan position indicator; position phototélégraphique internationale (f), international phototelegraph position [IPP]

PPID process program identification

PPL paquete de petición de llamada (s), call request packet [CRP]; polymorphic programming language (Harvard)

ppm pages per minute; parts per million, partes por millón (s); pulses per minute

PPM peak programme meter; periodic permanent magnet; periodic pulse metering; portable Pixelmap (Unix); presentation protocol machine; pulse position modulation

PPN project-programmer number

PPO primary program operator

P-POP plain-paper optimized printing (Canon)

ppp petición de página parcial (s)

PPP points par pouce (f), dots per inch (*ie* screen resolution) [DPI]; point-to-point protocol (RFC 1171), protocole de point à point (f)

PPPL Princeton Plasma Physics Laboratory

PPQN parts per quarter note (MIDI format)

PPR partial page request

PPRef process program reference

pps packets per second; pulses per second

PPS partial page signal; power personal system (IBM); preliminary product specification; pre-paid service; public packet switching

PPSN public packet-switched network

PPSU personal printer spooling utility

PPT preparado para transmitir (s), ready for sending [RFS]; process page table

PPTP point-to-point tunnelling protocol

PPV pay-per-view (TV), télévision à péage (f), Abonnementfernsehen (g)

PQET print quality (resolution) enhancement technology

PQFP plastic quad flat pack

PQS picture quality scale

pr pair

PR messages de passage sur canal sémaphore de secours et de retour sur canal sémaphore normal (f), change-over and change-back message [CHM]; packet radio; paging receiver; paging response receiver; per-recipient; performance rating; phase representation;

premature release; preparado para recibir (s), receive ready [RR]; prepare

PRA petición de rearranque (s), re-start request [RTR]; primary rate access (2 Mbit/s – ISDN; European term for PRI); punto de referencia auricular (s), earcap reference point [ECRP]

PRACSA Public Remote Access Computer Standards Association

pragma programming language statement

PRAM parallel random access machine (Macintosh)

prank virus Microsoft's preferred name for Word macro/virus

PRAS particle reactor analysis services

PRAT production reliability acceptance test

PrAz Prüfanrufziel (g) (test call destination)

PRB point de référence bouche (f), punto de referencia boca (s), mouth reference point [MRP]; pseudo-random binary

PRBS pseudo-random binary/bit sequence

PRC primary reference clock; primary routing centre; punto de referencia de calidad de servicio (s), quality of service reference point

PRCA Puerto Rico Communication Authority

PRCH packet radio channel

PRCS personal radio communications service

PRD información de petición de restauración por demanda [DR]

PRDMD private directory management domain, domaine de gestion d'annuaire privé (f), domaine de gestion privé d'annuaire (f)

PRE point de référence écouteur (f), earcap reference point [ECRP]; pre-formatted

PRECCX pre-C-compiler extended

Pref CUG preferential CUG

PREMO Presentation of Multimedia Objects

PreP Power PC reference platform (IBM/Motorola)

PRESS prediction error sum of squares

PRESTEL Press Telecommunications; processor request flag

PRF pulse repetition frequency

PRFD pulse recurrence frequency discrimination

PRFU processor ready for use

PRG message de progression d'appel (f), call progress message [CPG]

PrGt Prüfgerät (g) (test equipment)

PRI primary-rate interface (23B+D ISDN interface); pulse repetition interval; petición de reinicialización (s), restart request [RSR]

PRI-EOM procedure interrupt – end of message

PRI-EOP procedure interrupt – end of procedure

PRIMOS Prime OS

PRI-MPS procedure interrupt – multi-page signal

PRINT pre-edited interpreter (IBM)

PRIP priority indicator

PRISM photo-refractive information storage material

PRL mensaje de progresión de la llamada (s), call-progress message [CPG]; proof refinement logic

PRM prefijo interurbano mal marcado (s), mis-dialled trunk prefix [MPR]; premium (rate service); protocol reference model

PRMD private management domain

PRML partial response maximum likelihood

PRN printer (a system's first parallel port)

PRO point de référence oreille (f), punto de referencia oído (s), ear reference point [ERP]; punto de referencia óptica (s), optical reference point [ORP]

proc procedure

PROFS Professional Office System (IBM *tn*)

proglet a small program (as in piglet)

PROLOG logic programming language

PROM programmable ROM

PROMATS programmable magnetic tape system

PROMETHEUS (EU) Programme for European Traffic with Highest Efficiency and Unprecedented Safety (communications system between mobiles and road beacons)

PROSE programmable sequencer

prosign procedural signal

Proveedor SR proveedor del servicio de red (s), network service provider

PRQ photo reproduction quality (Epson *tn*)

PRR pulse repetition rate

PRRM pulse repetition rate modulation

PRS planning reference system; premium rate service; primary reference source; private radio systems; Prüfsatz (g) (test unit); prueba de conjunto de rutas de señalización (s), signalling route set test signal [RST]; pseudo-random sequence

PRSAN Anschalteprüfsatz (g) (test unit for line connections)

PRSC parametric response surface control

PRSIG Pacific Rim SMDS Interest Group

PRSL primary area switch locator

PRSTA Tischansteuerungsprüfsatz (g) (test set for desk access)

PRT pérdida de retornos transversal (s), transverse return loss [TRL]; Polonia Radio Television

PRTC Puerto Rico Telephone Company

PRTM printing response time monitor

PrtSc print screen (key)

PrvDN private data network

PRW Peachtree Report Writer

Ps location probability

PS packet-switched; paging system; permanent signal; phototelegraph station; physical sequential; proposed/(draft) standard, avant-projet de norme (f); point sémaphore (f), signalling point [SP]; punto de señalización (s), signalling point [SP]; porous silicon; port store; Postscript Level 2 (LZW) extension; product supplier; program store; programmed symbols

PS/2 Personal System/2 (IBM); Programming System 2

PSA port storage area; problem statement analyser; primitiva de servicio abstracto (s), abstract service primitive [ASP]; protocol-specific annex (Microsoft Windows Winsock 2)

PS-ALGOL persistent ALGOL

PSAP presentation (layer) service access point; public safety answering point – 911 (US)

PSB phase shifting blank; public service board

PSC Peachtree Support Center; picture start code; porous silicon capacitor; print server command; programmable sine controller; public service commission

PSCM process steering and control module

PSCS personal services communication space

PSD peripheral sharing device; permanent signal detection circuit; portable scheme debugger; port status display; power spectral density

PSDAU packet-switched data access unit

PSDC public-switched digital capability

PSDDS public switched digital data service

PSDN packet switched data network, réseau pour données à commutation par paquets (f), réseau public de commutation de données (f), red pública de datos conmutada (s), rede pública de dados conmutada (p), Telefonnetzverbund (g)

PSDS public/packet-switched data/digital service

PSDTN packet-switched data transmission network

PSDTS packet-switched data transmission services

PSDU presentation service data unit

PSE packet/(public) switch(-ed) exchange; programming support environment; project SE

PSEC punto de señalización del extremo cercano (s), near-end signalling point [NESP]

PSEN program store enable

PSERVE print server (NetWare)

PSF permanent signal finder; permanent swap file; petición de subsistema fuera de servicio (s), petición subsistema fuera de servicio (s), subsystem out of service request [SOR]; provisional system feature

PSG phase-structure grammar; phosphosilicate glass; product-shipping guide; protocolo de sistema de guía (s), directory system protocol [DSP]

PSGE severely-errored pseudo-seconds

PSI packet switching interface; paid service indication; pérdida de inserción para la sonoridad (s), loudness insertion point [LIL]; planned start installation; point sémaphore international (f), punto de señalización internacional (s), international signalling point [ISP]; portable scheme interpreter; protocolo de sesión interactivo (s), interactive session protocol [ISP]

PSID PostScript image data

PSII plasma source ion implantation

Psion Potter Scientific Instruments *tn* (Peter Potter, founder)

PSIU packet switched/switching interface unit

PSK phase-shift keying

PSL polystyrene latex; portable standard Lisp; power and signal list; power sum loss; process simulation language

PSL/PSA problem statement language/problem statement analyser

PSLS polystyrene latex sphere

PSM packet service module; phase shift/shifting mask; prestataire de service de maintenance (f); proveedor(es) de servicio(s) de mantenimiento (s), maintenance service provider [MSP]; printing systems manager

PSMA punto de sincronización mayor (s), major sync point [MAP]; punto de sincronización menor (s), minor sync point [MIP]

PSN öffentliches Wählnetz (g), public switched network; packet switch/switched network; packet switch node; people with special needs (RACE); point sémaphore national (f), punto de señalización nacional (s), national signalling point [NSP]; processor serial number (Intel Pentium III); public-switched (telephone) network, réseau public commuté (f)

PSNP partial sequence number packet (NetWare)

PSO private service operator, opérateur privé (f), privater Dienstanbieter (g)

PSP parámetro de serie de pruebas (s), test series parameter [TSP]; personal software products group (IBM); program segment prefix

PSPACE polynomial space

PSPDN packet-switched public data network

PSR parte servicio de red (s), network service part [NSP]; perfectly stirred reactor; point sémaphore faisant fonction de relais dans le SSCS (f), punto de señalización con funciones de relevo (s), signalling point with relay function [SPR]

PsRAM pseudo-static RAM

PSRI intelligent network signalling point (f)

PSS Packet Switch Stream (BT *tn*); packet-switching service; product support services

PSSIC phonetic speech synthesis integrated circuit

PS-SL Philips Semiconductors – Systems Laboratories

PSSP phone center staffing and sizing program

p-static precipitation static

PSTC public switched telephone circuits

PSTN public switched telephone network, réseau téléphonique public commuté (f), red telefónica pública conmutada (s), öffentliches Fernsprechnetz (g)

PSU port storage utility; power supply unit; primary switching unit

PS-user presentation service user

PSV closed-user-group validation check messages (s) [CVM]

PSVC point-to-point signalling virtual circuit

PSW processor/program status word

p-switch software system for UCSD Pascal; printer switch

PT parity data; pattern transfer, transferencia de patrón (s); pay tone; payload type; port number; poste de travail (f), workstation [WS]; program time; protocolo de transporte (s), transport protocol [TP]

PTAB Project Technical Advisory Board

P-TAC parallel three-address code

PTAT private trans-Atlantic Telecommunications

PTB Physikalisch-Technische Bundesanstalt (Braunschweig) (g)

ptc Pascal-to-C translator

PTC Pacific Telecommunications Council; Part Time Crackers; police tracking computer (UK); positive temperature coefficient; pre- and post-process treatment chambers

PTD parallel transfer disk drive; parte transferencia de datos (s), data transfer part [DTP]

PTE path terminating equipment (SONET)

PTEP pérdida del trayecto de eco ponderada (s), weighted listener echo path loss [WEPL]

PTF patch and test facility; problem temporary/trouble fix (IBM)

PTG precise tone generator

PTI packet type identifier (X.25); palabra de trama invertida (s), inverted frame word [IFW]; party identity; payload type identifier/indicator (ATM); portable tool interface; posición telefotográfica internacional (s), international telephotographic position [IPP]; public tool interface

PTIME polynomial time

PTL perfil de terminal lógico (s), logical terminal profile [LTP]; Praise the Lord; Pirates That Live

PTLXAU public telex access unit

PTM packet transport mode; parte (de) transferencia de mensajes (s), message transfer part [MTP]; portable traffic monitor

PTMPT point-to-multipoint

PTN plant test number; private telecommunication network; public telephone network

PTO Post and Telecommunications Organization (Denmark); public telecommunications operator

PTP peer-to-peer (networking)

PTR UAD puntero de unidad administrativa (s), administrative unit pointer [AU PTR]

PTR paper tape reader; pointer; poor transmission; printer; (señal de) prohibición de transferencia (s), transfer-prohibited signal [TFP]

PTS point de transfert sémaphore (f), punto de transferencia de señalización (s), signalling transfer point [STP]; presentation time stamp; proceed to select; public telephone system

PTSE PNNI topology state element (ATM Forum)

PTSP PNNI topology state packet; proceed-to-select protocol, protocolo de invitación a marcar (s)

PTSU protection transmission switching unit

PTT post telephone and telegraph; postes, télégraphes et téléphones (f), correos teléfonos y telégrafos (s), Post Telefon Telegraf (g); push-to-talk operation

PTTA Postal Telegraph and Telephone Administration

PTTC paper tape transmission code

PTTI precise time and time interval

PTTs post telegraph and telephone administrations (network operators)

PTTXAU public telex access unit

PTX charging parameter; parallel texts; textos paralelos (s)

pty party

PU messages de passage d'urgence sur canal sémaphore de secours (f), emergency change-over message [ECM]; parte de usuario (telefonía, datos, etc) (s), user part [UP]; physical unit (SNA); processor utility

PUA signal d'accusé de réception de passage d'urgence sur canal sémaphore de secours (f), emergency change-over acknowledgement signal [ECA]

PUBPOL-L Public Policy e-mail List (US)

PUC peripheral unit controller; Public Utilities Commission (US)

PUCP Physical Unit Control Point (SNA)

PUCT price-per-unit currency table

PUD parte (de) usuario de datos (s), data user part [DUP]

PUDU preferred user role data user

PUFFT Purdue University fast FORTRAN translator

PUI personal user identification

PUM processor utility monitor

PUN physical unit number

PUO signal d'ordre de passage d'urgence sur canal sémaphore de secours (f), emergency change-over order signal [ECO]

PUP PARC universal packet; peripheral unit processor; pick-up point

PU-RDSI parte usuario de RDSI (s), integrated services digital network user part [ISUP]

PUS processor upgrade socket; processor utility subsystem

PUSHF push flags

PUSI parte usuario de RDSI (s); öffentliches Wählnetz (g) (public switched network) [PSN]

PUT parte de usuario de telefonía (s), telephone user part [TUP]

PUTB put-back

PV parameter value

PVA pruebas de validación (s), validation test [VAT]

PVC permanent virtual circuit/connection; pruebas de validación de circuitos (s), circuit validation test [CVT]

PVCC permanent virtual channel connection

PVCP permanent virtual path connection

PVD physical vapour deposition

PVEM prueba de verificación de encaminamiento por la PTM (s), MTP routing verification test [MRVT]

PVES prueba de verificación del encaminamiento PCCS (s), SCCP routing verification test [SRVT]

PVLR previous visitor location register [VLR]

PVM parallel virtual machine; pass-through virtual machine (IBM)

PVN permanent virtual network; private virtual network

PVP parallel vector processing; permanent virtual path

PVQ pyramid vector quantization (codification)

PVQM perceptual video quality measure

PVR personal video recorder

PVS parallel visualization server

PVT perte de verrouillage de trame (f), loss of frame alignment [LFA]

pw password

pW picowatt

PWA Pirates with Attitude

PWB printed wiring board; programmers' workbench (Microsoft)

PWBA printed wiring board assembly

PWD print working directory (UNIX)

PWFG primary waveform generator

pwledit password list editor (Microsoft Windows)

PWM pulse-width modulation

PWP particles per wafer pass

pWp picowatt psophometrically-weighted

pWp0 picowatts relative to 0TLP

pwr power

PWS programmable workstation

PWSCS programmable workstation communication services (IBM)

pX peak envelope power in dB

PX peak envelope power in watts

pY mean power in dB

PY mean power in watts

Python scripting language

pZ carrier power in dB

PZ carrier power in watts

Q

Q quantity of electricity (coulombs)

Q channel one of eight channels on CD audio

Q1 Querleitung (tie line)

Q1P wait one (minute) for micturition (radio operator's signal)

QA Q (interface) adaptor; quality assurance

QAF Q-adaptor function

QAGC quiet automatic gain control

QAM quadrature amplitude modulation; queued-access method

QASPR QUALCOMM automatic satellite position reporting

Q-ball super-symmetric particle cluster

Qbasic quick Basic (*v* MS-BASIC)

QBE query by example

QBF query by form

QBIC Query by Image Content (IBM *tn*)

Q-bit qualified data bit (X.25)

Q-Browser Quetzal interface to intranet/Internet (Atlantik-Market)

QC quality control; quaternary centre

QCA quantum-dot cellular automata (University of Notre Dame project)

QCIF quarter(-inch) common intermediate format

QCM quartz crystal micro-balances

QCSE quantum confined stark effect

qd quad; quantizing distortion unit

QD queuing delay; qwerty/Dvorak (keyboard layouts)

QDOS Sinclair QL operating system *tn*

QDR quick dump rinse

QDS qualité de service (f), quality of service [QOS]

QDSS Q-bus Dragon subsystem *tn*

qdu quantizing distortion unit

QEMM Quarterdeck expanded memory manager (Quarterdeck Corp *tn*)

QEPC quad-exchange power controller

QET qualité d'écoulement du trafic (f), grade of service [GOS]

QFA quick file access

QFD quality function deployment

QFE Quick Fix Engineering (Microsoft)

QFM quantized frequency modulation

QFP quad flat pack

QH query handler/handling

QIC quality information using cycle time (Hewlett-Packard); Quarter-Inch Committee (O2/cartridge)

QiCA Qualification in Computer Auditing

QL Quantum Leap (Sinclair computer *tn*); query language; queue length

QLLC qualified logical link control (SNA)

Qlo Querleitung des Ortsdienstes (g) (local traffic tie-line)

QM queue manager

QMF quadrature mirror filters

QMIS Quality Management Information System (EU)

QMR qualitative material requirement

QMS quadrupole mass spectrometer

QN quarter-bit number

QOS quality of service

QPG quantum phase gate

QPM quantized pulse modulation

QPP quiescent push-pull

QPP amp quiescent push-pull amplifier

QPR-3 quadrature partial response with three amplitude levels

QPR-7 quadrature partial response with seven-level partial response

QPRS quadrature partial response signalling/system

QPSK quadrature phase-shift keying

QPSX queued packet synchronous exchange

QRC quick reaction capability

QRP QOS reference point

QRS quasi-random signal

QRSS quasi-random signal source

QS quality of service

QSAM quadrature sideband amplitude modulation

QSC Quetzal Support Centre (Atlantik-Market *tn*)

Q-signal narrow-band axis of chrominance signal in NTSC colour scheme

QSR quality system review

QSS quasi-stellar radio source (quasar)

QSTAG Quadripartite Standardization Agreement

QT queuing time

QTAM queued telecommunications access method (IBM)

QTC QuickTime conferencing (Apple Macintosh *tn*)

QTD degraded quality

QTI intolerable quality

QTML QuickTime Media Layer (Apple Macintosh *tn*)

QTN normal quality

QTVR QuickTime Virtual Reality (Apple Macintosh *tn*)

QTW QuickTime for Microsoft Windows (Apple Macintosh *tn*)

qty quantity

QU connection charge

quad quadruple

quadbit set of 4 bits (showing one of 16 possible combinations)

QUAM quantized amplitude modulation

QUARK quantizer, analyser, and record keeper

qubit quantum bit (of the future)

Quel a query language used by a DBMS (*also* QUEL)

queue *v* FIFO

QUIL quad-in-line

QUME peripherals manufacturer

QVE Quantisierungsverzerrung (g) (quantization distortion)

QVL Querverbindungsleitung (g) (tie-line)

QVSS Q-bus Viper sub-system *tn*

QWERTY (Anglo) keyboard layout

QWIP quantum-well infrared photo-detector

R

R persistence time; reception; red (s), network [N]; release; route beacon; symbol of resistivity

R i/f R interface

R&M reliability and maintenance

R&R rate and route

R(B)DS radio (broadcast) data signals

R/A recorded announcement

R/M read/mostly

R/O read-only

R/S relay set

R/T radiotelephony, radiotéléphonie (f), radiotelefonía (s), Funksprechverkehr (g); real time; receive/transmit; register translator

R/W read/write; rights of way

R1 signalling system (CCITT)

R2R run-to-run

R2RS revised RRS

Ra output resistance, Ausgangswiderstand (g)

RA Radiocommunications Agency (UK); random access; ready access; reanudación de actividad (s), activity resume [AR]; repeat attempt; re-synchronize acknowledgement; return authorization; routing arbiter (Internet); Rufabschaltung (g), ringing current disconnection; velocidad de adaptación (s), rate adaptation

RA1-3 signal de nouvelle réponse Nº 1 à Nº 3 (f), re-answer signal [RAN]

RAB random access burst

RABID Rebellion against Big Irrepressible Dweebs

RAC reflective array compressor; remote access and control; retorno a control (s), return to control [RTC]; Rolm Analysis Center; signal de raccrochage (du demandé) (f), clear-back signal [CBK]

RACAL Raymond-Calder (electronics company)

RACE R&D in Advanced Communications for Europe (Esprit); requirements acquisition and controlled evolution

RACF Resource Access Control Facility (IBM)

RACH random access channel

Racon radar responder beacon, baliza radar respondedora (s)

rad radian; radial; radiation absorbed dose

RAD (signal de) raccrochage du demandeur (f), calling-party clear signal [CCL]; random access device; rapid application development; remote antenna driver

RAD/RASP RAD/remote antenna signal processor

Rada random-access discrete address

RADAG Radar Area Guidance System (Pershing II)

radar radio detection and ranging/radio angle direction and ranging; appareil de radiodétection (f), aparato de radiodetección (s), Funkmeßgerät (g), Radargerät (g)

RADB routing arbiter database

RADHAZ electromagnetic radiation hazards

RADINT radar intelligence

RADIUS remote authentication dial-in user service/(security)

RADSL rate adaptive digital subscriber line

RAE residue after evaporation

RAF remote access facility; resource access facility

RAFAD radar automatic failure and alignment detection, système radar automatique de détection des pannes et des erreurs d'alignement (f)

RAG row address generator

RAI Radio Televisione Italiana; remote alarm indication

RAID redundant array of inexpensive disks/drives; relais automatique d'information 'digitale' (f)

RAIRS reflection-absorption infrared spectroscopy

RAIS redundant arrays of inexpensive systems

RAISE rigorous approach to industrial software engineering (language)

RAJ receiving ability jeopardized

RAL red de área local (s), local area network [LAN]; respuesta del abonado llamada (s), called party answer [CPA]; retardo de aceptación de la llamada (s), call acceptance delay [CAD]

RALA registered automatic line adaptor (RACAL-MILGO)

RALU register arithmetic logic unit

RAM random access memory; rarely adequate memory; reliability availability and maintainability; remote asset management (Compaq)

Rambus Rambus Inc (proprietary RAM architecture)

RAMDAC RAM digital-to-analogue converter (Sierra *tn*)

RAMIS Rapid Access Management Information System (On-line Software International *tn*)

RAMP reliability analysis and modelling program; remote access maintenance protocol

RAMS RAM store; random access measurement system

RAMSH reliability – availability – maintainability – safety – human factor

RAN reacondicionamiento negativo (s); re-answer signal

RANO réacheminement d'appel en cas de numéro occupé (f), call-forwarding busy [CFB]

RANR réacheminement d'appel en cas de non-réponse (f), renvoi d'appel sur non-réponse (f), call-forwarding no reply [CFNR]

RAO renvoi d'appel sur occupation (f), call-forwarding busy [CFB]

RAP rapid application prototyping; reacondicionamiento positivo (s)

RAPCON radar approach controls (US)

RAPDOC rapid document delivery (Holland)

RARE Réseaux Associés pour la Recherche Européenne (f)

RARP reverse address resolution protocol (RFC 903)

RAS random access storage; reliability – availability – serviceability (IBM); remote access server; remote access services (Microsoft Windows NT); replenishment at sea; row address/access strobe/(select)

RAS 1000 Radio Access System *tn*

RASAPI remote access service API (Microsoft Windows NT)

RASC renvoi d'appel sans condition (f), réacheminement d'appel sans condition (f), call-forwarding unconditional [CFU]

RASCOM Regional African Satellite Communications Organization, Organisation régionale africaine de communications par satellite (f)

RASP Radar Applications Specialist Panel, Groupe d'experts en applications radar (f); remote antenna signal processor

RAT signal de réponse avec taxation (f), answer signal – charge [ANC]

RATCC radar air traffic control centre, centre de contrôle radar (f)

RATCF Radar Air Traffic Control Facility (US)

RATEL Raytheon automatic test equipment language *tn*

RATFOR rational FORTRAN

RATP Paris transport authority (bus/metro)

RATS remote analogue testing system

RATT radio automatic teletype; radio teletypewriter

RAVE rendering acceleration virtual engine (Apple Macintosh *tn*)

RAZ remise à zéro (f), reset circuit [RSC]

Razz remote access (*v* DUN)

RB roll-back; Rückhörbezugsdämpfung (g), sidetone reference equivalent

RBA reset-band-acknowledgement message, signal d'accusé de réception de réinitialisation de bande (f); residential broadband

RBBS remote BBS; residential broadband service

RBCS remote bar code system

RBD Rückhörbezugsdämpfung (g), sidetone reference equivalent

RBDF recepción de bloqueo y desbloqueo de grupo de circuitos para el fallo del soporte físico (s), hardware failure-oriented circuit-group blocking and unblocking reception [HBUR]

RBDL recepción de bloqueo y desbloqueo de grupo de circuitos generados por el soporte lógico (s), software-generated circuit-group blocking and unblocking reception [SBUR]

RBI radar blip identification (message), message d'identification de plot radar (f); radar beacon interrogator; reset-band acknowledgement all circuits idle signal

RBL réception de signal de blocage (f), blocking and unblocking signal reception [BLR]

RBOC Regional Bell Operating Company (*aka* Baby Bells, US)

RBP radar by-pass, système de mise en dérivation radar (f)

RBS radar beacon system, radar secondaire de surveillance (f); refractive back-scattering; rule-based system; Rutherford back-scattering spectroscopy

RBT radar beacon transponder, transpondeur SSR (f); (bloque de) rechazo de bloque de transporte (s), transport block reject [TBR]; remote batch terminal; ring-back tone

RBV return beam Vidicon

RBX radio beacon transponder

RbXO rubidium-crystal oscillator

RC receive common (CCT 102B); reception control; receiver signal element timing (EIA-232-E); recording completing (trunk); re-drive counter, contador de reexcitaciones (s); reference clock; reflection coefficient; release complete; remote control; resin-coated (paper for photo-typesetting output); resistance-capacitance (coupling); resource controller; retransmission counter; ringing code; Rivest Cipher

RCA raccordement de la clientèle d'affaire (f); Radio Communications Agency (UK); RCA connector (audio/video devices attached to PC); signal d'accusé de réception de retour sur canal sémaphore normal (f), changeback acknowledgement signal [CBA]

RCAC Radio Corporation of America Communications; remote computer access communications service

RCAN recorded announcement

RCAT signalling route set congestion test control

RCB re-drive counter busy, contador de reexcitaciónes ocupado (s); reset-band signal, señal de reinicialización de banda (s)

RCC radio common carrier; remote cluster controller; reverse control/command channel; respuesta a continuar para corregir (s), continue to correct [CTC]; (Russian) Regional Commonwealth in the Field of Communications

RCCB residual current circuit-breaker

RCD receiver-carrier detector; réseau de communication de données (f), red de comunicación(es) de datos (s), data communications network [DCN]; route control digit

RCE remote channel extender

RCF radio communication failure message, message d'interruption des communications (f); remote call forwarding; routing control field

RCH rechazo (s), reject [REJ]

RCI mensaje (señal) de reinicialización de circuito (s), reset circuit message/signal [RSC]; routing control indicator; réseau de communication local (f), local communication network [LCN]; retardo de confirmación de liberación (s), clear confirmation delay [CLCD]; rotate carry left

RCLR retardo de conexión de la llamada dependiente de la red (s), network-dependent call connection delay [NCCD]

RCLU retardo de conexión de la llamada dependiente del usuario (s), user-dependent call connection delay [UCCD]

RCM reject - connect - modify signal; remote carrier module

RCO receiver cuts out; signal d'ordre de retour sur canal sémaphore normal (f), changeback declaration signal [CBD]

rcp remote copy (Unix)

RCP méthode avec retransmission cyclique préventive (f), método por retransmisión cíclica preventiva (s), preventive cyclic retransmission [PCR]; remote communications processor (IBM); remote control panel; restoration control point; restore cursor position; restoration control point

RCR rotate carry right

RCRE receiving corrected reference equivalent

RCRU Radio Communications Research Unit (UK)

RCS radar cross-section; records communications switching system; remote computer service/system; revision control system

RCSC remote spooling communications subsystem

RCT reference cluster tool; remote control terminal; señal de respuesta con tasación (s), answer signal – charge [ANC]; signalling-route-set-congestion-test message

RCTC rewritable consumer time-code

RCU radio channel unit

RCVR receiver (sequence number)

RD Radiocommunications Division – DTI (UK); received data (EIA-232-E); remove directory (DOS command);

referencia de destino (s), destination reference [DST-REF]; réseau de destination (f), red de destino (s), destination network [DN]; request disconnect; ringdown; routing domain

RD&T research development and technology

RDA remote database access

RDB receive data buffer; relational database

RDBMS relational database management system

RDC remote data concentrator

RDCC réseau pour données à commutation de circuits (f), red de datos con conmutación de circuitos (s), circuit-switched data network [CSDN]

RDCLP response document capability list positive

RDCP réseau pour données à commutation par paquets (f), red de datos con conmutación de paquetes (s), packet-switched data network [PSDN]

RDD requisition due date

RDDP response document discard positive

RDEP response document end positive, réponse positive à une commande de fin de document (f)

RDES remote data entry system

RDF radio direction finder, radiogonióometro (s), Peilfunkgerät (g); rate decrease factor; repeater distribution frame; resource description framework; route designator field

RDG closed user group selection and validation

RDGR response document general reject, réponse à une commande de rejet total de document (f)

RDI receiver data interface; red digital integrada (s), integrated digital network [IDN]; remote defect indicator; restricted digital information; route digit indicator

RDI-P remote defect indicator – path level (SONET)

RDL remote digital loop(-back)

RdM reste du monde

RDM remote display manager

RDN relative digital information; relative distinguished name, nom distinctif relatif (f), nom spécifique relatif (f)

RDO remote data object (Microsoft Visual Basic)

RDOS Realtime DOS (Data General *tn*)

RDP RACE definition phase; Radio Difusão Portuguesa (p); reliable datagram protocol

RDPBN response document page boundary negative, réponse négative à une commande de limite de page de document (f) [RPBDN]

RDPBP response document page boundary positive, réponse positive à une commande de limite de page de document (f) [RPBDP]

RDpriv red de datos privada (s), private data network [PvtDN]

RDPS radar data processing system, système de traitement automatique des données radar (f)

RDR retardo diferencial restringido (s), restricted differential time delay [RDTD]; rotating disk reactor

RDRAM Rambus dynamic RAM *tn*

RDRP response document re-synchronize positive, réponse positive à une commande de resynchronisation de document (f)

RDRPN response document recovery point negative, réponse négative à une commande de point de rétablissement du document (f)

RDRPR response document recovery point restart, réponse de reprise au point de rétablissement du document

RDS radio digital system; radio data/decoder system, système de décodage d'informations routières (f), sistema de codificación de información viaria (s), Radiodekodiersystem für Verkehrsinformationen (g); radiodiffusion numérique (f), radiodifusión digital sonora (s), digital audio broadcast [DAB]; random dot stereogram; Relational Database Systems Inc

RDSI red digital de servicios integrados (s), integrated service digital network [ISDN]

RDSI-BA red digital de servicios integrados de banda ancha (s), broadband ISDN [B-ISDN]

RDSN regional digital switched network

RDSS radio determination satellite services, Système radioélectrique de détresse et de sécurité en mer (f); Rapid Deployable Surveillance System (US)

RDS-TMC RDS-Traffic Message Channel

RDT radio digital terminal; restricted data transmission

RDTD restricted differential time delay

RDTO receive data transfer offset (IBM)

RE rapport d'état (f), status report [SRPT]; reference equivalent; regular expression

REA mensaje de reanudación (s), resume message [RES]; reencaminamiento automático (s), automatic re-routing [ARR]

REAA unidad de datos de protocolo de aplicación de RESPUESTA A-ASOCIACION (s) [AARE], A-ASSOCIATE RESPONSE application-protocol-data unit

READI Rights to Electronic Access to and Delivery of Information

README user information file attached to software

REC commande de la réception (f), reception control [RC]; receiver, récepteur (f), receptor (s); recepción (s); reception; recommendation; regular expression converter

Rec-DTT recepción de evento DTT (s), event reception transport data

RECS reseller electronic communication system

rect rectifier

RED random early discard; réponse de l'abonné demandeur (f)

Red Book NCSC publication: security aspects of trusted computer systems; protocol reference model for ISDN

redex reducible expression

REDIRIS Red de Interconnexión de Recursos Informáticos (s) (Spain)

REE Radio Exterior de España (s)

REF-DST referencia de destino (s), destination reference [DST-REF]

REG range extender with gain

regexp regular expression [Unix]

REI radar echo identification

REI-L remote error indicator – line level (SONET)

REIN REINICIACION (s), restart [SCCP]

REI-P remote error indicator – path level (SONET)

REJ reject, rejet (f), rechazo (s)

REL recommended exposure limit; release (message)

REL English rapidly-extensible language English

RELP regular-pulse excited LPC; residually-excited linear predictive

RELP-LTP regular-pulse excitation – long-term prediction

REM Remark (BASIC statement adding comment); ring error monitor

REMC rapport de réjection en mode commun (s), common-mode rejection ratio [CMR]

REMOB remote observation

REMOBS remote observation system

REMOTEREG remote registry service (Microsoft Windows 98)

REN remote enable (IEEE 488); ringer equivalency number/ring equivalent number

RENATER Réseau National de télécommunications pour la technologie, l'enseignement et la recherche (f)

REP message de réponse (f), answer message [ANM]; repeat; replication (function); routing exchange protocol

REPE repeat while equal

REPL restricted experimental programming language (EPL)

REPNE repeat while not equal

REPNZ repeat while not zero

REPROM re-programmable PROM

REPZ repeat while zero

REQ request

RER residual error rate

RES radio environment statistics; raster encoding standard; remote execution service; reset; resolution; restablecimiento de enlaces de señalización (s), signalling link restoration [LSLR]; resume message

RES4 Radio Equipment Specifications Sub-working Group 4

RESCU remote emergency satellite cellular unit

resis resistance

RESP reference equivalent speaking position; response

RESSFOX recessed sealed sidewall field oxidation

RET return/returning

ret resolution enhancement technology (Hewlett Packard *tn*)

RETAT Réseau de transport pour les applications de France Telecom (f)

RETEN retención de llamadas (s) [HOLD]

RETINAT Réseau de transport des informations numérisées de l'armée de terre (f)

RETMA Radio Electronics Television Manufacturers Association

REV reverse charging, cobro revertido (s); rectifier enclosure unit

REX re-locatable executable

REXEC remote executable/(execution)

REXX restructured extended executor (IBM)

REXXWARE REXX for Novell NetWare *tn*

RF radio-frequency; refuse; resonance frequency

RFA mensaje facilidad rechazada (s),

facility reject message [FRJ]; ready for acceptance; recurrent fault analysis

RFB reason for backlog

RFC radio-frequency choke coil; refus de connexion (s), connection refused [CREF]; Request For Comments: Internet protocols/procedures documents

RFCH radio-frequency channel

RFD request for discussion

RFE Radio Free Ethernet; Radio Free Europe

RFEP radar front-end processor, processeur frontal de données radar (f)

RFG roam-free gateway

RFH random-frequency hopping; remote frame handler (ISDN)

RFI radio-frequency interface; radio-frequency interference, interferencias de radiofrecuencia (s); ready for integration/installation; request for information

RFI/EMI radio-frequency interference/electromagnetic interference

RFID radio-frequency identification (device)

RFIP request for inclusion in program/plan

RFM radio-frequency monitoring

RFN reduced TDMA frame number

RFNM request for next message

RFO reason for outage; request for plan

RFP radio fixed part; radio-frequency probe; request for proposal

RFQ request for quote/quotation

RFR respuesta a fin de retransmisión (s), end of re-transmission [EOR]

RFS ready for sending; ready for service; remote file service/system/sharing

RFT radio-frequency terminal; ready for traffic; request for technology (OSF); revisable form text

RFTDCA revisable-form-text data communications architecture

RFTL radio-frequency test loop

RFU referencia de fuente (s), source reference [SRC-REF]; reserved for future use

RG release guard; ring generator; ring ground; ringing generator

RGA residual gas analysis

RGB signals signal correspondence to red green blue

RGB red – green – blue colour model

RGBI red – green – blue intensity

RGC mensaje de reinicialización de grupo de circuitos (s), circuit group reset message [GRS]

RGG Rufgenerator/Zählimpulsgenerator (g) (ringing and metering pulse generator)

RGPO range-gate pull-off

RGSL ring generator supervision unit

RGT réseau de gestion des télécommunications (f), red de gestión de las telecomunicaciones (s), telecommunications management network [TMN]

RH receive hub (telegraph); relative humidity; réseau hertzien (f), radio-relay system; request/response header

RHC Regional Holding Company (Bell)

RHCP right-hand circular polarization

rheo rheostat

RHM man-machine communication/operator command

RHR radio horizon range

Ri Innenwiderstand (g), internal resistance; número de referencia (s), reference identifier [RI]; referential integrity; related information; reliability improvement; réseau intelligent (f), intelligent network [IN]; repeat indication; response identifier; ring in (token-ring MAU); ring indicator; routing indicator

RI-2G second-generation intelligent network (f)

RIAA Recording Industries Association of America

RIAA curve RIAA standards for recording and equalization

RIC RACE industrial consortium; real-time interface co-processor; reinicialización de circuito (s), circuit reset [CRS]; repeating interface controller, contrôleur d'interface répéteur (f), controlador de interfaz repetidor (s), Schnittstellen-Kontrollbaustein mit Wiederholungsmodus (g); radio identity code, code d'identité de l'unité de radiomessagerie unilatérale (f), código de identidad de la unidad de radiobúsqueda (s), Aufbau des Empfängercodes (g), codice radio di identificazione (i)

RICS radio interface clock synchronization

RIE reactive ion etch; registro de identidad de equipo (s), equipment identity register [EIR]

RIF rate increase factor; routing information field (source route bridging)

RIFF raster image file format; resource interchange file format (Microsoft)

RIG mensaje de respuestas a indagación sobre grupo de circuitos (s), circuit group query response message [CQR]

RII routing information indicator (source route bridging)

RIK Radiofoniko Idrama Kyprou (Cyprus radio)

RILA restriction d'identification de la ligne appelante (f), calling line identification restriction [CLIR]

RILC restriction d'identification de la ligne connectée (f), restricción de la identificación de la línea conectada (s), connected line identification restriction [COLR]

RILL restricción de identificación de la línea llamante (s), calling line identification restriction [CLIR]

RILLR network-dependent call clear indication delay (s) [NCCID]

RILLU user-dependent call clear indication delay (s) [UCCID]

RILPR retardo de indicación de liberación por el tramo de red (s), network portion clear indication delay [NPCID]

RILR retardo de indicación de liberación

por la red, network clear indication delay [NCID]

RILT radio interface line terminal

RIM read-in mode; receiver intermodulation; remote installation and maintenance (Microsoft); request initialization mode

RIME RelayNet international message exchange

RIMM Rambus inline memory module (DRDRAM module)

RIN link-inhibit-denied signal (f) [LID]

RIOT Revolution in our Time

RIP radar input processing, traitement des données radar entrantes (f); raster image processor, procesador de imágenes en trama (s); remote imaging protocol; rest in proportion (instruction to printer); routing information protocol (RFC 1388)

RIPE Réseaux IP européennes (f)

RIPEM Riordan's Internet privacy-enhanced mail [IPEM]

RIPS raster image processing system

RIR radio interface data rate; reloj independiente de la red (s), network-independent clock [NIC]

RIS re-transmission identity signal

RISC reduced instruction set computer/(computing), microprocesseur à jeu d'instructions réduit (f), ordinateur à jeu d'instructions réduit (f), ordenador de juego de instrucciones reducido (s)

RISC OS RISC operating system (for Acorn's Archimedes computer)

RISE RD&T in Integrated Service Engineering

RISLU remote integrated services line unit

RIST rule induction and statistical testing

RIT raw input thread (Microsoft); repetitive information type

RITS radio interface time switch

RIUS respuesta a (una instrucción de) información de usuario de sesión (s), response session user information [RSUI]

RJ reject; remote/(registered) jack (*ie* phone connector)

RJ TPDU reject TPDU

RJ11 remote jack 11 – four-wire modular connector

RJ45 remote jack 45 – eight-pin modular connector

RJE remote job entry

RKM ROM kernel manual (Amiga)

RL real life; receive leg; reference loudness; reflection loss; return loss

RLAN radio LAN

RLC red local de comunicaciones (s), local communication network [LCN]; regional logistics centre; release-complete message

RLCU reference link control unit

RLD received line detect

RLE réseau local d'entreprise/d'établissement (f), local area network [LAN]; run-length encoding

RLF reverse line feed, cambio de renglón inverso (s); re-use library framework

RLG release-guard (signal), signal de libération de garde (f)

RLI resist lithography; response length indicator

RLL radio in local loop; run-length limited (encoding); reflexión de llamadas (s)

RLLAR reenvio de llamada en caso de ausencia de respuesta (s), call forwarding no reply [CFNR]

RLLI reenvio de llamadas incondicional (s), call-forwarding – unconditional [CFU]

RLLO reenvio de llamada en caso (de) ocupado (s), call forwarding busy [CFB]

RLM release mode

RLN remote LAN node

RLO restoration liaison officer

rlogin remote log-in (Unix)

RLP radio link protocol

RLR receive/(receiving) loudness rating

RLRQ A-RELEASE-REQUEST-application-protocol-data-unit

rls release

RLS remote line switch

RLSD received line signal detect/detector/detected; released

RLST release timer

RLT radio link transmission; remote line test

RLU remote line unit

rm remove empty (sub-)directory (UNIX command)

RM Reed-Müller (code); reference model replication manager; reset mode; resource management; roll-back module

RM code Reed-Müller code

RM ODP reference model for ODP (ISO)

RM386 expanded memory manager

RMA reject connect modify signal (f) [RCM]; return-merchandise authorization (order number)

RMAG recursive macro-actuated generator (IBM *tn*)

RMATS-1 remote maintenance, administration and traffic system-1

RMC RACE management committee; Radio Monte Carlo; réjection en mode commun (f), rechazo de modo común (s), common mode rejection ratio [CMR]

RMCS remote maintenance control system

RMDIR remove directory (DOS command)

RMDR remainder

RMF Radio Music Facts FM (Poland); rich music format

RMI radio magnetic indicator, indicateur radio magnétique (f), indicador radiomagnético (s); remote method invocation (Java library); route monitoring information

RML mensaje de rechazo de modificación de llamada (s), call modification reject message [CMRJ]; remote maintenance loopback

RMLA registered manual line adaptor (RACAL-MILGO)

RMON remote monitor/monitoring; remote network monitoring

RMOS refractory metal-oxide semiconductor

RMP remote maintenance processor (IBM); resource module platform

RMPS radar message processing system, système de traitement des messages radar (f)

RMS record management system (DEC *tn*); remote measurement subsystem; root mean square

RMSE root mean square error

RMSEP root mean square error of prediction

RMSM remote mass storage magazine

RMTF Recipe Management Task Force

RMTP réseau mobile terrestre public (f), red móvil terrestre pública (s), public land mobile network [PLMN]; réseau mobile terrestre public de rattachement (f), home public land mobile network [HPLMN]

RMTPP red móvil terrestre pública propia (s), home public land mobile network [HPLMN]

RMTPV red móvil terrestre pública visitada (s), visited public land mobile network [VPLMN]

RMU remote measurement unit

RMW read-modify-write

RN acuse de recibo negativo (s), negative acknowledgement [NAK]; receipt status notification; receive not ready, nicht empfangsbereit (g); red nacional, national network [NN]; reference noise; routing number

rnaapp remote network access (Microsoft Windows dial-up networking application)

RNC radio network centre; señal de respuesta no calificada (s), answer signal unqualified [ANU]

RNCS respuesta negativa a (una instrucción de) comienzo de sesión (s), response session start negative [RSSN]

RND random

RNE Radio Nacional de España (s)

RNF root normal form

RNG random number generator

RNI radio network investigation; réseaux numériques intégrés (f), integrated digital networks [IDN]

RNIS réseau numérique à intégration de services (f), réseau numérique avec intégration des services (f), integrated services digital network [ISDN]

RNLPD respuesta negativa a (una instrucción de) limite de página de documento (s), response document page boundary negative [RDPBN]

RNM radio network measurement

RNN recurrent neural network

RNP Rede Nacional de Pesquisa (Brazil); regional network provider

RNR radio network recording; receive not ready, non prêt à recevoir (f), nicht empfangsbereit (g)

RNSP radio network support product

RNTABLE routing number 128 integer table (in network hopping sequence)

RO read only; receive only; remote operation; reverse osmosis; ring out (token-ring MAU); route origin; routine order

ROA recognized operating agency

ROB re-order buffer (Intel)

ROC completion of calls to busy subscriber (f) [CCBS]; regional operating company (Bell); remote object communications; required operational capability

ROCA raccordement optique de la clientèle d'affaire (f)

ROE return on equity

ROE/ROS standing-wave ratio (f)

ROER RO-ERROR application-protocol-data-unit

ROF raccordement optique flexible (f), flexible optical connection

roff Unix text formatting language

ROH receiver off-hook

ROI return/rate on investment

ROIV RO-INVOKE application-protocol-data-unit

ROL rotate left

ROLAP relational on-line analytical processing

ROLR receiving objective loudness rating

ROM read only memory, mémoire inaltérable (f)

ROM OD ROM optical disk

ROMP remote operations microprocessor

ROOM real-time object-oriented modelling

ROP raster operation; result output period; RISC operation

ROPM remote operations protocol machine

ROR rotate right

RORJ RO-REJECT application-protocol-data-unit

RORS RO-RESULT application-protocol-data-unit

RORSAT Radar Ocean Reconnaissance Satellite

ROS radar operating system, système d'exploitation radar (f); radio operation and maintenance subsystem; read-only storage (v ROM), mémoire à lecture seule (f); remote operations service

Rosa recognition of open system achievement

ROSE remote operations service element (OSI/ITU-T); Research Open Systems in Europe (project within ESPRIT)

ROT Reign of Terror; running object table

rot13 rotate (letters of alphabet) 13 places (simple cipher)

ROTL remote office test line

ROTR receive-only typing reperforation

ROTS rotary out trunk switch

ROTT re-order tone trunks

ROU reference oscillator unit

ROW 'rights of way'

ROX run of experiments

RP radio part; regional processor; regular pulse; remise physique (f), physical delivery [PD]; renter's premises; répartiteur principal (f), repartidor principal (s), group distribution frame [GDF]; reply paid (telegram); restoration priority; roll-back process

RP3 Research Parallel Processor Project (IBM)

RPA re-entrant process allocator; regional processor adaptor; remote pass-phrase authentication

RPAS respuesta positiva a (una instrucción de) aborto de sesión (s), response negative abort positive [RSAP]

RPB rapport puissance bruit (f), noise-power ratio [NPR]; regional processor bus; registro de posición(es) base (s), home location register [HLR]

RPBC regional processor bus converter

RPC registered protective circuit; remote procedure call

RPCCS respuesta positiva a (una instrucción de) cambio de control de sesión (s), response session change control positive [RSCCP]

RPCPP remote procedure call print provider (Microsoft Windows)

RPCS respuesta positiva a (una instrucción de) comienzo de sesión (s), response session start positive [RSSP]

RPD réseau public pour données (f), red pública de datos (s), public data network [PDN]; regional processor device

RPDCC réseau public pour données à commutation de circuits (f), red pública de datos con conmutación de circuitos (s), circuit switched data network [CSPDN]

RPDCP réseau public pour données à commutation par paquets (f), red pública de datos con conmutación de paquetes, packet switched public data network [PSPDN]

RPDD respuesta positiva a (una instrucción de) descarte de documento (s), response document discard positive [RDDP]

RPDI red pública de datos internacional

(s), international public data network [IPDN]

RPE remote peripheral equipment

RPFD respuesta positiva a (una instrucción de) fin de documento (s), response document end positive [RDEP]

RPFS respuesta positiva a (una instrucción de) fin de sesión (s), response session end positive [RSEP]

RPFT remote power feed terminal

RPG report program generator (programming language)

RPH regional processor handler

RPI repetir la instrucción (s), repeat [RPT]; Rockwell Protocol Interface (for modems) *tn*

RPID réseau public international de données (f), international public data network [IPDN]

RPL repetición de prueba de continuidad-llegada (s), continuity re-check incoming [CRI]; requested privilege level; resident programming language; retardo de petición de liberación (s), clear request delay [CLRD]; reverse Polish LISP

RPLCD respuesta positiva a (una instrucción de) lista de capacidades de documento (s), response document capability list positive [RDCLP]

RPLPD respuesta positiva a (una instrucción de) limite de página de documento (s), response document page boundary positive [RDPBP]

rpm RedHat Package Manager (RedHat for Linux *tn*); revolutions/rotations per minute

RPM rate per minute; Relentless Pursuit of Magnificence

RPN real page number; reverse Polish notation

RPOA re-organized/(recognized) private operating agency

RPP registro de posiciones propio (s), home location register [HLR]

RPPROM re-programmable PROM

RPR message de reprise (s), resume

message [RES]; relación de potencia de ruido (s), noise-power ratio [NPR]; reprise (f), resume (message) [RES]

RPRINTR remote printer (NetWare)

RPRSD respuesta positiva a (una instrucción de) resincronización de documento (s), response document re-synchronize positive [RDRP]

RPS regional processor subsystem; relative performance score; repetición de prueba de continuidad-salida (s), continuity re-check outgoing [CRO]; revolutions per second; ring parameter server

RPT repeat, répétition (f), repetición (s); repetición de trama (s)

RPU radio power unit

RPV registro de posiciones visitado (s), visitor location register [VLR]; remotely-piloted vehicle; route processor connected to VME; virtual private network (s)

RQ repeat request; reportable quantity

RQBE Relational Query By Example (Fox Pro *tn*)

RR radio resource (management); radio-relay system; real reality; receive ready, prêt à recevoir (f), preparado para recibir (s), empfangsbereit (g); relay rack; release record; removal rate; repetition rate; re-route; resource reservation (protocol)

RR25E receiving R.25 equivalent

RRAS routing and remote access server (Microsoft Windows NT)

RRCR radio-related call release

RRD route-route destination

RRE receive reference equivalent; señal de repetición de respuesta (s), re-answer signal [RAN]

RRE1-3 señal de repetición de respuesta Nº 1-Nº 3 (s), re-answer signal Nº 1-Nº 3 [RAN1-3]

RRGC recepción de reinicialización de grupo de circuitos (s), circuit group reset receipt [CGRR]

RRGD respuesta a (una instrucción de)

rechazo general de documento (s),
response document general reject
[RDGR]

RRI re-route inhibit

RRL remote reference layer (Java)

RRLE A-RELEASE-RESPONSE-
application-protocol-data-unit

RRMSEP relative root mean square error
of prediction

RRO regional reporting office;
responsible repair organization

RRQ return request

RRRS route relief requirements system

RRS retransmission request signal

RRT ring-ring trip

RRUN re-run

RRX radio receiver

RS random splice; Recommended
Standard (EIA); record separator
(ASCII); Reed-Salomon (code);
reference store; remote single-layer;
répartiteur de groupe secondaire (f),
repartidor de grupo secundario (s),
supergroup distribution frame [SDF];
request to send; re-synchronize; ringing
and signalling set, Ruf- und
Signaleinrichtung (g); route switching

RS code Reed-Salomon code

RS PPDU re-synchronize PPDU

RS232C interface standard developed by
EIA

RSA reference system architecture; repair
service attendant; Rivest, Shamir and
Adelman encryption method

RSA PPDU re-synchronize acknowledge
PPDU

RSAC Recreational Software Advisory
Council (US)

RSAP response session abort positive

RSAT reliability and system architecture
testing

RSB signal de réinitialisation de bande
(f), señal de reinicialización de banda (s),
reset band signal

RSC remote switching center; renvoi de
données sans connexion (f), unitdata
service [UDTS]; reserva selectiva de

circuitos (s), selective service reservation
[SCR]; reset circuit message/signal,
signal de réinitialisation de circuit (f);
reset confirm/(confirmation); resource
service centre

RSCC red de señalización por canal
común (s), common-channel signalling
network [CSSN]

RSCCP response session change control
positive, réponse positive à une
commande de contrôle de changement
d'échange (f)

RSCE restoration switching control
equipment

RSCS remote source control system;
remote spooling communication
system/subsystem; réseau sémaphore par
canal commun (s), common-channel
signalling network [CSSN]

RSCV route selection control vector

RSE radio system entity; reactive sputter
etch; remote single-layer embedded;
removable storage elements; restoration
switching equipment; ringing and
signalling set, Ruf- und
Signaleinrichtung (g)

RSEP response session end positive,
réponse positive à une commande de fin
d'échange (f)

RSEP(r) response session suspend
positive

RSEU remote scanner and encoder unit

RSEXEC resource sharing executive

RSFG route server functional group

RS-flip-flop reset-set flip-flop

RSH remote shell

RSI answer signal – unqualified (f)
[ANU]; register sender inward; repetitive
strain injury

RSIG required type of signalling

RSL radio signalling link; RAISE
specification language (ESPRIT);
received signal level; request and status
link; specification language for time-
critical systems

RSLE remote subscriber line equipment

RSLM remote subscriber line module

RSM remote subscriber multiplexer; remote switching module; response surface methodology/matrix; (signalling) route set-test messages

RSN real soon now

RSO register sender outward

RSOT relación de señales de ocupado a tomas (s), busy-flash seizure ratio [BFSR]

RSP Radio Broadcasting Centre (Prague); Radiotelephony Speech Panel, Groupe d'experts sur le langage radiotéléphonique (f); rationalized software production; realización sometida a prueba (realización sujeta a pruebas) (s), implementation under test [IUT]; reliable stream protocol; replication synchronization process; restoration priority

RSR signalling-route-set-test signal for restricted destinations; reset request

RSRE Royal Signals and Radar Establishment

RSRT signalling route set test control

rss root-sum-square

RSS remote switch subsystem; reset synchronization signal, señales de reiniciación sincronización (s)

RSSI received signal strength indicator

RSSN response session start negative

RSSN(r) response suspended session re-activate negative

RSSP Radar System Specialist Panel, Groupe d'experts en systèmes radar (f); response session start positive

RSSP(r) response suspended session re-activate positive

RST mensaje de respuesta (s), answer message [ANM]; reset/restart; signal de réponse sans taxation (f), señal de respuesta sin tasación (f), answer signal – no charge [ANN]; signalling-route set test signal

RSTS resource sharing-time sharing (DEC)

RSTWN response session TWS negative, la réponse négative à une commande d'échange bidirectionnel simultané (f)

RSTWP response session TWS positive, la réponse positive à une commande d'échange bidirectionnel simultané

RSU remote subscriber/switching unit; remote switching units

RSUI response session user information, réponse à une commande d'information usager de l'échange (f)

RSV closed user group selection and validation response message (s) [CRM]

RSVG closed user group selection and validation response (s) [CSVC]

RSVP resource reservation protocol; Revolutionary Surrealist Vandal Party

RSX real-time resource sharing executive

RSZI regional subscription zone identity

(rt) insert with red marking, Kapsel mit roter Markierung (g)

RT radio-telephone; real time; register traffic; remote terminal; referencia temporal (s), time reference [TR]; reliable transfer; response time; ring trip; ringing tone; rot (g); rotation time; route treatment; routing type; run time

RTA Radio Telegraphy Act (UK); rapid thermal anneal; real-time accelerator; remote test access; retardo de tránsito acumulado (retardo de tránsito acumulativo) (s), cumulative transit delay [CTD]; redémarrage de trafic admis (f), traffic re-start allowed [TRA]; remote trunk arrangement

RTAC transfer-allowed control

RTAG real-time asynchronous grammar

RTAM remote terminal access method

RTB real-time backplane; retransmission buffer; basic telephone network (s)

RTBM read the blasted/bloody manual

RTC real-time clock; remote transcoder; réseau téléphonique commuté (f), public-switched telephone network [PSTN]; re-start confirmation; return to control

RTC/RTS regional tele-centre/regional tele-server

RTCC transfer-controlled control

RTCE real-time channel evaluation

RTCL retardo total de conexión de la llamada (s), total call connection delay [TCCD]

RTCP real-time transport control protocol

R-TCR receive TCR event

RTCVD rapid thermal chemical vapour deposition

RTD resistance temperature detector; retardo de tránsito deseado (o pretendido) (s)

RTD in IT Research And Technological Development in IT

RTDM real-time data migration; remote terminal digital magazine

R-TDT receive TDT event

RTE Radio Telefis Eireann; remote terminal emulator/emulation; route

RTF rich text format (Microsoft text interchange format)

RTFAQ read the FAQ

RTFM read the festering/fucking manual

RTG radio-isotope thermo-electric generator; radio-telegraphy, radiotélégraphie (f), radiotelegrafía (s), Funktelegraphieverkehr (g)

RTGC red telefónica general conmutada (s), general switched telephone network [GSTN]

RTH regional telecommunications hub; remote transcoder handler

RTI return from interrupt; run-time type identification

RTILL retardo total de indicación de liberación de la llamada (s), total call clear indication delay [TCCID]

RTL Radio Tele Luxembourg; register transfer language/level; resistor-transistor logic

RTLL retardo total de liberación de llamada (s), total call clearing delay [TCLD]

RTLP reference transmission level point

RTM Radio Television Maroc; rapid thermal multiprocessing; read the manual; real-time monitor; reference test

method, méthode de mesure de référence (f); remote terminal magazine; remote test module; response time monitor; run-time manager (Borland)

RTMA retardo de tránsito máximo aceptable (s), maximum acceptable transit delay [MATD]

RTMM remote terminal modem magazine

RTMP routing table maintenance protocol (AppleTalk *tn*)

RTN re-train negative

RTNR ring-tone no reply

RTO rapid thermal oxidation; regenerative thermal oxidizer

RTOS real-time OS (generic, not specific)

RTP Radio Television de Portuguesa; rapid transport protocol; rapid thermal processing/processor; real-time protocol; redundant transmitter power; remote test port; re-train positive; routing update protocol

RTPC réseau téléphonique public commuté (f), red telefónica pública conmutada (s), public switched telephone network [PSTN]

RT-PC RISC technology PC (IBM)

RTPG réseau téléphonique général avec commutation (f), general switched telephone network [GSTN]

RTPM reliable-transfer-protocol-machine

RTPP remote test port panel

RTPR network-dependent data-packet transfer delay (s) [NPTD]

RTPU user-dependent data-packet transfer delay (s) [UPTD]

RTR real-time reporting; re-start request

RTRC transfer restricted control

RTS Radio Televisia Serbia; radio transmission and support subsystem; reliable transfer server; remote take-over system; remote test/testing system; request to send; return to service; run-time support

RTS-5A remote test system 5A (AT&T)

RTSA real-time structured analysis

RTSC Regional Telematic Service Centre
RTSE reliable transfer service element; remote transfer service element
RTSP real-time streaming protocol
RTSPC real-time statistical process control
RTT radio teletypewriter; Radio Television Tunisie; radio transmission technology; remote transceiver terminal; round-trip time
RTTI run-time type information (C++)
RTTPR retardo total de transferencia de paquete por la red (s), total data packet network transfer delay [TPTD]
RTTY radio teletypewriter
RTU reference time unit; remote terminal/test unit; right to use
RTV real-time video; room temperature vulcanizing
RTX charge category
RTXPB radio transmitter power booster
RTXPF radio transmitter power filter
RTY re-try limit
RU recovery unit; replacement unit, unité de correction (f); request/response unit
runcom run command [Unix]
RURAX rural automatic exchange
RV Regelverkabelung (g), standard cabling
RVA receive volt-ampere; relative virtual address
RVC reverse voice channel
RVCH recording of voice channel handling
RVD radar viewing display
RVEM resultado de verificación de encaminamiento por la PTM (s), MTP routing verification result [MRVR]
RVI reverse interrupt
RVL reference vocal level
RVP reverse pulse (dialling)
RVWG Reliability and Vulnerability Working Group
RW receiver window; Richtungswähler (g), route selector
RWI radio and wire integration
RWIN receive Window

RWM read-write memory
RWP remote write protocol (Internet)
RWR radar warning receiver; receptor de alarma radar (s)
RX receive data; receive/receiver; remote exchange; through toll operator
RXBP receiver bandpass filter
RXD receive data; receiver divider
RXDA receiver divider amplifier
RXLEV received signal level
RXMC receiver multi-coupler
RXQUAL receiver signal quality
R-Y signal component of colour TV chrominance signal
RZ rechazo (s), reject; return to zero
RZA message d'accusé de réception de remise à zéro de groupe (f), circuit group reset acknowledgement message [GRA]
RZC message de remise à zéro de circuit (f), reset circuit message /signal [RSC]
RZG message de remise à zéro de groupe de circuits (f), circuit group reset message [GRS]
RZGE remise à zero de groupe de circuits à l'émission (f), circuit group reset sending [CGRS]
RZGR remise à zero de groupe de circuits à la réception (f), circuit group reset receipt [CGRR]

S

S Geräteschnittselle (g) (equipment interface) B+B+DO 144kb/s net; Satz (g) (units, assemblies); sending; session, sesión (s); set-up; Signalerde (g), signal ground CCT 102 AB; Speicherteil (g) (memory unit); Spitze-Spitze p-p (g) (peak-peak as suffix); Sprechkapsel (g), microphone insert; Steuerfeld – Daten (g) (control field – data); sulphur; supervisory; supervisory function bit, bit de función de supervisión (s); supplier, suministrador (s); switchboard

S/A sensor/actuator

S/B rapport signal/bruit [SNR]

S/D source/drain

S/DMS SONET/digital multiplex system

S/H sample and hold

S/MIME secure multipurpose Internet mail extensions

s/n serial number

S/N signal-to-noise ratio, relación señal ruido (s) [SNR]

S+N/N signal-plus-noise to-noise ratio

S/PDIP Sony/Philips digital interface protocol

S/R send/receive

S/SYS storage system, système de mémorisation (f), sistema de almacenamiento (s)

S/TK sectors per tracks

S Ts Sprechkapsel mit Transistor (g) (microphone isert with transistor)

S ZB Sprechkapsel für ZB-Betrieb (g) (microphone insert for CB working)

S-100 bus early backplane 100-pin bus (IEEE 696)

S1DN stage 1 distribution network (part of SMAS)

S2M Teilnehmer-Netz Schnittstelle (g) (subscriber-network interface 30B+1D)

S3TC S3 texture compression

SA alignment signal, signal d'alignement (f) [AS], señal de

alineación/alineamiento (s) [AS]; message de supervision de l'appel (f), call supervision message [CSM]; selective availability; service administrator; service alarm; service application; service area; set-up acknowledge; single armour; slow-acting; source address; storage allocator; storage array; structured analysis; structured (systems) analyst; sub-area; sub-resolution assist; surface area

SA/RT structured systems analysis for real time

SAA segmented addressing architecture; service aspects and applications; stateless address auto-configuration; static automated analysis; systems applications architecture (IBM)

SAAL signalling ATM adaptation layer

SAB señal de asignación de velocidad binaria (s), bit-rate allocation signal [BAS]; sensor/actuator bus

SABM set asynchronous balanced mode, paso al modo equilibrado/simétrico asíncrono (s)

SABME set asynchronous balanced mode extended, paso al modo equilibrado/simétrico asíncrono ampliado (s)

SABRE store access bus recording equipment

SAC secret area code; secure authentication centre; señalización asociada al canal (s), channel associated signalling [CAS]; serving area concept; single attachment concentrator; small area coverage; special area code

SACCH slow associated control channel

SACCH/T slow associated control channel/traffic channel

SACCH/TF slow associated control channel/traffic channel full-rate

SACCH/TH slow associated control channel/traffic channel half-rate

SACS set additional character separation

SACU single application computer user

SAD space antennae diversity;

STREAMS administrative driver; systems analysis definition

SADT structured analysis and design technique

SAE Signalanpaßeinrichtung (g) (signal matching equipment); Society of Automotive Engineers

SAFE Security and Freedom Through Encryption; Security Awareness From Education; signature analysis using functional analysis; store and forward element

SAFFI special assembly for fast installation

SAF-TE SCSI accessed fault-tolerant enclosure

SAGE Security Algorithm Group Experts; Strategic Air Ground Environment (NORAD); Systems Administrators Guild (within USENIX Association)

SAIC Science Applications International Corporation (US)

SAID security association ID (IP v6)

SAIL Stanford Artificial Intelligence Laboratory

SAITLM servicio abstracto de interfuncionamiento telemático (s) [TIAS]

SAK secret authentication key

SAL señales de acondicionamiento de la línea (s), line-conditioning signals [LCS]; shift arithmetic left; single assignment language; SPARK annotation language (ICL); SQL application language; SQL-Windows application language; symbolic assembler language

SALT subscriber's apparatus line tester

SAM scanning auger microscopy; security accounts manager (Microsoft); señal de alineación de multítrama (s), multiframe alignment signal; sequential access method; serial access model; service administration management; service attitude measurement; single application mode (Microsoft); subsequent address message, message

d'addresse subséquent (f), mensaje subsiguiente de dirección (s)

SAM 1-7 subsequent address message Nos 1-7, message d'addresse subséquent (No 1 à No 7) (f)

SAMC subscriber access maintenance centre

SAME Standard ANSI module language with extensions; subscriber access maintenance entity; system(s) management application entity

SAMeDL SQL Ada module description language

SAMIP servicio abstracto de mensajería interpersonal (s), interpersonal messaging abstract [IPMAS]

SAMSARS Satellite-based Maritime Search and Rescue System

SAN satellite access node; storage area network; system area network

SANCHO sector abbreviations and definitions for telecommunications thesaurus-oriented database (ITU-T)

Sandra System analyser diagnostic and reporting assistant (SiSoft *tn*)

SANE standard Apple numeric environment *tn*

SAO Signatory Affairs Office(UK), junta de asuntos del signatario (s); subsequent address message with one signal

SAP second audio programme; señal de acceso prohibido (s), access barred signal [ACB]; service access point (DEC); Service Advertising Protocol (Novell *tn*); symbolic assembler point; symbolic assembler program (IBM)

SAP AG Systeme Anwendungen Produkte in der Daten Verarbeitung (g), system application products

SAPI scheduling application programming interface; service access point (SAP) indicator/identifier/interface; speech application programming interface (Microsoft)

SAPS SpartaCom Asynchronous Port Sharing (Icon Technology *tn*)

SAR search and rescue; segmentation and

reassembly (Sun chip functionality *tn*); service analysis request; shift arithmetic right; spare at relief (number); store/street address register; successive approximation register; synthetic aperture radar

SARC Symantec Anti-virus Research Center

SARIE selective automatic radar identification equipment

SARM set asynchronous response mode

SARTS switched access remote test system (AT&T)

SAS segmented address space; single attachment station/system; Small Astronomical Satellite; statistical analysis system; subscriber administration system; switched access system

SASD structured analysis – structured design

SASE specific applications service element

SASI Shugart Associates System Interface (*now* SCSI)

SASL St Andrews (University) static language

SAT satellite; (Japan) Satellite System; site acceptance tests; señal de alineación de trama (s), frame alignment signal [FAS]; special access termination; spray acid tool; stepped atomic time; supervisory audio tone

SATAN Security Administrator Tool for Analysing Networks

SATCC Southern Africa Transport and Communications Commission

SATNET Satellite Network

SATRM servicio abstracto de transferencia de mensajes (s), message transfer abstract service [MTAS]

SatStream BT satellite digital leased circuit service *tn*

SATT Strowger automatic toll ticketing

SAV service après-ventes (f), after-sales service

SAVDM single application virtual DOS machine

SAW surface acoustic wave

SAWO surface acoustic wave oscillator

SAX Simple API for XML; small automatic exchange

Sb antimony (used for hardening alloys in semiconductors)

SB signal/signalling battery; simultaneous bulletin; sound board; station de base (f), base station [BS]; strong base ion exchange; synchronization burst

SBA scene balance algorithms (Kodak *tn*); software-generated group blocking-acknowledgement message

SB-ADPCM sub-band ADPCM

S-band microwave band 2-3 GHz

SBB service building block; subtract with borrow; system building block

SBC mensaje de sobrecarga (s), overload message [OLM]; single-board computer; sub-band codec; switching and bridging control; sub-sample control

SBCS single-byte character set (IBM)

SBD Sendebezugsdämpfung (g) (sending reference equivalent)

SBDN Southern Bell Data Network

SBH strip-buried heterostructure

SBI spare bit interface; speaker box interface; storage bus interconnect; streamer-buffered interface; synchronous backplane interconnect/interface

SBIRS space-based infrared system

SBL Super Basic Language

SBLN South Bristol Learning Network (UK)

SBM secuencia de bits M (s), M-bit sequence [MBS]; solution-based modelling; subscription management; successful backward-set-up information message; super-bit mapping

SBMS Southwestern Bell Mobile Service

SBO select-before-operation

S-box transformation process in cipher systems

SBPA séquence binaire pseudo-aléatoire (f), pseudo-random bit sequence [PRBS]

SB-Prolog Stony Brook Prolog

SBr Speisbrücke (g) (feed current bridge)

SBR standby-ready signal; statistical bit rate; Storage Business Review

SBS satellite business systems

SBSA secuencia binaria (de bits) seudoaleatorio (s), pseudo-random bit sequence [PRBS]

SBSM single-bearer state model

SBSVC selective broadcast signalling virtual channel (B-ISDN)

SBT Software Business Technologies

SBUR software-generated circuit group blocking and unblocking receipt

SBUS software-generated circuit group blocking and unblocking sending

SBUV/TOMS solar back-scatter ultra-violet/total ozone mapper system

SC messages de supervision de circuit (f), circuit supervision message [CCM]; scanner controller; send common (CCT 102A); secondary centre; sending complete; service centre; service code; service channel; service customer; sub-committee; subscriber categories

SC1/2 Standard Clean 1/2

SCA subsidiary carrier authorization; surface charge analysis; Swiss Cracking Association; synchronous concurrent algorithm; system communication architecture

SCADA supervisory control and data acquisition

SCAF service control access function

SCAI switch-to-computer application interface

SCAM SCSI configure/configuration automatically

SCAMS scanning microwave spectrometer

SCAN switched circuit automatic network

Scan-EDF scan earliest deadline first

ScanRegW registry checker Windows interface (Microsoft Windows 98)

SCART single-channel asynchronous receiver/transmitter; Syndicat des Constructeurs d'Appareils Radio Recepteurs et Televiseurs (f)

SCAS scan string

SCB special clear back; string control byte; string control data; subsystem control block (IBM)

SCBE spatially-coupled bipolar electrochemistry

SCC satellite communications controller; satellite control centre; señalización por canal común (s), common-channel signalling [CCS]; serial communications controller; serial controller chip; signal converse circuit; single cotton-covered; specialized common carrier; strategic cell controller; strongly-connected component; supervisory audio-tone colour code; switching control centre; synchronous channel check (IBM)

SCCF Satellite Communications Control Facility

SCCP signalling connection control part

SCCS source code control system (Unix); switching control centre system

SCD Software Chronicle Digest; SPARC compliance definition; standard colour display; system configuration data

SCDMA synchronous code-division multiple access

S-CDR serving GRPS support node – call detail record

SCE saturated calomel electrode; security configuration editor (Microsoft Windows NT); service creation environment (IN entity); signal conversion electronics

SCEF service creation environment function

SCEG speech coder expert group

SCENIC Siemens-Nixdorf PC

SCF señal de conexión fructuosa (s), connection-successful signal [CSS]; selective call forwarding; service control function/facility; super-critical fluid

SCFA service control function area

SCFM sub-carrier frequency modulation

SCH seizures per circuit per hour;

signalling channel; synchronization channel

schem schematic

SCHEME dialect of LISP

Schottky TTL Schottky transistor-transistor logic

SCI scalable coherent interface (IEEE 1596 – 1 Gbit/s); serial communications interface; subscriber controlled input; surface charge imaging

SCIC semiconductor integrated circuit

SCIM speech communications index meter

SCL System Control Language (ICL *tn*)

SCLM software configuration and library management (IBM)

SCLog security log

SCM scanning capacitance microscopy; ScreenCam file (Lotus); security configuration manager (Microsoft Windows NT); select coding method, choix de la méthode de codage (f), selección de método de codificación (s); semiconductor mode; service configuration management; single-channel modem; station class mark; sub-carrier multiplex; subscriber's carrier/concentration module

SCMM software capability and monitoring model

SCN self-compensating network; sub-channel number

SCNA sudden cosmic noise absorption

SCO Santa Cruz Operation (XENIX Corp); select character orientation, selección de orientación de caracteres (s)

SCODL scan conversion object description language

SCOFI signal processor CODEC/filter

SCOOP structured concurrent object-oriented Prolog

SCOOPS scheme object-oriented programming system (Texas Instruments *tn*)

SCOPE simple communications programming environment (Hayes); Software evaluation and certification programme Europe (ESPRIT)

SCOSTEP Special Committee on Solar-Terrestrial Physics

SCOTICE Scotland to Iceland submarine cable system

SCOTS surveillance and control of transmission systems

SCP save cursor position; service control point; session control protocol; signal control point; single chip package; subscriber categories pointer; subsystem control port; SunLink communications processor (Sun Microsystems *tn*); system control programme

SCP(S) subscriber's call processing (subsystem)

SCPC single channel per carrier, (sistemas de) un solo canal por portadora (s)

SCPI standard commands for programmable instruments (HP)

SCR selective chopper radiometer; selective circuit reservation; semiconductor-controlled rectifier; sequence control register; signal conversion relay; silicon-controlled rectifier; station de coordination du réseau (f), network coordination station [NCS]; sustainable cell rate (UNI 3.0); system change request

SCRA single-channel radio access

SCRC SCCP routing control

SCRCE severe cyclic redundancy check error

SCRE send corrected reference equivalent

SCRI Supercomputer Computation Research Centre (US)

SCRN screen

SCRS scalable cluster of RISC systems

SCRT selected channel rate and type; subscriber's circuit routine tester

SCS satellite control satellite; sélection du canal sémaphore (f), signalling link selection code [SLS]; signalisation sur voie commune (f), common channel signalling [CCS]; self-configuring system; signalling channel selection; silicon-controlled switch; SNA character string; suppressed-carrier system

SCSA signal computing system architecture (Dialogic)

SCSI small computer systems interface

SCSP server cache synchronization protocol (RFC 2334)

SCSR single-channel signalling rate

SCSSI Service Central de la Sécurité des Systèmes d'Information (f)

SCT simulador de cabeza y tronco (s), head and torso simulator [HATS]; subscriber carrier terminal

SCTE serial clock transmit external; Society of Cable and Telecommunications Engineers

ScTP screened twisted-pair

SCTR system conformance test report, rapport de test de conformité du système (f)

SCTS secondary clear to send

SCU servicing control unit; signalling system control unit, unité de signalisation pour la commande du système de signalisation (f), unidad de señalización para el control del sistema (de señalización) (s); subscribers' concentrator unit; system configuration utility (Microsoft Windows 98); system control signal unit

SCUAF service control user agent function

SCUG start of closed user group sequence

SCV simulador de carga vocal (s), voice load simulator [VLS]

SCVFT single-channel voice-frequency telegraphy

SCVL speech coder version list

SCX specialized communications exchange

SD referencia de fuente (s), source reference [SRC-REF]; schematic drawing; sección digital (s), digital section [DS]; send data/ digits;

Sendebezugsdämpfung (g) (sending reference equivalent); service data/design; single density (disk); small dual in-line package; starting delimiter (LAN); structured design; super density; surface-mounted device, oberflächenmontierte Bauelemente (g)

SDA sélection directe à l'arrivée (f), direct dialling-in [DDI]; señalización digital de abonado No 1 (s), digital subscriber signalling (system) [DSS 1]; Silicon Dream Artists; software design automation; software disk array; source data automation; system display architecture (Digital)

SDC select dot composition, choix de la composition des points (f); self-correction distributed computation (technique)

SDCCH stand-alone dedicated control channel

SDD subscriber direct dialling

SDDC security description definition language (Microsoft Windows NT)

SDDI Sony digital data interface *tn*

SDE selección directa de las extensiones (s), direct dialling-in [DDI]; señal de desinhibición de enlace (s), link uninhibit signal [LUN]; serveur de données externe (f), external data server; small digital exchange (Siemens); software development environment; submission and delivery entity

SDEF señal de desinhibición del enlace forzada (s), link forced uninhibit signal [LFU]

SDEG signalling dimensioning expert group

SDF satellite distribution frame; service data function; software development facilities; space delimited file; supergroup distribution frame; system description file

SDFL Schottky-diode FET logic

SDFR sección digital ficticia de referencia [HRDS]

SDH synchronous digital hierarchy (ITU-T 155 Mbit/s)

SDI selective dissemination of information; señal/mensaje de dirección incompleta (s), address-incomplete signal [ADI]; serial digital interface; single document interface (Microsoft Word 2000); sintaxis de datos de interfuncionamiento (s), interworking data syntax [IDS]; software development interface (Mosaic); standard disk interconnect; standard drive interface; (US) Strategic Defense Initiative (*ie* 'Star Wars'); system data interface

SDIA sensibilidad diferencial de intervención adaptativa (s), adaptive break-in differential sensitivity [ABDS]

SDIS switched digital integrated service

SDL shielded data link; (functional) specification and description (design) language (ITU-T); structure definition language (DEC)

SDL/GR specification and description language–graphic representation

SDL/PR specification and description language–phase representation

SDLC synchronous data link control (IBM network protocol)

SDM site data mediation; space-division multiplex/multiplexing; specific device model (for sensor/actuator bus); system development multi-tasking

SDMC sistema digital de multiplicación de circuitos (s) [DCMS]

SDMI Secure Digital Music Initiative (EU consortium)

SDML signified document mark-up language

SDMS SCSI device management system (NCR)

SDN software-defined network (AT&T); subscribers' directory number; switched digital network; synchronized digital network

SDNS secure data network service; software-defined network service (AT&T)

SDO secuencia de detección de originador (s), originator detection pattern [ODP]

SDOC selective dynamic overload controls

SDOU site domain and organizational unit (Microsoft Windows NT)

SDP semiconductor die processing; service data point; session description protocol (RFC 2327); signal despatch point

SDPM software development process model

SDPS sélection directe d'un poste supplémentaire [DDI]

SDR secuencia de detección de respuesta (s), answerer detection pattern [ADP]; speaker-dependent recognition; special drawing rights; streaming data request; Sud Deutscher Rundfunk (g)

SDRAM synchronous dynamic RAM

SD-ROM super-density ROM

SDRP source demand routing protocol

SDS Scientific Data Systems; smart distributed system; Sondersatz (g) (special unit); switched data service; switched digital service; synchronous data set; sysops distribution system

SDSC San Diego Supercomputer Centre

SDSF system (spool) display and search facility

SDSI synchronous data-link control

SDSL single-line digital subscriber line; symmetric digital subscriber line

SDSN Secure Data System Network

SDSU SMDS data service unit

SDT SDL development tool; select domain in mobile terminal; serial data transfer; serveur de tarif (f), rate server; structured data transfer; switching terminal wander; synchronous data transport

SDTCP servicios de transmisión de datos con conmutación de paquetes (s) [PSDTS]

SDTR serial data transmitter/receiver

SDTU servicio de dato unidad (s), unitdata service [UDTS]

SD-TV standard definition TV

SDU mensaje subsiguiente de dirección con una señal (s), subsequent address message with one signal [SAO]; service data unit

SDV switched digital video (AT&T)

SDXC synchronous digital cross-connect

Se selenium (used in photo-electric cells and as a semiconductor)

SE message de succès de l'établissement émis vers l'arrière (f), successful backward set-up information message [SBM]; secondary electron; secondes erronées (f), segundos con error (s), errored seconds; service expansion; servicio especial (s), special service; Signalanpaßeinrichtung (g) (signal matching equipment); système d'exploitation (f), sistema de explotación (s), operating system [OS]; Society Elite; software engineering; south-east; spectroscopic ellipsometry; status enquiry; structure element; support entity; suppression d'écho (f), echo suppressor; switching element; systems engineer

SEA self-extracting archive; standard extended attribute (IBM OS/2); sudden enhancement of atmospherics; System Enhancement Associates

SEAC Standards Eastern Automatic Computer (1950)

SEACOM South-East Asia Commonwealth submarine cable system

SEAJ Semiconductor Equipment Association of Japan

SEAL screening external access link (DEC); Secure Electronic Authorization Laboratory; segmentation and re-assembly layer; simple and efficient adaptation layer

SEAM scalable and efficient ATM multicast

sec secant; second/s

SEC secondary; secondary emission control; section; security; single-edged contact/connect; single error correction; size exclusion chromatography; sufijo de punto extremo de conexión (s), connection endpoint suffix [CES]; switching-equipment-congestion (signal)

SECA semi-permanent connection

SECAM séquence de couleurs avec mémoire (f)

SECB severely-errored cell block

SECBR severely-errored cell block ratio

SECD stack environment control dump machine

SECDED single-error correction double/dual error correction

sech hyperbolic secant

SECH service channel

SECO station engineering control office

SECORD secure voice cord board

SECS Semiconductor Equipment Communications Standard

SECTEL secure telephone

SED selección de enlace de datos de reservada (s), standby data link selection [LSDS]; servicio de enlace de datos (s), data link service [DLS]; status entry devices; stream editor

SEDCC servicio de enlace de datos con conexión (s), connection-mode data link service [CODLS]

SEDIT systems engineering (for network) debugging integration and test

SEDM status entry device multiplexer

SEE software engineering environment; system equipment engineer

SEF severely-errored frame; simultaneous engineering environment; sistema de entrega física (s), physical delivery system [PDS]; support entity function

SEFS severely-errored framing second

SEG security experts' group; segment; selective epitaxial growth; special effects generator

Sega Service Games *tn*

segfault segmentation fault

segv segmentation violation

SEH structured exception handling (Visual Basic)

SEI Software Engineering Institute (Carnegie Mellon University)

SEIM software engineering improvement method

SEIRT selección e indicación del retardo de tránsito (facilidad) (s), transit delay selection and indication [TDSAI]

SEL select/selector; Space Environment Laboratory

SELCAL selective calling

SEM scanning electron microscope/microscopy; sous-entité de maintenance (f), subentidad de mantenimiento (s), maintenance sub-entity [MSE]; space environmental monitor; specific equipment model maintenance sub-entity [MSE]; standard electronic module; station engineering manual

SEMI Semiconductor Equipment and Materials International

SemiSPIN Semiconductor Software Process Improvement Network

SEMPER Secure Electronic Marketplace (IBM *tn*)

SEMS solar environment monitor subsystem

SE-NS système d'exploitation – noeud de service (f), operating system – service node

SE-ODP support environment for open distributed processing

SEOS Synchronous Earth Observatory Satellite

SEP signalling end point; software engineering process

SEPG Software Engineering Process Group

SEPIA standard ECRC Prolog integrating applications

SEPP secure encryption payment protocol

Seq x sequence number x

SEQDB Semiconductor Equipment Database

SEQUEL structured English query language (SQL precursor)

SEQUIN international quality evaluation system

SER octet de service (f), service information (octet) [SIO]

SERC Science and Engineering Research Council

SERCnet precursor of JANET

SERI intelligent network operating system (s)

Servlet small Java (applet) program on a Web server

SES satellite Earth stations; selección de enlaces de señalización (s), signalling link selection (code) [SLS]; service provision (sub-)system; severely-errored second; ship-earth station; Société Européenne des Satellites (f) (Switzerland); source end station; special exchange service

SESDL ship-earth station low-speed data

SESE système d'essai de suppresseurs d'écho (f), echo suppressor testing system [ESTS]

SESRP ship-earth station response

SESRQ ship-earth station request

SEST Sender-/Empfänger-Steuerung (g) (send/receive control); ship-earth station telex

SET secure electronic transaction (online-buying protocol); select environment terminal; standard d'échange et de transfert (f); supervision de l'état du canal sémaphore (f), line signalling channel [LSC]

SETA SouthEastern Telecommunications Association

SETAB working standard with subscriber's equipment

SETAMS System engineering technical assistance and management services

SETEC Semiconductor Equipment Technology Center

SETED working standard having an electro-dynamic microphone and receiver

SEU smallest executable unit
SEVAS secure voice access system
SEW Satellitenbilder der europäischen Wetterlage (g)
SEX software exchange
S-F secure and forward
SF service function; servicio físico (s), physical service [PhS]; sign flag; single frequency (signalling); spare frame; spécification fonctionnelle (f), functional specification [FS]; spreading factor; status field; straightforward; super-framed (D4 format); system function
SFA sales force automation
SFBI shared frame buffer interconnect (Intel)
SFC sensitivity/frequency characteristics; supercritical fluid chromatography; system file checker (Microsoft Windows 98)
SFCS I/F shop floor control system interface
SFD simple formattable document; start-of-frame delimiter
SFE severe framing error
SFERT (European) master reference system for telephone transmission, système fondamental européen de référence pour la transmission téléphonique (f)
SFET synchronous frequency encoding technique
SFF características de sensibilidad frecuencia (s), sensitivity frequency characteristics [SFC]; service switching function; small form factor
SFH slow frequency hopping
SFI Society for the Freedom of Information
SFM service management function
SFMC Satellite Facility Management Center
SFMT San Francisco-Moscow Teleport (US corp)
SFN señal de fin de numeración (s), string terminator [ST] (end-of-pulsing signal)

SFQL structured full-text query language
SFR Société Française de Radiotéléphonie (f); système de référence fondamental (f), sistema fundamental de referencia (s), fundamental reference system [FRS]
SFS single frequency signalling; single function software; software standards and facilities; start (of) frame sequence; system file server
SFT system fault tolerance
SFU status fill-in unit; store and forward unit
SFX sound effects
SG referencia de fuente (s), source reference [SRC-REF]; Signalerde (g), signal/signalling ground; CCT 102 AB); signalling category; sistema de gestión (s), management system; Steuerung (g) (control/controller); SuperGroup (submarine cable transmission system); (messages de) supervision de groupe de circuits (f), circuit group supervision messages [GRM]
SGA Sysops Guild Association
S-gate ternary threshold gate
SGB software-generated group blocking message
SGBD système de gestion de bases de données (f), sistema de gestión de bases de datos (s), database management system [DBMS]
SGBDOO système de gestion de base de donné orientée objet (f), object-oriented database management system [OODBMS]
SGBDR système de gestion de bases de données relationelles (f), relational database management system [RDBMS]
SGC signalling group channel
SGCP Simple Gateway Control Protocol
SGD signalling ground
SGDF super-group distribution frame
SGDT store global descriptor table; système de gestion de base de données techniques (f), technical database management system

SGE secondes gravement erronées (f), severely-errored seconds [SES]; signal generator; sistema (de) gestión (s), management system [MGMT]

SGF structured graph format (XML format language)

SGI Silicon Graphics Inc

sgl single

SGM segmentation; señal de gestión de red y mantenimiento (s), network management and maintenance system [NMM]; shaded graphics modelling

SGML standard generalized mark-up language (method)

SGMM Semiconductor Generic Manufacturing Model

SGMP simple gateway management protocol

SGMRS Semiconductor Generic Manufacturing Requirements Specification

SGOD système de gestion d'objets distribué (f), distributed object(s) management system

sgp supergroup

SGR select/set graphic rendition, choix de la reproduction graphique (f), selección de reproducción (s), select graphic rendition; signaux de gestion du réseau (f), señales de gestión de red (s), network management signals [NMS]

SGRAM synchronous graphics RAM

SGRS señal de gestión de la red de señalización (s), signalling network management signal [SNM]

SGS système de gestion de service (f), service management system

SGSN serving GRPS support node

SGU single management system (s); software-generated group unblocking message

SGV closed user group selection and validation (f)

SGVD closed user group selection and validation request (f) [CSVR]

SGVR closed user group selection and validation response (f) [CSVC]

SGX selector group matrix

SH send hub (telegraphy); speech handler; switch handler

SHA secure hash algorithm (NSA); sidereal hour angle; Swedish Hackers' Association

shar shell archive (Unix)

SHARE (early) IBM users' group

SHARP self-healing alternate route protection

SHASE shareware search engine

S-HDSL single-pair high-speed digital subscriber line

SHE safety health and ergonomics

SHED segmented hypergraphic editor

SHF super-high frequency

SHG segmented hypergraphics

SHH service handler home

SHIFT scalable heterogeneous integrated facility testbed (CERN)

SHL shift logical left

SHM session handler mobile terminal

SHN self-healing network

SHNS self-healing network service

SHO session handler originating network

SHR special handover request

SHRN special handover request – network

SHRN/U special handover request network/user

SHRU special handover request – user

SHS select character spacing; select horizontal spacing; short message service subscriber

SHSCC satellite high-speed computer communications

sht short

SHT session handler terminating network

SHTML server-parsed HTML

S-HTTP secure hypertext transfer protocol

SHUG Scottish hypermedia users' group

SHV session handler in visited network; standard high volume (servers)

SHV/H session handler in visited/home network

SHWY Super Highway

SHY soft hyphen

Si silicon

SI screening indicator; service identifier/indicator, identificateur de SPDU (f); service interworking; shift-in; special instruction; SPDU identifier; speech interpolation; status indicator; supplementary information; system information; système d'information; Système International (d'Unités) (f), SI units

SI/SO serial-in/serial-out

SIA Satellite Industry Association; Semiconductor Industry Association (US); serial interface adaptor; signal d'indication d'alarme (f), señal de indicación de alarma (s), alarm indication signal [AIS]; supplementary information A

SIAD señal de indicación de alarma distante (s) [RAI]; système d'aide à la décision (f), decision-making support system

SIAF service indicator associated field

SIB service-independent building block; status indication 'busy'; system interbus

SIBIL Système Informatisé pour Bibliotheque (f) (Switzerland)

SIC silicon integrated circuit; software integrated circuit; specific inductive capacity; standard industrial code; subscriber-line integrated circuit; système d'information commercial (f), commercial information system; Teilnehmerschaltung (g) (subscriber-line integrated circuit)

SICL standard instrument control library (IEEE)

SICOFI signal processor CODEC/filter

SICP service international de commutation par paquets (f), servicio internacional de conmutación por paquetes (s), international packet switch stream [IPSS]

SICRA servicio internacional de cobro revertido automática (s)

SICS Swedish Institute of Computer Science

SID Satanic Incarnate of Doom; scheduled issue date; security identifier; serial input data; señal de instrucción digital (s), digital control signal [DCS]; silence description (frame); silence descriptor; silence information descriptor; sistema de interconexión digital (s), digital interconnection signal; site identifier; Society for Information Display (US); station identification (AT&T); subscriber/subscription identity device; sudden ionospheric disturbance; SWIFT interface device; symbolic interactive debugger; system ID (number)

SIDD señal de identificación digital (s), digital identification signal [DIS]

SIDF system-independent data format

SIDH system identification for home systems

SIDP sputter-ion depth profiling

SIDT store interrupt descriptor table

SIE señal de inhibición de enlace (s), link-inhibit signal [LIN]; standard interface equipment; status indication 'emergency terminal status'; status indication 'E' ('emergency alignment')

SIEC signalling information element confidentiality

SIED señal de inhibición de enlace denegada (s), link inhibit denied (signal) [LID]

SIF signalling information field; SONET Interoperability Forum

SIFT share internal FORTRAN translator; Stanford (University) information filtering tool

SI-FT Système d'information de FT (f), France Télécom information system

sig signal/signalling

SIG SMDS Interest Group; special interest group

SIGARCH Special Interest Group for Computer Architecture

SIGCAT SIG on CD-ROM applications and technology

SIGGRAPH Special Interest Group on Computer Graphics

SIGINT signals intelligence

SIGLA Sigma Language (for industrial robots by Olivetti *tn*)

SIGPLAN Special Interest Group on Programming Languages

SIGR sistema integrado de gestión de la red (s), integrated network management system

SIGVERIF signature verification tool (Microsoft Windows 98)

SIL single in-line device (flat-pack); SNOBOL implementation language; speech interference level

SILD (Division du) Support informatique linguistique et documentaire (f), Data-processing Language and Documentation Support Division (EU), División de Apoyo Informático Lingüístico y Documental (s), Abteilung Unterstützung in den Bereichen Informatik Linguistik und Dokumentation (g)

Silent 700 portable data terminal (Texas Instruments Inc)

SILS standard for interoperable LAN security (IEEE 802.10)

SIM service instruction message; set initialization mode; simulator; single address message; single identification module; subscriber identity module (smart card); switch interface module

SIMD single instruction (stream) multiple data (stream)

SIME subscriber installation maintenance entity

SIMM single in-line memory module

SIMO single-input multi-output

SIMOX separation by implantation of oxygen

SIMP satellite information message protocol; sistema de mensajería interpersonal (s)

simplex transmission of data in one direction only

SIMS secondary-ion mass spectroscopy

SIMTEL simulation and teleprocessing

SIMULA Algol 60 based language for simulation programmes

sin sine

SIN status indication 'normal terminal status'; status indication 'N' ('normal alignment'); status indication 'normal terminal status'

SINAD signal interference noise and distortion

S-Indik Scheifenschluß-indikator (g) (loop indicator)

SingTel Singapore Telecom

sinh hyperbolic sine

SIO serial input/output; service information octet; Scientific and Industrial Organizations (ITU-T); status indication 'out of alignment'

SIOS status indication 'out of service'

SIP señal de interrupción del procedimiento (s), procedure interrupt signal [PIS]; service independent platform; single in-line package; SMDS interface protocol; subscriber interrogation point; supervision de l'isolement des processeurs (f), processor outage control [POC]

SIPC simply interactive PC (Microsoft)

SIPO serial input/parallel output; status indication 'processor outage'

SIPP simple internet protocol plus; single in-line pin package; SMDS interface protocol

SIR serial infra-red (Hewlett Packard); sistema intermedio de referencia (s), intermediate reference system [IRS]; sustained information rate; système d'information réseau (f), network information system [NIS]

SIRDS single image random dot stereogram

SIREN Système d'identification au Répertoire des Entreprises (f)

SIRET Système d'Identification au Répertoire des Établissements (f)

SIRIJ Semiconductor Industry Research Institute of Japan

SIRS satellite infrared spectrometer

SIRT selección e indicación de retardo de tránsito (f) [TDSAI]

SIS Satellite Information Services; signalling interworking subsystem; sound-in-sync, son dans la synchronisation (f), sonido en sincronización (s); Standardiseringskommissionen I Sverige (Swedish standards organization); superconductor-insulator-superconductor; subscriber identity security

SISD single instruction/(instance) stream single data stream

SIS-MS SIS mobile station

SISO single input single output; serial in serial out; Software and Intelligent Systems Office (US DoD)

SIT Signalisierungsumsetzer (g) (signalling converter); special information tone; static indication transistor; système d'information technique (f), technical information system; Système Interbancaire de Télécompensation (f), French inter-bank clearing system

SITA Société internationale de télécommunications aéronautiques (f), Society of International Aeronautical Telecommunications

SITC Satellite International Television Center

SITLM sistema de interfuncionamiento telemático (s), telematic interworking system [TIS]

SITM simple Internet transition mechanism

SITTA system integrated transmission test analogue

SITU situation

SIU slide-in unit; synchronization interface unit; system interface unit

SIV sensors in vacuum

SJF shortest job first

SKTs Sprechkapsel mit Transistor (g) (microphone insert with transistor)

SKZB Sprechkapsel für ZB-Betrieb (g) (microphone insert for CB)

SK Sofdox Krackers; Sprechkapsel (g) (microphone insert)

skinnable a program allowing its interface to be changed

SKIP simple key management for Internet protocol

SKR single key response

SKTDH stop kicking that dead horse

Skynet satellite digital service (AT&T)

SL send leg (telegraphy); service logic; signalling link; sincronización de línea (s), line synchronization [LS]; slate; specification limit; stability loss; subscriber loop

SLA señal de liberación por abonado llamante (s), calling party clear signal [CCL]; service level agreement; Standard Linux Association; synchronous line adaptor

SLAC Stanford Linear Accelerator Laboratory; subscriber line audio-processing circuit

SLAI service de libre-appel international (f), international freephone service [IFS]

SLAM scanning laser acoustic microscopy; single layer alumina metallization

SLANG structured language

SLAT system line-up and test

slc selon les cotes (f), not to scale; straight line capacitance

SLC service level contract; signalling link code; software license configurator; surface laminar circuit; subscriber loop carrier

SLCA subscriber line circuit analogue

SLD señal de liberación diferida (s), delayed release signal [DRS]; single-line data/display; straight line depreciation; super-luminescent diode

SLDRAM SyncLink DRAM (SyncLink Consortium *tn*)

SLDT store local descriptor table

SLE small local exchange
SLED single large expensive disk
SLEE service logic execution environment
SLF Smurf Liberation Front; symbolic link file
SLI scan-line interleave/interleaving; señal de llamada infructuosa (s), call-failed signal [CFL]; service logic interpreter
SLIC stereo line-in connector; subscriber line integrated circuit; subscriber loop multiplex; system level integration; Teilnehmerschaltung (g) (subscriber line integrated circuit)
SLICS SunLab's Interactive Collaborative Systems (Sun)
SLIM subscriber line interface module
SLIP serial line Internet protocol
SLIP/PPP serial line Internet protocol/point-to-point protocol
SLL semi-loop loss
SLM selective level meter; shared local memory; signalling link management; silicon-layer metal (gate array); single longitudinal mode; spatial light modulator
SLMA subscriber line module analogue
SLMCP subscriber line module coordination processor
SLMR silly little mail reader
SLMT signalling link test message
SLOC source lines of code; stereo line-out connector
SLOTH suppressing line operands and translating to hexadecimal
SLP service location protocol; service logic program/procedure; single-link procedures, procedimientos monoenlace (s)
SLPI service logic program instance
SLR send/sending loudness rating, índice de force des sons à l'émission (f)
SLRN source local reference number
SLS second line support; set line spacing, establecimiento de espaciamiento (s);

signalling link selection; storage library system
SLSI super-large-scale integration
SLSS systems library subscription service (IBM)
SLT service level target (Quetzal); switchman's local test
SLTA signalling link test (message) acknowledgement
SLTC signalling link test control
SLTM signalling link test message
SLU secondary logical unit; serial line unit; spare line unit; subscriber line use/unit
SLUS subscriber line use system
slw straight line wavelength
Sm samarium (used in solid-state and superconductor technologies)
SM security management; service management layer; service monitoring; set mode; shared memory; short message; signalling module; signal de mesure (f), señal de medida (s), measurement signal [MS]; single mode (optic fibre; *aka* monomode); station mobile (f), mobile station [MS]; stress migration; switching module (AT&T); system management (entity)
SMA sub-miniature A connector
SMAE system(s) management application entity, entité d'application de gestion du système (f) [SAME]
SMAF service management access function
SM-AL short message – application layer
SMAP system management application part/process
SMART self-monitoring analysis and reporting technology; Strategy for Mobile Advanced Radio Telecommunications (DG XIII)
SOG-ITS Senior Officials Group for IT Standards
SMAS service management application area in TMOS; supplementary main store; switched maintenance access system (AT&T)

SMASH step-by-step monitor and selector hold

SMATV satellite master antenna TV; single-mast antenna TV

SMB server message block; single-mask bumping; unidad de señalización de sincronización de multibloque (s), multi-block synchronization signal unit [MSB]

SMC short message control; système de multiplication de circuits (f), sistema de multiplicación de circuitos (s), circuit multiplication system [CMS]; surface-mounted components; switch maintenance centre

SMCC Sun Microsystems Computer Corp

SMCD sistema de multiplicación de circuitos digitales (s), digital circuit multiplication system [DCMS]

SMCN système de multiplication de circuit numérique (f), digital circuit multiplication system [DCMS]

SM-CP short message – control protocol

SMD surface-mounted device, oberflächenmontierte Bauelemente (g)

SMDF subscriber main distribution frame

SMDI storage module disk interconnect

SMDL standard music description language

SMDR station message detail recording

SMDS switched multi-megabit data service (Bellcore)

SME segundos con muchos errores (s), severely-errored seconds [SES]; service modification environment; short message entity; small and medium-sized enterprises; Society of Manufacturing Engineering; software maintenance engineer; subject matter expert

SMETDS standard message trunk design system

SMF service management function; single-mode fibre (interface); software maintenance function; system manager facility (Compaq); sub-multiframe

SMFA service management function

area; simplified modular frame assignment system

SMG screen management guidelines (DEC); software message generator; special mobile group (ETSI)

SMH signalling message handling

SMI structure of management information (RFC 1155); Sun Microsystems Inc; system management interrupt (Intel)

S-MIC signal transfer point – medium interface connector

SMIF standard mechanical interface

SMIL synchronized multimedia integration language

smiley :- colon dash left/right bracket: image of pleasure ☺ or pain ☹

S-MIME secure MIME

SMIP sistema de mensajería interpersonal (s), interpersonal message/messaging system [IPMS]

SMIT system management interface tool (IBM)

SMJ smartjack

SMK software migration kit (Microsoft)

SML software master library; standard functional language; standard meta-language

SMM system management mode (Intel)

SMMC system maintenance monitor console

SMMR scanning multi-channel microwave radiometer

SMP SiMetrical processing standard *tn*; señal de multipágina [MPS]; señal máxima permitida, signal maximal permis (f), permitted maximum signal [PMS]; service management point/process; simple management protocol; switching module processor; symbolic manipulation program; symmetric multiprocessing

SMPC shared memory parallel computer

SMPDU service message protocol data unit

SMPM SECS message protocol machine

SMPS switching mode power supply

SMPTE Society of Motion Picture and TV Engineers (standard time code)

SMR specialized mobile radio; semiconductor mask representation; short message relay

SMRC service management reference configuration

SMRT single message rate timing

SMS Satellite Multiservice System; SECS message service; self-maintenance services; service management system (IN entity); servicio marítimo por satélite (s), maritime satellite service [MSS]; short message service (for mobile phones); sistema del servicio móvil por satélite (s), mobile satellite service system [MSS]; standard management system; storage management services (NetWare); synchronous meteorological satellite; system management server (Microsoft Windows NT)

SMS/PP short message service/point-to-point

SMSA standard metropolitan statistical area

SMSC short message service centre

SMSCB short message service cell broadcast

SMS-GMSC short message service – gateway mobile services centre

SMSI systems management service interface

SMS-IWMSC short message service – interworking mobile switching centre

SMS-PP short message service – point to point

SMSS switching and management subsystem

SMS-SC short message service – service centre

SMSW store machine status word

Smt short message terminal

SMT sub-multiframe, sous-multitrame (f), submultitrama (s); sistema de medidas de tráfico (s); station management; sub-multiframe (interface); surface-mount technology

SMTE super-mastergroup translating equipment, équipement de traduction de groupe quaternaire (f)

SM-TL short message – transfer layer

SMTP simple mail transfer protocol (RFC 822)

SMTS Strategic Material Transport System

SMU scaled measurement unit, unité de mesure pondérée (f); super-module unit

SMWG synchronized multimedia working group

SMXU service message transfer unit

SN sequence number; serial number; service network/note; slot number; subscriber number; switching network

SNA systems network architecture (WAN-only protocols)

SNA/DS SNA distribution services

SNACP subnetwork access protocol

SNAP service node application program; single number access plan; standard network access protocol; sub-network access protocol; sub-network attachment point (IEEE 802.1a); sustained necessary applied pressure

SNAPS standard notes and parts selection

SNB subscriber number

SNC sub-network connection; synchronous network clock

SND send

SNDCF sub-network dependent convergence functions

SNDCP sub-network dependent convergence protocol

SNDR sender

SNDS status NTE diagnostic state

SNE/FTF strategic network environment – file transfer facility

SNET Southern New England Telephone

SNFR section numérique fictive de référence (f) [HRDS]

SNG satellite news-gathering

SNH signal-to-noise handshake

SNI serial network interface; Siemens Nixdorf Informationsisteme AG; subscriber network interface

SNIA Storage Networking Industry Association (US)

SNICF sub-network independent convergence functions

SNIRR selected negotiation of intermediate rate requested

SNL selected nodes list; signalling network layer

SNM signalling-network-management signal, signal de gestion du réseau de signalisation (f)

SNMP simple network management protocol (RFC 1157)

SNMP/IP simple network management protocol/Internet protocol

SNMPv2 simple network management protocol version 2

SNMS sputtered neutral mass spectroscopy

SNOBOL string-oriented symbolic language

SNOM scanning near-field optical microscopy

SNP sequence number packet (Novell NetWare *tn*); sequence number protection; serial number/password (Open Technology)

SNPA sub-network point of attachment

SNPC Software New Products Committee

SNPP simple network paging protocol (RFC 1568)

SNR séquence numérique de référence (f), digital reference sequence [DRS]; serial number; signal-to-noise ratio; signal-to-quantization noise ratio

SNRM set normal response mode, paso a modo de respuesta normal (s)

SNRME set normal response mode extended

SNS service node subsystem

SNT switching national terminal

SO service order; shift-out, hors code (f), fuera de código (s); sistema(s) de operaciones (s), operating system [OS]; slow to operate; small outline (packing density on PCBs)

SOA start of address, comienzo de dirección (s); semiconductor optical amplifier; suppress outgoing access (CUG SS)

SOAP symbolic optimal assembly program (IBM)

SOAR Smalltalk on an RISC (computer); state operator and result (Common Lisp)

SOC statement of compliance

SOCC standard optical cable code

SOD sales object drawing; spin-on-dielectric; servicio de operación a distancia (s), remote operation service [ROS]; statement of direction

SODA symbolic optimum DEUCE assembly program (English Electric)

SODCF subscription option data communication facility

SO-DIMM small outline DIMM

SOE standard operating environment; standards of excellence

SOF satisfactory operation factor; service order form

SOFRECOM Société Française d'études et de réalisations d'équipements de télécommunications (f)

SOG service order gateway; spin-on glass; subsystem out-of-service grant

SOGA SNA-open gateway architecture

SOGT senior officials group for telecommunications

SOH section overhead; start of header/heading

SOHF sense of humour failure

SoHo small office – home office

SOHO télétravail/travail à domicile (f), teleworking

SOI service data point – open interface; silicon on insulator (technology); single-operand instruction

SOIC small-outline integrated circuit (SMD)

SOIF summary object interchange format

SOJ small-outline J-lead (SMD)

sol soluble; solenoid

SOL simulation-oriented language; small-outline leaded; Sons of Liberty

soly solubility

SOM scanning optical microscope/microscopy; service order mediator; start of message; sulphuric acid-ozone mixture; system object model (IBM)

SOM/DSOM system object model/distributed system object model

SOMA software machine

SONAD speech-operated noise-adjusting device

SONDS small office network data system

SONET Synchronous Optical Network (fibre-optic transmission)

SOP standard operating procedure; start of record; sum of products (expression)

SOR subsystem out of service request; support of optional routing

SOS service order system; silicon on sapphire; Son of Stopgap; speed of service; start of string

SOSDCF subscription option supplementary – default call forwarding

SOSE silicon-on-something-else

SOSIG social science information gateway

SOST special operator service traffic

SoT the Sea of Tranquillity

SOT selectable on test (of discrete component value); state of termination; subscriber originating trunk

SOTA state of the art

SOW statement of work, énoncé de travaux (f) [EDT]

SOX sound exchange

sp gr specific gravity

SP space, espace (f); semi-public; service pack; service plane (TINA); service provider; servicio de presentación (s), presentation service; signalling end-point; signalling point; signal processor; single-pole; Southern pine (telephone pole); speech packet; Speicherteil (g) (memory unit); stack pointer; station phototélégraphique (f), phototelegraph station [PS]; supervisory process; support processor; system product

SP/2 scalable POWER parallel 2 (IBM *tn*)

SP/DIF Sony Philips/digital interface

SPA satellite paging area; semi-permanently associated; signalling point allowed; Software Publishers' Association; specific poll address

Spacenet GTE Corp Business Systems network

SPADE single-channel per carrier (SCPC) PCM multiple access demand assignment equipment

SPAG Standards Promotion and Application Group (BISYNC)

SPAI specified alarm interface

Spam stupid person's advertisement; unsolicited e-mail

SPAN Space Physics Analysis Network

SPARC Scalable Processor Architecture (Sun Microsystems *tn*)

SpBr Speisbrücke (g) (feed current bridge)

SPC secondary point code; semi-permanent circuit (Telstra); signalling point code; signalling point congested; small peripheral controller; Software Productivity Centre (Canada); Southern Pacific Communications; statistical process control; stored program control, commande par programme enregistré (9); suppress preferential CUG

SPC 1 sucesión de protocolos candidata Nº 1 (s), candidate protocol suite Nº 1 [CPS 1]

SPC exchange stored-program control exchange

SPCC Southern Pacific Communications Corp

SPCE sistema de pruebas de compensadores de eco (s), echo-canceller testing system [ECTS]

SPCL spectrum cellular

S-PCN satellite PCN

S-PCS satellite PCS

SPCS stored program-control system/switch

SPD secuencia de prueba digital (s),

digital test sequence [DTS]; selección de la dirección de presentación (s), select presentation direction; serial presence detect; software product description

SPDIF Sony/Philips Digital Interface *tn*

SPDL standard page description language

SPDT single-pole double-throw

SPDTDB single-pole double-throw double-break

SPDTNCDB single-pole double-throw normally-closed double-break

SPDTNO single-pole double-throw normally-open

SPDTNODB single-pole double-throw normally-open double-break

SPDU (receipt of) session protocol data unit, unité de données de protocole de session (f)

SPE signal processing extension; Software Practice and Experience (journal); SONET synchronous payload envelope; sporadic E (layer of ionosphere)

Spec specification

SPEC speech-predictive encoded communications; Standard Performance Evaluation Corporation (US)

SPECmark Systems Performance Evaluation Cooperative benchmark

Spellstar Microsoft International Corporation

SPESS stored program electronic switching system

SPF shortest path first; system performance factor; system productivity facility

SPG support processor group; synchronized pulse generator

SPGA staggered pin-grid array (Intel)

sph spherical

SPI safeguarding proprietary information; security parameters index; serial peripheral interface; service provider identity; service provider interface (Winsock 2); Software Process Improvement; station program identification

SPICE simulation program with integrated circuit emphasis; scientific personal interactive computing environment

SPID service protocol identifier (ISDN); service provider identity

SPIN Software Process Improvement Network

SPIRES Stanford (University) public information retrieval system

SPIRS Silver Platter Information Retrieval System *tn*

SPITE switching processing interface telephone events

SPJ self-protective jammer; self-provided telecom systems (UK); service provider link; set priority level (Unix); software parts list; sound pressure level; Space Program Language (USAF); splice; spooler; system programming language (Hewlett Packard)

SPLaT Society for the Preservation of Lasting Anarchy and Terrorism

SPLM sound pressure level meter

SPM scanning probe microscopy; sequential Paralog machine; service performance management; service protocol machine, machine protocole de session (f); space switch module; strategic product management; system performance monitor (IBM)

SPMT señal de prueba multitono (s), multi-tone testing signal [MTTS]

SPN single personal number; software performance monitor; solar proton monitor; subscriber premises network

SPOC single point of command; single point of contact

SPOF single point of failure

SPOM self-programmable one-chip microcomputer (Bull, France *tn*)

SPOOL simultaneous peripheral operation on-line

SPOT shared product object tree (IBM)

SPP señal de página parcial (s), partial page signal [PPS]; sequenced packet protocol (NetWare); signalling point prohibited; signal processing part; single-phase printing

SPPAY semi-postpay pay-station

SPPS scalable power parallel system (IBM)

SPR semiconductor process representation; señal de paso a enlace de reserva (s), change-over signal [COV]; send priority and route digit; signalling point relay; signalling point with SCCP relay function; sistema probado (s), system under test [SUT]; spare; special purpose register; statistical pattern recognition

SPRC signalling procedure control

SPRINT Southern Pacific Communications' switched long-distance service; special police radio display inquiry unit

sprite moveable pattern of pixels (*eg* cursor)

SPS Self-Preservation Society; standby power supply/system; support processor subsystem; surface preparation system; symbolic programming system (IBM); system performance score

SPSE echo suppressor testing system (s) [ESTS]

SPST single-pole single-throw (switch)

SPSTNC single-pole single-throw normally-closed

SPSTNO single-pole single-throw normally-open

SPT sectors per track; shortest processing time; system page table

SPTN single protocol transport network

SPTS single program transport stream

SPU signal processing unit

SPV surface photovoltage

SPVC soft permanent virtual circuit

SpVSt speisende Vermittlungsstelle (g) (switching centre supplying feed current)

SPX sequenced packet exchange (Novell); simplex

sq square

SQ signal quality (detector)

SQA software quality assurance

SQC statistical quality control

SQCIF sub-quarter (inch) common intermediate format

SQD signal quality detect/detector

SQE signal quality error (IEEE)

SQF surveillance de la qualité de fonctionnement (f), performance monitoring [PM]

SQID super-conductive quantum interference device

SQL structured query language

SQL/DS structured query language/data system (IBM)

SQPMM software quality and process maturity model

SQUID super-conducting quantum interference device (Fujitsu)

SQWID Search query-weighted information display

sr steradian (SI unit of solid angular measure)

SR scanning radiometer; service de réseau (f), servicio de red (s), network service [NS]; service release/routing; set-reset; shift register; signal-strength receiver; slow release; sonoridad de referencia (s), reference loudness [RL]; source reference; status response; supervisor

SR25E sending R.25 equivalent

SRA signal d'accusé de réception de liaison de réserve prête (f), standby ready acknowledgement signal

SRAC Supplier Relations Action Council

SRAM sideways RAM; static RAM

SRAPI speech recognition application program interface (Novell)

SRATS Solar Radiation and Thermospheric Structure Satellite (Japan)

SRB source route bridge/bridging

SRBP statistical report basic package

SRC Science Research Council (UK); Semiconductor Research Corporation; Seymour Roger Cray (supercomputer designer)

SRCC servicio de red con conexión (s),

connection mode network service [CONS]

SRC-REF source reference (field)

SRCS set reduced character separation, establecimiento de espaciamiento de caracteres reducido (s)

SRD satellite receiver decoder; screen reader system; secondary received data; secuencia de referencia digital (s), digital reference sequence [DRS]

SRDC sub-rate digital cross-connect

SRDE señal de acuse de (recibo de) desinhibición de enlace (s), link uninhibit acknowledgement signal [LUA]

SRDF servicio de recolección de datos de facturación (s), billing data collection service

SRDM sub-rate digital multiplexer (part of DDS)

SRDRAM self-refreshed DRAM

SRE send reference equivalent; signalling range extender; site resident engineer; surveillance radar

SREJ rechazo selectivo (s), selective reject

SRES signed response (authentication)

SRF Science Research Foundation (UK); specialized resource function; specifically routed frame; surface roughness factor; système de référence fondamental (f), fundamental reference system [FRS]

sRGB simple red green blue (colour palette)

SRI SIP relay interface; Société du Réseau International (Bull Marketing); standard run-time interface (Siemens); Stanford Research Institute (US); Swiss Radio International; système de référence intermédiaire (f), intermediate reference system [IRS]

SRIE señal de acuse (de recibo) de inhibición de enlace (s), link inhibit acknowledgement signal [LIA]

SRL stability/structural return loss

SRM security reference monitor; señal de rechazo de mensaje (s), message refusal

signal [MRF]; signal strength receiver module; signalling route management

SRMA split-channel reservation multiple access

SRMC servicio de red en modo conexión (s), connection mode network service [CONS]

SRO shareable and read-only; sous-répartition optique (f), optical cross-connect

SRP spreading resistance probe; statistical report package; système de référence de planification (f), sistema de referencia de planificación (s), planning reference system [PRS]; système de remise physique (f), physical delivery system [PDS]

SRPI server-requester programming interface

SRPT shortest remaining processing time; status report

SRQ service request (IEEE-488); status request

SRR serially re-usable resource; signal-strength receiver redundant

SRS select reverse spacing, comienzo de cadena inversa (s); señal de respuesta sin calificación (s), answer signal – unqualified [ANU]; software requirements specification; sound retrieval system; sub-rate switch; synchronous relay satellite

SR-SA service region – service area

SRSM sub-rate switch module

SRT selección de retardo de tránsito (s), transit delay selection [TDS]; select resources terminal; source route transport (token ring); source transparent routing (IEEE standard); subscriber radio terminal

SR-TB source routing-transparent bridging (IEEE)

SRTP sequenced routing table protocol

SRTS synchronous residual time stamp

SR-UAPDU status report user-agent protocol data unit

SRUP sequenced routing update protocol

SRVT SCCP routing verification test

ss Spitze-Spitze (g) (p-p: peak-peak as suffix)

SS semi-final splice; service subscriber; servicio de sesión (s), session service [SS]; signalling service; signal strength (maximum); single sideband; single-sided; soft-sectored; solid state; Sondersatz (g), special unit; space switch; special service; spread spectrum; stainless steel; station-to-station; subscriber switching; subsystem, sous-système (f), subsistema (s); supplementary service; switching system; system simulator

SS Nº 6/7 signalling system Nº 6/7, sistema de señalización Nº 6/7 (s) [SS Nº 6/7]

SS/W space switch

SS2/3 single shift, inversion unique (f), cambio individual a G2/3 (s) [SS2]

SS6/7 signalling system 6/7 (control of office switches and processors)

SS7/CCS signalling system 7/common-channel signalling

SSA secuencia seudoaleatoria (s), pseudo-random sequence [PRS]; Semiconductor Industry Association; serial storage architecture (IBM); station select address; subsystem allowed, sous-système autorisé (f), subsistema autorizado/admitido (s); subsystem available

SSAC AC signalling system

SSACC signal strength access

SSADM structured systems analysis (and) design method

SSAEM sous-système d'application d'exploitation et de maintenance (f), operation and maintenance subsystem application

SSAM sous-système application mobile (f), mobile application subsystem

SSAN système de signalisation à accès numérique (f), digital access signalling system [DASS]

SSAP session service access point; source service access point

SSAS station signalling and announcement subsystem

SSB signal strength blocking; single-sideband (transmission); subscriber busy signal, signal (électrique) d'abonné occupé (f); subconjunto sincronizado de base (s), basic synchronized subset [BSS]

SSBA suite synthétique des benchmarks de l'AFUU (f)

SSBAM single-sideband amplitude modulation

SSB-SC single sideband suppressed carrier (transmission)

SSC sector switching center; single silk-covered; SNA simulation code; special services centre; specialized common carrier; subsystem congested, subsistema congestionado (s); supplementary service control string

SSCC common channel signalling system (f)

SSCCN7 Nº 7 common-channel signalling system (f)

SSCF service specific coordination function; service switching and control function; signalling specific coordination function

SSCOP service specific connection-oriented protocol (ITU-T Q.2110)

SSCP service switching control point; systems service control point (IBM)

SSCS service specific convergence sub-layer; sous-système de commande des connections sémaphores (f), signalling connection control subsystem [SCCP]; utilisateur terminal (f), end-user [SCCP]

SSD Scalable Server Division (Intel); signal strength decrease; single-sided disk; solid state disk; sous-système disponible (f), subsistema disponible (s), subsystem available [SSA]

SSDC signalling system directional coupler

SSDU session service data unit

sse single silk covering over enamel insulation

SSE scanning supervisory equipment; secondes sans erreur (f), segundos sin error (s), error-free seconds [EFS]; sous-système encombré (f), subsystem congested [SSC]; sum squared error

SSEM Stepper Specific Equipment Model; sous-système pour l'exploitation la maintenance et la gestion (f), operation and maintenance subsystem [OMAP]

SSF service specification framework; service switching function; single sheet feed (printer); single-sided frame; sub-service field

SSFC sequential single frequency code system

SSFD solid state floppy disk

SSFS special services forecasting system

SSG Suburban Survival Guide

SSGA system support gate array

SSGT sous-système de gestion de capacité de transaction (f), transaction capability application part (f) [TCAP]

ssh secure shell (Unix)

SSH signal strength hand-off

SSI server-side include (Web); service script interpreter; signal strength increase; single-system image; small-scale integration; start signal indicator; sous-système interdit (f), subsystem prohibited [SSP]; switch site implementation

SSID Service spécialisé d'informatique documentaire (f), specialized service for documentary data processing; subsystem identifier

SSII Sociétés de Service d'Ingénierie Informatique (f)

SSL secure-server layers; service script logic; semi-loop loss; software slave library; synthesizer special language

SSM service session manager (TINA); signal strength measurement; single segment message; special safeguarding measures; strategic sourcing methodology

SSMA spread spectrum multiple access

SSMB special services management bureau

SSMF multifrequency signalling system (f)

SSMIN signal strength minimum

SSN station serial number; subsystem number; switched services network

SSO signal strength overlaid

SSODB Secret Society of Dark Birds

SSOP shrink small outline package

SSP service switching point; Siemens switching processor *tn*; signalling service point; sistema sometido a prueba (s), system under test [SUT]; special services protection; sub-satellite point; subsystem prohibited, subsistema prohibido (s); switch-to-switch protocol

SSPA solid state power amplifier

SSPS satellite solar power station

SSPU signal and speech processing unit

SSQA Standardized Supplier Quality Assessment (US)

SSR secondary surveillance radar; signal strength receiver; solid state relay; station source routing; strict source route (RFC 791); switching selector repeater

SSRA spread spectrum random access system

SSREG signal strength registration

SSS service subsystem; signal strength supervision; simetría de la señal de salida (s), output signal balance [OSB]; single sideband system; subscriber switching subsystem; suspended substrate stripline

SSSI sous-système services intermédiares (f), intermediate service part [ISP]

SSSP station-to-station sent paid (direct distance dialling)

SST secure server technology (Netscape); send-special-information tone signal, (signal) envoyez la tonalité spéciale d'information (f), señal de envio de tono especial de información (s); Sender-

/Empfänger-Steuerung (g), send/receive control; single sideband transmission; spread spectrum transmission; subscriber transferred (signal); subsystem status test; synchronous service transport

SSTM (sous-système) transport de messages (f), message transfer/transport part

SSTV slow-scan television

SSU session support utility (DEC); single signalling unit; signal strength underlain; (sous-système) utilisateur (f), user part [UP]; stratospheric sounding unit; subscriber switching unit; subsequent signal unit, unité de signalisation subséquente (f), unidad subsiguiente de señalización (s)

SSU RNIS sous-système utilisateur du réseau numérique à intégration de services (f), ISDN user part [ISUP]

SSUD (sous-système) utilisateur données (f), data user part [DUP]

SSUR (sous-système) utilisateur pour le RNIS (f), ISDN user part [ISDN-UP]

SSURI international ISDN user part (f)

SSURN national ISDN user part (f)

SSURT ISDN user part for third-party networks (f)

SSUT (sous-système) utilisateur téléphonie (f), telephone user part [TUP]

SSW set space width, establecimiento de anchura de ESPACIO (s)

SSWO special service work order

St Steuerfeld – Daten (g) (control field – data)

ST key pulsing (*ie* Start); satellite de communications (f), satélite de comunicaciones (s), communications satellite [CS]; service de transport (f), servicio de transporte (s), transport service [TS]; string terminator, terminador de cadena (s); end of pulsing signal (*ie* stop), signal de fin de numérotation (f) [ST]; segment type (DQDB); sidetone; servicio de telecomunicación (s), telecommunications service [TS];

Signalisierungsumsetzer (g), signalling converter; signalling terminal; signalling tone; supervisor tester

ST/SYS storage and transfer system, système de mémorisation et de transfert (f), sistema de almacenamiento y transferencia (s)

sta station

STA servicios de tráfico aéreo (s), air traffic services [ATS]; spanning-tree algorithm; station terrienne d'aéronef (f), aircraft earth station [AES]; supervision du traitement d'appel (f), call processing control [CPC]

STAB selective tabulation, tabulación selectiva (s)

STAC storage allocation and coding program (DEUCE/English Electric)

STACK start acknowledgement

STAD start address

STAIRS document retrieval system (IBM *tn*)

STALO stabilized local oscillator

STANAG Standardization Agreement – NATO

STAPLE St Andrews (University) Applicative Persistent Language

STAR Second Time Around (ICL recycling of PC hardware); self-defining text archive; simultaneous transmitted and reflected; Special Telecommunications Actions for Regional Development (in Europe)

Starlan AT&T local area network *tn*

STARS software technology for adaptable reliable software (US DoD)

STARTS UK initiative promoting best practice in software development

STARTUP.CMD batch file for initializing OS/2

STATLIB statistical library

STAX standardization (taxonomy) reference database (RACE I)

STB set-top box (*ie* satellite signal decoder); station de travail banalisé (f), common/shared workstation

STC señal de transferencia de la carga

(de tráfico) (s), load transfer signal [LTR]; servicio de teleconferencia internacional (s), (international) teleconference system [TCS]; serving test centre; signal terminal centre; Standard Telephone and Cable Ltd (UK); station technical control; sub-technical committee of ETSI; station terrienne côtière (f), coast earth station [CES]; switching and testing centre; system time clock

S-TCA send TCA action

STCM signal terminal centre magazine

STD secondary transmitted data; sistema de transmisión digital (s), digital transmission system [DTS]; state transition diagram; subscriber trunk dialling (UK); synchronous time division

STD1 Standard 1 (Internet Official Protocol Standards)

STDA StreetTalk directory assistance (Banyan)

STDAUX standard auxiliary

STDbus standard bus

stderr standard error (Unix library routine)

STDIN standard input

stdio standard input/output (Unix library routine)

STDM statistical time-division multiplexing (*aka* statMUX)

STDn IAB standards list (n = number)

STDN space flight tracking and data network

STDOUT standard output

STDPRN standard printer

STE section terminating equipment (SONET); signalling terminal equipment; spanning-tree explorer; span terminating equipment; super-group translating equipment

STEA surveillance du taux d'erreur pendant la procédure d'alignement (f), alignment error rate monitoring [AERM]

STEL short-term exposure limit

STEM scanning transmission electron microscope/microscopy; strategic

telecommunication evaluation model (RACE I); Standard for the Exchange of Product Model Data (ISO)

STET simulateur de tête et de torse (f), head and torso simulator [HATS]

STF elemento de servicio transferencia fiable (s), reliable transfer signal [RTS]

STFS standard time and frequency signal (service)

STG Steuerung (g) (controller); Strategic Technologies Group; sub-technical group

STI set interrupt flag; statistics time interval

STIR service de transport de base indépendant du réseau (f), network-independent basic transport service

StKr Stromkreis (g) (circuit - general)

STL salto de línea (s), line skip [LSK]; sistema telefónico local (s), système téléphonique local (f), local telephone system [LTS]; standard telegraph level; studio-to-transmitter link

STLC signalling link test control

STLR sidetone loudness rating

STM scanning/standard tunnelling microscopy; selective traffic management; signalling traffic management; système de messagerie (f), servicio/sistema de tratamiento de mensajes (s), message handling system [MHS]; synchronous transfer mode; synchronous transport/(transfer) module

STM-1 synchronous transport module 1

STM-N synchronous transport module level N

STM-n Synchronous transport module n

STM-nc synchronous transport module n concatenated

STMR sidetone masking rating

STN station terrienne de navire (f), ship earth station [SES]; super-twist nematic

STO sistema telefónico de operadora (s), operator telephone system [OTS]; system test objectives

STOIC string-oriented interactive compiler (Smithsonian Observatory)

STONE structured and open environment (BMFT)

STOS store string

STOVSt Steuerende ortsvermittlungsstelle (g) (controlling local switching centre)

STP secure transfer protocol; service traffic position; services transaction program (IBM); shielded twisted pair (cable); short-term prediction; signal traffic point; signal transfer point; spanning tree protocol (IEEE); standard temperature and pressure; synchronized transaction processing; system test plan

STPI services transaction program interface (IBM)

STPL sidetone path loss

STPR sidetone power rating

STPST stop-start

STR sidetone reduction; signalling terminal regional; store task register; synchronous transmit/receive (IBM)

STRATSAT Strategic Satellite System (US Air Force)

STRE sidetone reference equivalent

STREAMS kernel support for network data/communications drivers (Sun *tn*)

STRESS structural engineering system solver (language)

Stretch very large computer (IBM 7030, 1950s)

STRM signalling terminal regional micro; sistema de transferencia de mensajes (s), message transfer system [MTS]

STROBES shared time repair of electronic systems

STRP signalling terminal remote processor

STRUDL structural design language

STS Satellite Transmission Systems (US corp); shared tenant service; station terrienne au sol aéronautique (f), aeronautical ground earth station [GES]; statistics and traffic management subsystem; synchronous transport signal (SONET)

STS3c SONET specification

STSC Scientific Time-Sharing Corp

STSK Scandinavian committee for satellite communications

STS-nc synchronous transport signal 'n' concatenated

STT Secure Transaction Technology (Microsoft *tn*)

STTC start of transfer through-connect signal

STTS surveillance du taux d'erreur sur les trames sémaphore (f), signal unit error-rate monitor [SUERM]

STU secure telephone unit; signal transmission unit; store unit; subscribers' trunk unit

STVUR station de television uniquement réceptrice (f), television receive-only station [TVRO]

STX start of text (control character 02 – 02H), début de texte (f), comienzo de texto (s)

su substitute user (Unix command)

SU service user; signal unit, unité de signalisation (f), unidad de señalización (s); spectrum utilization; sub-resolution attenuated; supervisory unit

SUA software-generated group unblocking-acknowledgement; stored upstream address

SUAC message d'acceptation de service supplémentaire (f) [FAA]

sub subscriber; substitute (character)

SUB sub-addressing; subroutine; substitute character, carácter de sustitución/(sustitutivo, sustituto) (s)

SUBLIB subroutine library

subNMS sub-network management system

substa sub-station

SUD session user data

SUDM message de demande de service supplémentaire (f), facility request message [FAR]

SUDS service supplémentaire désactivé (f), facility de-activated [FAD]

SUE stupid user error

SUERM signal unit error rate monitor

SUH suppression time for hand-off attempts

SUIF Stanford University intermediate format

SUIN information de service supplémentaire (f), facility information [FAI]

SUM set-up (control) module

SUNet Stanford University Network

SUNIST serveur universelle national de l'information Scientifique & Technologique (f)

sup supply

SUP support part

SuperCalc spreadsheet and financial modelling software (Sorcim Corp *tn*)

SuperJANET enhanced JANET (successor) via optical fibre links

SuperPLL super-phased locked loop

SUPS solid-state UPS

SUR message de surcharge (f), overload message [OLM]

SURF message de refus de service supplémentaire (f), facility reject message [FRJ]

SURFnet Dutch academic network

SUS single UNIX specification; single user system; subscriber service subsystem; suspend message, message de suspension (f), mensaje de suspensión (s)

SUT (call) set-up time; service usage trial; system under test, système à tester

SUU signalisation d'usager à usager (f), señalización de usuario a usuario (s), user-to-user signalling (s) [UUS]; supervision unit

SV speaker verification; speech voltmeter

SVA service à valeur ajoutée (f), value-added service

svc service

SVC signalling virtual channel; still video camera; supervisor (supervisory) call; switched virtual circuit/connection (frame relay use)

SVCC switched virtual channel connection

SVD simultaneous voice and data (modem); singular value decomposition

SvDo Steckverbindungsdose (g) (socket box)

SVE SAP (service access point) vector element; system and vendor evaluation

SvE Steckverbindungsdoseneinsatz (g) (socket box insert)

SVF simple vector format

SVG scalable vector graphics

SVGA super-video graphics adaptor

S-VHS super VHS

SVI service interception; sub-vector identifier

SVID system V interface definition (Unix)

SVLog servicing log

SVLUG Silicon Valley Linux Users' Group

SVM Science & Vie Micro (magazine)

SvN Stromverteilungsnetz-Nachbildung (g) (balancing circuit for current distribution)

SVN software version number; switched virtual network

SVP surge voltage protector; switched virtual path

SVPC switched virtual path connection

SVR server

SVR4 System V Release 4 OS (AT&T)

SVRC Software Verification Research Centre (Australia)

SVS select line spacing; select vertical spacing, choix de l'espacement vertical (f), selección del espaciamiento vertical/(de lineas) (s); single virtual storage; switched voice service

SVSS small voice switching system (US)

SVSt speisende Vermittlungsstelle (g) (switching centre supplying feed current)

SVT signal de verrouillage de trame (f), frame alignment signal [FAS]; secuencia de verificación de trama (s), frame check/checking sequence [FCS]

SVV sub-vector value

SW switch
SWAG scientific wild ass guess
SWaT Special Warez acquisition Team
SWAT Software Action Team
SWB Southwestern Bell
swbd switchboard
swc switching centre
SWC surge-withstand capability
SWD self-wiring data
SWE spherical wave expression
SWEAT standard wafer-level electro-migration accelerated test
SWEDAC Swedish Board For Technical Accreditation
SWEPL scaled weighted echo path loss
SWF short wave fade-out; Sudwest Deutscher Fernsehen (g)
SWFD Selbstwählferndienst (g) (subscriber trunk dialling)
SWFG secondary waveform generator
SWG standard wire gauge
SWI special word interval; static walkthrough inspection
SWIFT Society for Worldwide Inter-bank Financial Telecommunications; Streamlined Worldwide Information Today
SWIG Software Writers' International Guild
SWIM semiconductor workbench for integrated modelling; see what I mean; super WOZ integrated machine
SWISH simple web indexing system for humans
SWM Semiconductor Worldwide Management
SWOP specifications for web offset publication
SWP simple Web printing (Microsoft/HP secure printing proposal); single-wafer processing
SWPV software production and verification
SWR semiconductor wafer representation; standing wave ratio
SWV square wave voltammetry
SX simplex (signalling)

SXS step-by-step switching system
SxS Strowger Switch
SYBASE DBMS relational database system
SYLK symbolic link (application program format)
SYN synchronous idle (character), sincronización (s)
SyncLink private-circuit data service using SDH (Mercury *tn*)
SYNTRAN synchronous transmission
SYSADMIN system administrator
sysgen system generator/(generation)
SYSLOG system log
SYSOM system O&M
sysop system operator (of a BBS)
SysRq system request (key)
System 3000 local area network (Sytek Inc)
System U version of UNIX
System X BT computerized exchange switching system
SYSTRAN EC machine translation system
SYU synchronization signal unit, unité de signalisation de synchronisation (f), unidad de señalización de sincronización; synchronization unit/utility
Sz Schaltungszeichung (g) (circuit diagram)
SZ seizure

T

T dialect of LISP; ISDN reference point; signal unit error rate monitor [SUERM] threshold; tera, 10^{12}, one million million; tesla (unit of magnetic flux); timer; transparent; transport, transporte (s); type only

T&C time and charges

T&L terminate and leave

T/B top and bottom

T/C thermo-compression

T/C/D timer, counter, data

T/E test and evaluation

T/L/V type/length/value

T/R transmit/receive

T/S thermo-sonic

T/SYS transfer system, système de transfert (f), sistema de transferencia (s)

T1 synchronous data rate: 1 544 Mbit/s (AT&T)/Bell digital standard DS1)

T1 tiempo de retransmisión local (s)

T1DM T1 digital multiplexer

T2 6.3 Mbit/s multiplex data rate (AT&T Bell standard DS2)

T²L TTL (time to live)

T3 45 Mbit/s high-speed multiplex data rate (AT&T Bell standard DS3)

Ta tantalum (used in capacitors)

TA tape-armoured; Telecommunications Act (UK, 1984); telegrafía armónica (s), voice frequency telegraphy [VFT]; terminal adaptor (ISDN); test access; test alert; test analyser; timing advance; traffic area; transferred account; transmission authenticator; type approval

TA-182 AN/TA-182 signal converter

TAA track average amplitude; transfer-allowed acknowledgement signal, signal d'accusé de réception de transfert autorisé (f)

tab table; tabulate; tab key

TAB tape automated bonding; technical advisory board

TAC technical assistance centre; Telnet access controller; test access control; translator assembler-compiler (Philco); type approval code

TACACS terminal access control access control system

Tacc type acceptance

TACI test access control interface

TACIS Technical Assistance to the Commonwealth of Independent States (ex-USSR)

TACL Tandem advanced command language (Tandem *tn*); Telecommunications Analysis Centre Library

TACS television automatic control system, installation de commande automatique par télévision (f); total access communication system (UK cellular radio system)

TAD telephone-answering device; test access digroup; thrust-augmented improved data; transmission automatique de données (f), automatic data transmission

TADIL tactical digital information link

TADS teletypewriter automatic despatch system; tactical automatic digital switching

TADSS tactical automatic digital switching system

TAE Telekommunikation Anschlußeinheit (g) (telecommunications line unit); transportable applications environment; tráfico aleatorio equivalente (s), equivalent random traffic [ERT]

TAED telex automatic emitting devices

TAF technologically-aware family; terminal access facility; terminal adaptation function; time alignment flag

TAG Technical Advisory Group (IEEE); Techniques d'Avant-Garde *tn* (f); terminal audio de grupo (s), group audio terminal [GAT]

TAGF transferencia, acceso y gestión de ficheros (s), file transfer access and management (protocol) [FTAM]

TAHC taux d'appels à l'heure chargée (f), busy-hour call attempts [BHCA]

TAI traitement automatique de l'information (f), automatic data processing; temps atomique international (f), international atomic time

TAL Tandem application language (Tandem *tn*)

TALTC test access termination circuit

TAM telephone answering machine, répondeur téléphonique (f), contestador telefónico (s), Telephonanrufbeantworter (g); test access multiplexer; timer-active monitor

TAMA telegrafía armónica por modulación de amplitud (s), amplitude-modulated voice-frequency telegraphy [AMVFT]

TAMC telegrafía armónica multicanal (s), multi-channel voice-frequency telegraphy [MCVFT]

TAMF telegrafía armónica por modulación de frecuencia (s), frequency-modulated voice frequency telegraphy [FMVFT]

tan tangent; tandem offer

TANE Telephone Association of New England

tan h hyperbolic tangent

TANJ 'There ain't no justice'

TAO object-oriented dialect of LISP; signal d'ordre de transfert autorisé (f), transfer-allowed signal [TFA]

TAOS technology for autonomous operation survivability

TAP technological assistance program; telecommunications application platform; telelocator alphanumeric protocol; telematics applications program; terminal access point; test access path; traitement automatique de la parole (f), automatic speech processing; tool application program

TAPCIS The Access Program for the CompuServe Information Service

TAPDU telematic(s) access protocol data unit

TAPI telephony applications programming interface

tar tape archive (compressed format)

TAR información de tarificación (s), charging [CRG]; technical action request; temporary alternative re-routing; trial ATN routers

TARDIS time and relative dimensions in space

TARP target identifier [TID] address resolution protocol

TARS turn around ranging station

TAS (mensaje de información de) tasación (s), charging [CRG]; Telecommunications Authority of Singapore; Telephone Access Server (WordPerfect *tn*); telephone answering service; test access selector; The Apostle Syndicate; transceiver administration subsystem; trace analysis system

TASA Telefónica de Argentina SA (s)

TASC telecommunications alarm surveillance and control

TASCC test access signalling conversion circuit

TASD The Association of Social Disorder

TASI time-assignment speech interpolation

TASM terminal access state model; turbo assembler (Borland *tn*)

TASO Television Allocations Study Organization

TASP Toll Alternatives Studies Program

Tass Telegraph Agency of the Soviet Union, Agencia Telegráfica de la Unión Soviética (s)

TASS trouble analysis system/subsystem

TAT theoretical arrival time (used in GCRA definition); trans-Atlantic telecommunications/telephone cable; turnaround time

TAU terminal adaptor unit; test access unit

TAWK tiny AWK

TAX taxation (f) (charging); (message de) taxation [CRG]

TAXI Transparent asynchronous transmitter/receiver interface

TAZ transient absorber

Tb terabit, 10^{12}; terbium (used in lasers and solid-state devices)

TB tail bits; terabyte, 10^{12}, one million million; terminal broadcast; transmission buffer; transparent bridging

TBA to be advised/announced

TBC time-based competitiveness; time-based corrector; tren de bits continuo (s), continuous bit stream-oriented [CBO]

T-BCSM terminating – basic call state model

TBD to be determined

TBDF trans-border data flow

TBE transient buffer exposure

TBF transfer on busy fixed

TBGA tape ball grid array

TBINKA Technische Betriebsleitung Internationale Kreditkarten (g), technical operations office international credit cards

TBK tool builder kit

tbl trouble (*ie* a fault)

TBP transmission break protection

TBR technical basis for regulation; transport block reject

TBRL terminal balance return loss; test balance return loss

TBS tape back-up system; to be specified

TBT technology-based training

TBU tape back-up unit; terminal buffer unit

TBV transfer on busy variable

Tc committed rate measurement interval

T_C sampling interval

TC T-carrier; technical committee; temperature coefficient; terminaison de commutateur (f); terminal de central (f), terminación de central (s); terminal de commutation (f), exchange terminal [ET]; terminal congestion; terrestrial channel; tertiary centre; test control; thermocouple; time constant; toll centre; traffic channel; transaction capabilities; transmission control; transmission convergence (sublayer); transmitted signal element timing; transport connection; trunk channel; trunk code; type de codage (f), encoded information type [EIT]

TC PPDU capability data PPDU

TC1 Telecom 1 (French telecommunications satellite)

TCA téléscripteur à commutation automatique (f), automatic teletype exchange; test calibration assembly; Telecentre Association (UK); Tele-Communications Association; time of closest approach; toll-completing (trunk); transport connection accept

TCAD technology computer-aided design

TCAM telecommunications access method

TCAP transaction capabilities application part

TCAS T-carrier administration system

TCB task control block; transceiver cabinet; 2-Weg Fernleitung (g) (trunk circuit bothways); trusted computing base

TCBC changeback control

TCBH time-consistent busy hour

TCC tactical cell controller; target cells and connections; libération de connexion de transport (f), transport connection clear; telecommunications centre; Telecommunications Coordinating Committee; Telecommunications Corporation (Jordan); telephone country code; terminating call control; The Computability Centre (UK); through-connected circuit; transmission control character

TCCB TC common box

TCCD total call connection delay, temps total d'établissement de la communication (f)

TCCF Tactical Communications Control Facility

TCCID total call clear indication delay

TCCN target cells and connections – network

TCCN/U target cells and connections network/user

TCCU target cells and connections – user

TCC PPDU capability data acknowledge PPDU

TCDE control de disponibilidad de los enlaces (s), link-availability control [TLAC]

TCDM tool development cost model

TCE technique de compensation d'écho (f), técnica de compensación de eco (s), echo-cancellation control technique [ECT]; telephone company engineered; temperature coefficient of expansion; transcoding equipment; transit connection element

TCER control del paso a enlace de reserva (s) [TCOC]

TCF TC-user function; technical control facility; training check

TCFA TC-user function area

TCG Teleport Communications Group; test call generator

TCGS Twente (University) compiler generator system

TCH tentatives de prise par circuit et par heure (f), tomas por circuito y por hora (s), seizures per circuit per hour [SCH], bits per circuit per hour [BCH]; traffic channel

TCH/F full-rate traffic channel

TCH/FS full-rate speech traffic channel

TCH/HS half-rate speech traffic channel

TCI trunk circuit incoming, Fernleitungkommend (g); Telecommunications Corporation (US); telewriting code interface; terminales de comunicación informatizados (s), telephone circuit [CCT]; transceiver control interface

TCIC transit centre identification code

TCIF Telecommunications Industry Forum

TCIP time code in picture

TCL temps de confirmation de libération (f), clear confirmation delay [CLCD]; toll circuit layout; tool command

language (Sun Microsystems); traffic class; transportable applications environment (TAE) command language; transverse conversion loss

TC-LAN tightly-coupled LAN

TCLD total call clearing delay

TCLR toll circuit layout record

TCM test call module; thermal control module; time-compression multiplexing/multiplexer (ping-pong transmission); trellis-coded/coding modulation; tunnelling current microscopy

TCMC The Complete Movie Channel (Benelux)

TCMF touch call multifrequency

TCMS toll centring and metropolitan sectoring

TCN throughput class negotiation; topology change notification

TCO message de transfert sous contrôle (f), transfer-controlled message [TFC]; Tjänstemännens Central Organization (monitor safety specifications) – Swedish Confederation of Professional Employees

TCOC change-over control

TCON test connection

TCOS trunk class of service

TCP tape carrier packaging, conditionnement sous film (f); test coordination procedure; transformer-coupled plasma; transmission/(transport) control protocol (RFC 793)

TCP/IP transmission control protocol/internet protocol

TCQAM trellis-coded quadrature amplitude modulation

TCR transport connection request, demande de connexion de transport (f); tagged cell rate; temperature coefficient of resistance

TCRC controlled re-routing control, control de reencaminamiento controlado (s)

TCRF control de reencaminamiento forzado (s), forced re-routing control

TCRF transit connection-related function

TCRS control de retorno al enlace de servicio (s), changeback control [TCBC]

TCS teleconference/(teleconferencing) service; The Criminal (Crime) Syndicate; traffic control subsystem; transmission convergence sub-layer; trusted computer system

TCSEC Tested Computer Systems Evaluation Criteria (US government)

T-CSI terminating CAMEL subscriber information

TCSM terminal communication session manager (TINA)

TCSP tandem cross-section program

TCT terminal control table; toll connecting trunks

TCTL transverse conversion transfer loss

TC-TR technical committee technical report

TCTS Trans-Canada Telephone System

TCU telecommunications control unit; teletypewriter control unit; timing control unit; transmission control unit; trunk coupling unit

TCVXO temperature-compensated voltage-controlled crystal oscillator

TCXO temperature-compensated crystal oscillator

TD table data; temporarily disconnected; terminal digit(s); test distributor; thermal desorption; time delay; transferencia de datos (s), data transfer [DT]; transmit/transmitted data; transmitter-distributor; typed data

TD PPDU presentation data PPDU

TDA toll dial assistance; tunnel-diode amplifier

TDAM technical documentation architecture model

TDAS traffic data administration system

TDB (clase) transferencia de documento en bloque (s), document manipulation class [DM]

TDC tape data controller; temps de demande de la communication (f) [T1]

TDCC Transportation Data Coordination Committee

TDD telecommunications device for the deaf; time-division duplexing; tipo de dirección (s), null-pointer indication [TOA]

TDD/IPN tipo de dirección/indicador de plan de numeración (s), numbering plan indicator/null-pointer indication [NPI/TOA]

TDDB time-dependent dielectric breakdown

TDE terminal display editor

TDEAT tetrakis (diethylamide) titanium

TDEV transport-level deviation

TDF technique de description formelle (f), técnica de descripción formal (s), formal description language [FDT]; telediafonía (s), far-end crosstalk [FEXT]; Télé-Diffusion Français (f); time differential factor; trunk distribution frame

TDFR trayecto digital ficticio de referencia (s), hypothetical reference digital path [HRDP]

TDG taux de gel (*also* blocage partiel) (f), freeze-out fraction [FOF]; traffic destination group

TDI transmission drop/insert board; transit delay indication; transport driver/device interface; two-wire direct interface

TDL temps de demande de libération (f), clear request delay [CLRD]; tono de llamada (s), calling tone [CNG]; tiempo de vida de UDSR distante-local

TDM time-division multiplex/multiplexing; transferencia de mensajes (s), frequency modulation transmitter [FMT]

TDMA time-division multiple access

TDMA/DSI TDMA/digital speech interpolation; time-division multiple access, Vielfachzugriff im Zeitmultiplex (g)

TDMS telegraph distortion measuring set; terminal data/display management system; thermal desorption mass spectrometry; transmission distortion measuring set

TDM-VDMA TDM-variable destination multiple access

TDN (señal de) trayecto digital no proporcionado (s), digital path not provided (signal) [DPN]; tipo de número (s), type of number [TON]

TDNS total data network system

TDP Telefónica de Perú; telelocator data protocol; trigger/triggering detection point

TDP-N trigger detection point – notification

TDP-R trigger detection point – request

TD-PSK time-differential PSK

TDR temporarily disconnected at subscriber's request; terminal digit(s) requested; time domain reflectometer/reflectometry; transmit data register

TDRE tracking and data relay experiment

TDRS tracking and data relay (geostationary) satellite

TDS tabular data stream; thermal desorption spectroscopy; tracking and data acquisition station; transaction-driven system; transit delay selection; tratamiento digital de la señal (s), digital signal processing [DSP]

TDSAI transit delay selection and indication

TDSR transmitter data service request

TDT transport data, données de transport (f)

TDU topology database update

TDX time delay to X; transferencia de datos tipificados (s), typed data transfer

TDY time delay to Y

Te tellurium (used in thermo-electric devices)

TE taux d'erreur (f), error rate; techno-economic evaluation; telematics engineering; terminal equipment (ISDN), équipement terminal (f); transit

exchange; transmitted electron; transverse electric (mode/wave); trunk expansion

TE PPDU expedited data PPDU [ED PPDU]

TE&DA techno-economic evaluation and demand analysis

TE/2 terminal emulator/2 (Oberon)

TE1 terminal equipment type 1 (ISDN capable)

TE2 terminal equipment type 2 (non-ISDN)

TEA transverse excited atmosphere; Telnet external access

TE-A terminal equipment agent (TINA)

TEAM Techno-Economic Ad-hoc Meeting (RACE tools working group)

TEB taux d'erreur sur les bits (f), tasa de errores en los bits (s), bit error rate [BER]; thread environment block

TEBL tasa de errores en los bloques (s), block error rate [BLER]

TEC teleservice code; temps d'établissement de la communication (f), call set-up time [SUT]; test and electrical characterization; thermal expansion coefficient; tiempo de establecimiento de la comunicación (s), call set-up time [SUT]; Tokyo Electronics Corp; total electron content

TECAP transistor electrical characterization and analysis program

Technion Israel Institute of Technology

TECO text editor and corrector (on early DEC equipment)

TED tenders electronic daily system; threshold extension device; transient enhanced diffusion; transmitted electron detection; trunk encryption device

TEDA Techno-Economic evaluation and IBC Demand Analysis (RACE)

TEDIS Trade Electronic Data Interchange Systems (EU)

TEF-A fixed terminal equipment agent (TINA)

TEF-M mobile terminal equipment agent (TINA)

TEG technical exchange group

TEGAS time generation and simulation

TEHO tail end hop off

TEI terminal endpoint identifier; Text Encoding Initiative (US academic project)

TeilVSt Teilvermittlungsstelle (g), partial switching centre

TEIU test equipment interface unit

TEK traffic encryption key

Tel telephony services, Telefondienst (g)

Telam Telenoticias Americanas (Argentina) (s)

telco telecommunications company, opérateur de télécommunications (f), operador de telecomunicaciones (s), operador de telecomunicações (p), Betreiber im Bereich der Telekommunikation (g)

TELCOR Instituto Nicaragüense de Telecomunicaciones y Correos (s)

Telebras Telecomunicacoes Brasilieras (p)

Telecel Comunicações Pessoais (p) (private Portuguese telco)

Telecities EU telematics project

Telecom Telecomunicações dos CTT (Portugal) (p)

telecon teletypewriter conference

Teledesic system of 288 communications satellites (consortium including Bill Gates)

Téléfi Télévision Française Internationale (f)

TELEHUILA Compañía Telefónica del Huila, SA (s)

Telematique EU data communications initiative (*aka* Telematica)

Telemig Telecomunicações de Minas Gerais (Brazil) (p)

TELENARIÑO Empresa de Telecomunicaciones de Nariño (Colombia) (s)

TELENET public switching network in US; Telenet Communication Corp

Telenet Transport and Services (GTE Business Systems)

TELEPAK telephone package

teleran sistema teleran (s), air traffic control

Telesp Telecomunicações de São Paulo (Brazil) (p)

Teletas Telekomunikasyon Endustri Ticaret (Turkey)

TELETEX télétraitement de texte (f), super-telex service

teletext computer-held information via TV signal (*eg* CEEFAX)

teleworking télétravail/travail à distance (f), trabajo a distancia (s), teletrabajo (s), Fernarbeit (g)

Telex CCITT-defined correspondence exchange via teleprinter/telegraph

telg telegram/telegraph

Telidon Canadian videotex system

TELINT telemetry/telecommunications intelligence

Tellytalk Lewisham Council residents' videoconferencing access points

Telmex Teléfonos de México (s)

Telnet teletype network providing terminal emulation

TelOp teleconference operator

TELPAK AT&T private line bulk rate tariff *tn*

TELR talker echo loudness rating

TELRIC total element long-run incremental cost

Telstar first telecommunications satellite, 1962

TEM tampon d'émission (f), transmission buffer [TB]; terminal d'essai maritime (f), maritime test terminal [MTT]; transmission electron microscopy; transverse electromagnetic (mode/wave)

TeMIP Telecom Information Management Platform *tn*

TEML turbo editor macro-language (Borland)

TEMPEST temporary emanation and spurious transmission; transient electromagnetic emanations standard

Tempo original name for Mac OS

TEMS telecommunications management system; test mobile system

TEN Television Education Network (UK); trans-European networks

TENEX time sharing OS (DEC)

TENS transcutaneous electronic nerve simulators

TEO telephone equipment order

TER message de test d'encombrement de faisceau de routes semaphore (f), signalling route set congestion test message [RCT]; tasa de error residual (s), residual error rate [RER]; thermal eclipse reading (Sony); tiempo de espera de una reasignación/resincronización (s), time to wait for re-assignment/re-synchronization [TWR]

tera 10^{12}, one million million

TERENA Trans-European Research and Education Networking Association

term terminal

termcaps terminal capability files

TER-TLM telematic(s) terminal, terminal telemático (s) [TLM-TER]

TES terminal de signalisation (f), terminal de señalización (s), signalling terminal [STE]; test d'état d'un sous-système (f), subsystem status test [SST]; time encoded speech; trials end system

TESA Telefónica de España (s)

TESUG The European Satellite User Group

TeT techno economic-oriented tools

TET text enhancement technology (Epson *tn*); traffic estimation tool

TETRA trans-European trunked radio

TE-wave transverse electric wave

TeX text formatter (macro based)

texel textured picture element (in 3D graphics)

TF messages de transfert interdit ou de transfert autorisé (f), transfer-prohibited and transfer allowed messages (TFM); test/trunk frame; thin film (hard disk drive heads); timing function; Trägerfrequenz (g), carrier frequency;

transferencia fiable (s), reliable transfer [RT]

TF1 Télévision France 1 (f)

TFA transfer-allowed (signal)

TFC mensaje de transferencia controlada (s), transfer allowed signal [TFA]; total fault coverage; traffic; transfer-controlled (signal); transmission fault control

TFDD text file device driver

TFE telephony front-end subsystem

TFEL thin-film electroluminescent (display)

Tflops teraflops, 10^{12} floating point operations per second

TFM transfer-prohibited and transfer-allowed messages

TFMD temps de fonctionnement moyen avant défaillance (f), mean time to failure [MTTF]

TFMS trunk and facilities maintenance system

TFP transfer-prohibited (signal)

TFR theoretical final route; transfer-restricted (signal); transformée de Fourier rapide (f), fast Fourier transform [FFT]

TFRC forced re-routing control

TFS thin-film switching; translucent file service; traffic flow security; traffic forecasting system; Trägerfrequenzsatz (g), carrier frequency unit; trunk forecasting system

TFSG Trägerfrequenzsatz gehend (g) (carrier frequency unit outgoing)

TFSK Trägerfrequenzsatz kommend (g) (carrier frequency unit incoming)

TFT thin-film transistor

TFTP television facility test position; trivial file transfer protocol (RFC 1350)

TFTS terrestrial flight telephone service/system

TFUe Trägerfrequenzsatz-Übertragung (g) (carrier frequency line unit)

TFX toxic effects; Telefaxdienst (g) (fax service)

tg telegraph

TG technical guide; thermo-gravimetry;

titulo global (s), global title [GT]; tone
generator; transceiver group;
transmission group; trunk group;
transcoding gain

TGA Targa graphics adaptor *tn*; thermal
gas analysis; thermal gravimetric
analysis

T-gate ternary selector gate

TGB termination barred; transceiver
group bus

TGC transmitter gain/(group) control

TGF through-group filter

TGID trunk group identification
(number)

TGITM The Ghost in the Machine

TGM titulo global de móvil (s), mobile
global title [MGT]; trunk group
multiplexer

TGMS third-generation mobile system

TGN trunk group number

TGV train à grande vitesse (f) (high-
speed train)

Th thorium (used in photo-electric cells)

TH transmission header

THA transaction handling

THD total harmonic distortion

THENET Texas Higher Education
Network

therm thermistor

THF tremendously high frequency (300
GHz-3 000 GHz)

THG The Humble Guys

thin client PC a PC without internal
data storage (*ie* not self-sufficient)

THIR temperature/humidity infrared
radiometer

THL trans-hybrid loss

THMF télégraphique harmonique à
fréquences vocales avec modulation de
fréquence (f), frequency-modulated
voice-frequency telegraph [FMVFT]

TH-NIX Thailand – National Internet
Exchange

Thomas The House Open Multimedia
Access System (House of
Representatives)

THOR Tandy high-performance optical
reading *tn*

THP terminal handler process; The Hill
People

THR transmit holding register

throbber animated icon in top corner of
Internet browser

THT token-holding timer

T$_{hu}$ temps de traitement par le sous-
système utilisateur téléphone (f), tiempo
de tratamiento de la parte de usuario de
telefonía (s), user handling time

thunk to call 16-bit DLL functions from
32-bit code (*or* vice versa)

THz terahertz

TI task individual; terminación interredes
(s), interwork termination [IT]; terrestrial
interference; Texas Instruments
Corporation; transaction
identifier/identity

TIA signal d'accusé de réception de
transfert interdit (f), transfer-prohibited
acknowledgement signal (TPA);
Telecommunications Industry
Association (US); telematics
interworking application; thanks in
advance; The Internet Adaptor; time
interval analyser

TIAS telematic(s) interworking abstract
service

TIB timing bus

TIC terminal international centre; test
interface controller; token-ring interface
coupler; tipo de información codificada
(s), encoded information type [EIT]

TICK terminating IN category key

TICTAC The Intermittent Connectivity
Technical Advisory Committee

TID target identifier; terminal
identifier/identification, identificateur de
terminal (f); travelling ionospheric
disturbance

TIDF trunk intermediate distribution
frame

TIE señal de envio de tono especial de
información (s), send-special
information-tone message [SST];

terminal interface equipment; time interval error

TIES time-independent escape sequence

TIF telematic(s) interworking facility; telephone interference/influence factor; text image format; text interchange format

TIF.0 text interchange format 0, formato de la imagen de texto 0 (s)

TIF.1 text interchange format 1, text image format 1

TIFF tagged image file format

TIG telegram identification group; telegraph identification group; Topology Information Group (ATM Forum)

TIGA Texas Instruments graphics architecture *tn*

TII technology independent interface

TIM TCP/IP inverse multiplexing protocol; technical information memo (Compaq); Telephone International Media (UK); transmitter inter-modulation; transistor interface module

time code time unit coding system for identification of frames

TIMI technology-independent machine interface (IBM)

TIMS telephone information management system; transmission impairment measuring set

TIN terminating intelligent network subscriber class supplementary service; test d'inactivité (f), inactivity test [IT]

TINA telecommunications information networking architecture

TINA-C Telecommunications Information Networking Architecture Consortium

TIO signal d'ordre de transfert interdit (f), transfer-prohibited signal (TFP)

TIP telecommunications interface processor (EU); terminal interface (message) processors; Texas Instruments Pascal *tn*; transition in progress

TIPHON Telecommunications and Internet Protocol Harmonization over Networks (ETSI) *tn*

TIPS telemetry impact prediction system

TIQ task input queue

TIR total indicator run-out; total internal reflection

TIRA Telematics in Rural Areas (UK)

TIRC Telecommunications Industry Research Centre (UK)

TIRI Telecom Information Resources on the Internet (US)

TIRKS trunks integrated record-keeping system (Bellcore)

TIROS Television and Infra-red Observation Satellite

TIS Tele-Info-Service; terminal interface subsystem; telematic(s) interworking system; tool-induced shift

TIU telematics interworking unit

TJR trunk and router/routing

T_k response timer

TK telecine; test desk; trunk equipment

TKAnl Telekommunikationsanlage (g), telecommunications installation

TKK The Kiwi Killers

TKO Telekommunikationsordnung (g) (telecommunications regulations); trunk offer

TKSyst Telekommunikationssystem (g) (telecommunication system)

Tl Teilvermittlungsleitung (g) (line on partial switching centre)

TL terminaison de ligne (f), terminación de línea (s), line termination [LT]; tie line; transfer layer; transmission level

TL1 transaction language 1 (SONET)

TL-4 trans-local node

TLA three-letter acronym

TLAC link availability control

TLAP TokenTalk Link Access Protocol (Apple *tn*)

TLB translation look-aside buffer

TLC terminal de ligne client; The Learning (TV) Channel; thin layer chromatography; transport layer class (0-4)

TLD telephone line doubler, Zweieranschlußeinheit (g); tiempo de vida de UDSR local-distante (s)

TLE tool loading elevator; two line elements (satellite tracking data)

TLF trunk line frame

TLH transport layer header

TLHC tentativas de llamada en hora cargada (s), busy hour call attempts [BHCA]

TLI transport layer/(level) interface

TLL total scanning line-length, longitud total de la línea de exploración (s); transferencia de llamadas (s), call transfer [CT]

TLM tape-laying machine; telemeter; transition line model; telematics

TLMA telematics agent

TLMAU telematics access unit

TLML theological mark-up language

TLM-TER telematics terminal

Tln Teilnehmer (g) (subscriber)

TLO terminal de ligne optique; thread language zero

TLP transmission level point

TLR toll line release key

TLS Television Library System; transparent LAN service

TLTP trunk line test panel

TLU table look-up; terminal logic unit

TLV threshold limit value; time – length – value (call record format variables)

TLV/TWA threshold limit value/time-weighted average

TLX telex; telex type

TLXAU telex access unit

TM système de transfert de messages (f), message transfer system [MTS]; technical manual; terminal maritime (f), terminal marítimo (s), maritime terminal [MT]; terminal mobility; terminal multiplexer; thermal mechanical analyser; time modulation/timing module; traffic management; traitement des messages (f), tratamiento de mensajes (s), message handling [MH]; transfer mode; transfert de messages (f), message transfer [MT]; transmission and multiplexing; transport module;

transverse magnetic wave/mode; Turing Machine

T-M time-to-market

TMA Telecommunications Managers Association (UK); tower-mounted amplifier

TMAI tiempo medio acumulado de indisponibilidad (s), mean accumulated down time [MADT]

TMAX maximum time

TMBF temps moyen de bon fonctionnement (f), mean time before failure [MTBF]

TMC telecommunication management centre; Telematics Management Committee; Télévision Monte Carlo (f); traffic measurement on cells; transfer module controller; transmission maintenance centre; transport module controller

TMCB timing module connection board

TMD temps moyen de disponibilité (f), tiempo medio de disponibilidad (s), mean up-time [MUT]; transferencia y manipulación de documentos (s), document transfer and manipulation [DTAM]

TMDB clase transferencia y manipulación de documento en bloque (s), document bulk transfer and manipulation class [DBM]

TMDF trunk main distribution frame

TMDL target machine description language

TMDR tiempo medio de reparación (s) [MTTR]

TMDS transition-minimized differential signalling

TME Tivoli Management Environment; trunk maintenance files

TMEF tiempo medio entre fallos (s), mean time between failures [MTBF]

TMF teclado multifrecuencia (s), multi-frequency push-button [MFPB]

TMG temps moyen de Greenwich (f), Greenwich Mean Time (GMT)

TMI Telemedia International (Italy);

temps moyen d'indisponibilité (f), tiempo medio de indisponibilidad (s), mean down-time [MDT]

TMIA tiempo medio de indisponibilidad acumulado (s), mean accumulated down-time [MADT]

TMIIA tiempo medio de indisponibilidad intrínseca acumulado (s), mean accumulated intrinsic down-time [MAIDT]

TMIN minimum time

TML tandem matching loss; teléfono manos libres (s), hands-free telephone [HFT]; total mass loss

TMMS telephone message management system

TMn traffic mix n(=1-4)

TMN telecommunications management network (ITU-TS M.3010)

TMOS Telecommunications Management and Operations Support *tn*; T metal-oxide semiconductor transistor

TMP teleprinter multiplexer, multiplexeur de téléimprimeur (f); terminal management processor (National Semiconductor); terminal marítimo de pruebas (s), maritime test terminal [MTT]; test management protocol; thermal magnetic duplication

TM-PDU test management PDU

TMPFS temporary file system

TMP-IL transmission maintenance point (international line)

TMPS Telecommunications Management and Planning Service (EU)

T$_{Mr}$ message transfer part receiving time, temps de réception de sous-système transport de messages (f), tiempo de recepción de la parte de transferencia de mensajes (s)

TMR terminal radar; test measurement receiver; transmission medium requirement

TMRS traffic measuring and recording system

T$_{Ms}$ message transfer part sending time, temps d'émission du sous-système

transport de messages (f), tiempo de emisión de la parte de transferencia de mensajes (s)

TMS telecommunications message switch; time-multiplexed switch/switching; traffic measurement subsystem; transmission measuring set; transport management system; tratamiento de mensajes de señalización (s), signalling message handling [SMH]

TMSC transit and combined MSC (mobile switching centre)

TMSCP tape mass storage control protocol

TMSF tracks, minutes, seconds, frames – time format audio CD

TMSI temporary mobile station/subscriber identity

TMT terrain modelling tool; testing methods and techniques

TMTA temporary mobile terminal address

TMTI temporary mobile terminal identifier/identity

TMU transmission medium used, transmission message unit

TMUX-P type P transmultiplexer, transmultiplexeur de type P (f), transmultiplexor de tipo P (s)

TMUX-S type S transmultiplexer, transmultiplexeur de type S (f), transmultiplexor de tipo S (s)

Tmx Temexdienst (TEMEX service)

Tn temporizador [Tn]

TN Telnet (protocol); time-slot number; tone; transit node; transport network; twisted nematic

TN3270 variation of Telnet connecting to an IBM mainframe

TNA/IDA Telematics Network for Administration/Interchange of Data between Administrations

TNAS Total NET Administration Suite

TNBT transmission numérique dans la bande téléphonique (f) [DIV]

TNC terminal national centre; terminal node controller

TNDS total network data system

TNE terminal numérique d'extrémité (f)

T-network two equal series arms linked by shunt arm

TNF transfer on no reply – variable

TNL terminal net loss; terminal numérique de ligne (f)

TNLO terminal numérique de ligne optique (f), optical line digital terminal

TNM mux-demux terminal (interface); telecommunications network management

TNOP total network operations plan

TNR terminaison numérique de réseau (f), digital network terminal

TNRN terminating network routing number

TNRO terminal numérique récepteur optique (f), optical receiver digital terminal

TNS telephone network simulator; Tolmes News Service; transaction network service; transit network selection

TNSV transmission numérique supravocale (f), data over voice [DOV]

TNT Turner Network Television

tnx thanks (also TKS)

To time-out

TO transistor outline package; technical order; telecommunications; operator traffic order

TOA take-off angle; time of arrival; type of address

TOAO The One and Only

Toast CD-R software (Adaptec *tn*)

TOC table of coincidences; table of contents; task-oriented costing; television operating centre; total organic/oxidizable carbon

TOCC technical and operational control/coordination centre

TOD time of departure

TODC time of day clock

TOEM technical original equipment manufacturer

TOF time of flight; time out factor; tone off

TOFF tagged object format file

TOH transport overhead (SONET)

TOL transverse output level

ToLC thread-of-life communications

TOLD telecoms on-line data (system)

TOLL Tariff On Line (TMA Ventures *tn*)

TOLR toll restricted; transmitting objective loudness rating

TOM transaction-oriented money-maker

TON tone off; type of number

TONLAR tone-operated net loss adjuster

T-Online Deutsche Telekom access service

TOOLS technology of object-oriented languages and systems (ISE)

TOP technical and office protocol (OSI TP-4); tube à ondes progressives (f), travelling-wave tube [TWT]

TOPES telephone office planning and engineering system

TOPP task-oriented plant practice

TOPS software allowing PCs and Macs to share files *tn*; tape and optical products; traffic operator position system

TOR telegraph on radio; teleprinting over radio circuits; term of reference; top of range

TORC (automatic) traffic overflow re-route control

TOS OS for Atari ST computer; taken out of service; temporarily out of service; time-ordered system; Tiros operational satellite; type of (IP) service

TOSD Telephone Operations And Standards Division (Rural Electrification Administration, US)

TOT Telephone Organization of Thailand; transfer of technology

TOVS Tiros operational vertical sounder

TOW tube-launched optically-tracked wire-guided (missile)

tp telephone

TP polling time (between ICMP bursts); tele-remote processing; téléinformatique (f); teleprocessing, télétraitement (f);

terminal pole/portable; terminal portability; test position; toll point/prefix; traitement à distance (f), teleprocessing; transaction processing/program; transport protocol (ISO); tratamiento de paquetes (s), packet handling [PH]; two-procedures modem (TP modem)

TP-4 transport protocol 4

TP monitor transaction processing machine

TPA toll pulse accepter; transfer-prohibited acknowledgement (signal)

T-PAD terminal packet assembler/disassembler

TPAU twisted-pair attachment unit

TPC telecom purpose computer; The Phoney Coders; Transaction-processing Performance Council; trans-Pacific cable

TPCC third-party call control

TPD temperature program desorption

TPDR temps de propagation différentiel restreint (f), tiempo de propagación diferencial restringido (s), restricted differential time delay [RDTD]

TPDU AK data acknowledgement TPDU (f) [AK TPDU]

TPDU CC connection confirm (f) [CC TPDU]

TPDU CR connection request (f) [CR TPDU]

TPDU DC disconnect confirm TPDU (f) [DC TPDU]

TPDU DT data TPDU (f) [DT TPDU]

TPDU EA expedited acknowledgement TPDU (f) [EA TPDU]

TPDU ED expedited data (f) [ED TPDU]

TPDU ER error TPDU (f) [ER TPDU]

TPDU RJ reject TPDU (f) [RJ TPDU]

TPDU transfer/(transport) protocol data unit

TPE terminal portable d'exploitation (f); transmission path endpoint

TPF text processable format; time prism filter; transaction processing facility (IBM *tn*); transmission priority field

TPG test pattern generation

tpi tracks per inch

TPI tone protection information; transmission performance index; transport provider interface

TPL table-producing language; template (Microsoft Mail Post Office); terminal per line; toll pole line; transaction processing language

TPM third-party maintenance; total productive maintenance/manufacturing; transactions per minute

TP-MIC twisted-pair media interface connector

TPNS terminal performance network simulation (IBM)

TPO telecommunications program objective; twisted-pair only

TPON telephony over a passive optical network (UK)

TPORT twisted-pair port transceiver (AT&T)

TPP telephony pre-processor

TP-PMD twisted-pair physical media dependent

TPR taux de prises avec réponse (f), answer seizure ratio [ASR]; terminal probado (s), terminal under test [TUT]

TPRS temperature-programmed reaction spectroscopy

TPS Telephone Pioneers of America; Telephone Preference Service (UK); three-party service (3PTY); transaction processing system; transactions per second

TPSO taux de prises avec signal d'occupation (f) [BFSR]

TPT test progress tone

TPTB the powers that be

TPTD total data packet network transfer delay

TPU text processing utility (DEC *tn*); thermal processing unit; transceiver power unit

TPW Turbo Pascal for Windows

TQ total quality

TQC technical quality control

TQM total quality management

TQPF thin-quad flat pack

TQS transmission quality supervision

tr transpose

TR message de test de faisceau de routes sémaphores (f), signalling route set test message [RSM]; telegram to be called for; terminal ready; terminaison de réseau (f), terminación de red (f), network termination [NT]; tip/ring, token ring; traffic route; transfer restricted; transmission report; transaction; trouble report

TR1 terminación de red (de tipo) 1 (s), network termination (type 1) [NT 1]

TR2 terminación de red (de tipo) 2 (s), network termination (type 2) [NT 2]

TRA señal de autorización de transferencia (s), transfer-allowed signal [TFA]; temps de réponse de l'abonné (f), subscriber response time [SRT]; traffic restart-allowed signal; traffic routing administration; transcoder rate adaptor; trouble report answer

TRA1 radar transfer of control message, message de transfert de contrôle radar (f)

TRAC Telecommunications Regulations Application Committee (EU); telefonía radio(-electrica) por aceso celular (s), cellular-access radiotelephony

TRAD traducteur (f), translator

TRAM tools for radio access management; transcoder rate adaptor magazine

TRAMPS traffic measurement and path search

TRAN Computer Transmission Corporation of California

Transdyn US corporation: network management and systems provider

TRANSEC transmission security

Transfix sub-rate point-to-point/multipoint digital private line service

TRANSMIC French digital leased lines service

TRANSPAC French packet-switched network

TRANSPLAN transaction network service planning model

transputer modular programmable VLSI device with communication links

transzorb transient absorber

TRAP tandem recursive algorithm process

TRAU transcoder/transceiver rate adaptor unit

TRAVIS traffic retrieval analysis validation and information system

TRC transcoder controller; transverse redundancy check; tubo de rayos catodicos (s), cathode ray tube [CRT]

TRCC signalling route set congestion control; 'T' carrier restoration control centre

TRF receiver tuned radio frequency receiver

TRF transformada rapida de Fourier [FFT]; tuned radio-frequency

trfr transfer

TRG Telecommunications Research Group (US)

TRH transceiver handler

TRHM transceiver handler magazine

TRI tableau de raccordement intérieur (f); The Remote Informer; transmission remote interface

TRIB transmission rate of information bits

triniscope colour display system (involving three separate CRTs)

Trinitron three-colour TV tube design (Sony *tn*)

TRI-TAC tri-services tactical (equipment)

TRL Telecom Research Laboratories (Australia); transistor-resistor logic; transverse return loss; trunk register link

TRLC tasa de reclamaciones/llamadas completadas [CTCR]

TRM traffic-restart-allowed message; transceiver module; transferencia de mensajes (s), message transfer [MT]

TRN transfer – no reply

TRO tail recursion optimization; terminal

de réseau optique (f), optical network terminal

trombone adjustable U-shaped waveguide transmission line

Tron The Real-time Operating system nucleus (Hitachi *tn*)

TROV telepresent remotely operated-vehicle

TRP TMOS running platform

TRPC transaction remote procedure call

TRS signal de test de faisceau de routes sémaphores (f), signalling route set test signal [RST]; term rewriting system; toll room switch; transceiver subsystem; triple signal strength receiver

trsp transpose

TRSS token ring subsystem

TRT tampon de retransmission (f), retransmission buffer [RTB]; text-retrieval technique; token rotation timer; tone receiving unit; traffic route testing; Turkish Radio Television Corp

TRU transceiver unit

TRUD transceiver unit digital part

TRUSIX Trusted Unix OS

TRV terminal de representación video (s), visual display terminal [VDT]

TRX transceiver (transmitter-receiver)

TR-x Travan standard Nº x

TRXC transceiver controller

TRXCONV transceiver (DC/CD) converter

TRXD transceiver digital

TRXT transceiver tester

TS tara de sección (s), section overhead [SOH]; technical specification/standards; telematic(s) service; teleservice; terminal de signalisation (f), signalling terminal equipment; time slot; time stamp; time switch; toll switch(-ing trunk); Toxic Shock; traffic shaping; traffic station; trame sémaphore (f), signal unit [SU]; transport service; transport stream; Typed Smalltalk

Ts1 Teilnehmerschaltung (g) (subscriber line unit on switching centre)

TS16 time slot 16

TSA telecommunications service analysis; time slot assignment, zeitliche Zuordnung DE-Datenkanäle (g)

TSAC time slot assignment circuit

TSAN The Sysops Association Network; TriState Area Sysops Network (US)

TSAP transport service access point

TSAPI telephony services application program interface

TSAP-ID transport service access point identifier

TSAT T-1 rate small aperture terminal

TSAU time slot access unit

TSB termination status block; twin sideband

TSC Telecommunications Security Council; telecommunications service code; test system controller; time slot code; training sequence code; transit switching centre

TSDI transceiver speech and data interface

TSDU transport service data unit

TSE state signalling frame; trame sémaphore d'état du canal sémaphore (f), link status signal unit [LSSU]

TSF telegraphie sans fil (French radio); through-supergroup filter

TSFA traffic simulation function area

TSFC signalling traffic flow control

TSFS trunk servicing forecasting system

TSG time-slot generator

TSI signal d'envoi d'une tonalité spéciale d'information (f), send-special-information-tone signal [SST]; time slot interchange; Tivoli Systems Inc (USA); transmitting subscriber identification

TSK transmission security key

TSL telecom services licences (UK); total service life

TSLRIC total service long-run incremental cost

TSM timer standby monitor; time switch module; trame sémaphore de message (f), message signal unit

TSO telecommunications service order; telephone service observations;

terminating screening office; time-sharing option

TSO/E time-sharing option/extensions

TSOP thin small outline package

TSORT transmission system optimum relief tool

TSP Telecommunications Service Priority system; temperature-sensitive parameter; traffic service position; travelling salesman problem; Trusted Service Provider (monitoring encryption keys, UK)

TSPS traffic service position system

TSPSCAP traffic service position system real-time capacity program

TSR telecommunications service request; terminate-and-stay-resident (program); trame sémaphore de remplissage (f), fill-in signal unit/frame [FISU]; triple signal-strength receiver

TSRC signalling routing control

TSS tangential signal sensitivity; task state segment; Telecommunication Standardization Sector (ITU-T); time-sharing system; toll switching system; trunk and signalling (sub-)system; trunk servicing system

TSS-C transmission surveillance system – cable

TSSDU typed data session service data unit

TSSI time-slot sequence integrity

TSSST time-space-space-space-time (division switching)

TSST time-space-space-time (division switching)

TST time switch time; time-space-time (division switching)

TSTA transmission signalling and test access

TSTE Texas Society of Telephone Engineers

TSTN triple super-twisted nematic

TSTOA time-slot trunking over ATM (asynchronous transfer mode)

TSTPAC transmission and signalling test plan and analysis concept

TSTS time-space-time-space

TSU tandem signal unit; tape search unit; telephone signal unit; test signal unit

TS-USER utilisateur de service de transport (f), transport service user

TSV tab-separated value; test status verification

TSW tele-software; time switch; time switch-board (extension module)

TT tara de trayecto (s), path overhead [POH]; technology transfer; telegraphic transfer; teletype; teletypewriter terminal; temporarily transferred; text transfer, transfert de texte (f), transferencia de texto (s); toll ticketing; traffic tester; translation type

TT&C tracking telemetry and control

TT/N test tone-to-noise ratio

TTA terminating transport area; traffic trunk administration; transport-triggered architecture

TTB TMOS tool box; toll test-board

TTC teletypewriter centre; terminating toll centre; time to custom; transaction capabilities; transmit through-connect signal, signal de connexion de transit (s)

TTCH tentativas de toma por circuito y por hora (s), bids per circuit per hour [BCH]

TTCN tree and tabular/tabulator combined notation

TTD signal de connexion des centres de transit (f), transit centres through-connected signal; télétransmission des données (f), data tele-transmission; temporary text delay; transit centres through-connected signal

TTE señal de transferencia de tráfico de emergencia (s), emergency load transfer signal [ELT]; talk time effect

TTF TrueType Font *tn*

TTFN ta-ta for now

TTFS Teilnehmer-Trägerfrequenz-system (g) (subscriber carrier frequency system)

TTL 'time to live' field (0-255 hops on Internet connections); transistor-transistor logic, logique transistor-

transistor (f), lógica transistor-transistor (s); transverse transfer loss

TTM test transmitter

TTNIC temporary telex network identification code

TTO traffic trunk order

TTP thermal-transfer printer/printing; timed token protocol; trusted third party (official encryption key custodian)

TTPR taux de tentatives de prise avec réponse (f), tasa de tomas con respuesta (s), answer-bid ratio [ABR]; time to try re-assignment/resynchronization, tiempo de tentativa de reasignación resincronización (s) [TTR]

TTRT target token-rotation time

TTS Telerate Trading System (UK); tele-typesetter; text-to-speech; Transaction Tracking System (Novell)

TTSD Messeinrichtung für digitale Übertragung (g) (transmission test system digital)

TTTN tandem tie trunk network

TTTR tasa de tentativas de toma con respuestas (s), answer-bid ratio [ABR]

TTV total thickness variation

Ttx Teletexdienst (g) (teletex service)

TTX teletext

TTXAU teletext access unit

TtxE Teletex-Endeinrichtung (g) (teletex terminal)

TtxFestVerb Ttx-Festverbindung (g) (teletex point-to-point link)

TTY teletypewriter (Teletype Corp *tn*), teleimpresora (s)

TTYL talk to you later

TTYS tele-typesetters

TU tape unit; test unit, Testeinheit (g) (timing unit)

TUA Telecommunications Users' Association (UK)

TUB Technische Universität Berlin

TUBA TCP and UDP with Bigger Addresses

TUC total user cell count

TUCC Triangle University Computing Centre

TUCD total user cell difference

TUCOWS The Ultimate Collection of Winsock Software (on Internet) *tn*

TUF Telecommunications Users' Foundation

TUFF The Underground Fone Foundation

TUG tape unit group; Telephone Users' Group (UK); TeX Users' Group

TUGOS The United Guild of Sysops

TUI text-based user interface (WordPerfect)

TUIMG Telecom/Utilities Industry Marketing Group

Tulip electronic format for technical journal publication (Elsevier Science *tn*)

TUMS table update and management system (Stanford University)

TUP telephone user part (SS7)

tuple set of related values (in database table)

TUPLE Toyohashi University Parallel Lisp Environment

TUPS technical user performance specifications (USITA)

TUR traffic usage recorder

Turingol language for programming a Turing Machine

turtle actual/virtual drawing device used by LOGO

TüV Technische überwachungs Verein (Germany) (g)

TV TeleVoting Service, Televotum; television, télévision (f), televisión (s); type and value

TVB Television Broadcasts (Hong Kong TV station)

TVC thermal voltage converter

TVE Television Trust for the Environment, (UK); Televisión de España (s)

TVFS Toronto virtual file system (IBM)

TVHD télévision haute définition (f), high-definition TV [HDTV]

TVI television interference; terminal virtual (s), virtual terminal [VT]

TVM time voltage minimum (ferro-electrics)

TVN Television Norge (Norway)

TVNZ Television New Zealand

TVOL television-on-line

TVR television recording

TVRO television receive-only (station), estación(es) de televisión con recepción únicamente (s)

TVS triangular voltage sweep; Televotum service

TVSt Teilnehmervermittlungsstelle (g) (subscriber-line switching centre)

TW transit working; transmitter window; travelling wave

TW antenna travelling wave antenna

TWA time-weighted average; The Warez Alliance (.alt presence on BBS); two-way alternate, mode d'échange bidirectionnel à l'alternat (f)

TWAIN technology without an interesting name

TwB Tastenwahl-Betrieb (g) (key-dialling operation); Tonwahl-Betrieb (g) (multi-frequency code dialling)

TWCB The Whacko Cracko Brothers

TWDD two-way delay dial

TWEB transcribed weather broadcast

TWERLE tropical wind energy conversion reference level experiment

TWG Technical Working Group

TWID two-way immediate dial

TWIMC to whom it may concern

TWR TransWorld Radio

TWS two-way simultaneous, bidirectionnel simultané (f)

TWT travelling-wave tube

TWTA travelling-wave-tube amplifier

TWWS two-way wink start

TWX1 Teletype Exchange Service, Téléscripteur à commutation automatique (Canada) *tn* (f)

Tx Telexdienst (g) (telex service); transmitter/transmission

TX terminating toll operator; transit exchange; transmit, transmitir (s); transmitter; transmission

TXBP transmitter bandpass filter

TX-BUS transmitter bus

TXC transmission control

TX-cell throughput accelerator

TXCMB transmitter combiner

TXD telephone exchange digital; transmit/transmitted data; transmitter divider

TXE telephone exchange electronic; transmitter empty

TxHAs Telex-Hauptanschluß (g) (telex exchange line)

TXK telephone exchange crossbar

TXn charging category based on source

TxNSt Telex-Nebenstelle (g) (telex extension)

TxNStAnl Telex-Nebenstellenanlage (g) (telex PBX)

TXPA transmitter power amplifier

TXPWR transmit/transmitter power

TXRF total X-ray fluorescence

TXS telephone exchange (Strowger)

TXT2ST text to structured file (Lotus)

Tymnet public data packet-switching network (McDonnell Douglas)

TYMNET Timeshare Inc Network (commercial network in US) *tn*

typ typical

TyP terminación y partida (s) [T and L]

TZ transmitter zone

TZS Ton-zweiersatz (g) (tone 2-party circuit)

U

U unclassified; unit; unnumbered; up-link; U reference point (ISDN); utility

U(t) instantaneous availability, indisponibilité (instantanée) (f), indisponibilidad instantánea (s)

U/L universal/local; up-link

U/S ultrasonic

UA unbalanced asynchronous; unnumbered acknowledgement; user agent, agente de usuario (s); Underground America; unité d'accés (f), unidad de acceso (s), access unit [AU]

UAA United Anarchists of America

UACB user adaptor control block

UAD unidad administrativa (s), administrative unit [AU]

UAE unrecoverable application error; user agent entity

UAEF unidad de acceso de entrega física (s) [PADU]

UAHM user authentication handler mobile terminal

UAHN user authentication handler network

UAI Union Astronomique Internationale (f)

UAL unité arithmétique et logique (f) [ALU]; user agent layer

UAMPT African and Malagasy Postal and Telecommunications Union

UAN universal access number service; user access node

UAP user application

UAPDU user agent protocol data unit

UAPDU IM IP-message user agent protocol data unity

UAPDU SR status report user-agent protocol data unit

UAR unidad de almacenamiento y retransmisión (s), store-and-forward unit [SFU]

UARMS unidad de almacenamiento y retransmisión marítima por satélite (s), maritime satellite store-and-forward unit [MSSFU]

UARP unité d'accès de remise physique (f), physical delivery access unit [PDAU]

UART universal asynchronous receiver/transmitter

UARTO United Arab Republic Telecoms Organization (Egypt)

UAS unavailable seconds; unidad aislada de señalización (s), lone signal unit [LSU]

UATLM unidad de acceso telemático (s), telematic access unit [TLMAU]

UATLX unidad de acceso al télex (s), telex access unit [TLXAU]

UATLXP unité d'accès télex public (f), unidad de acceso al telex público (s), public telex access unit [PTLXAU]

UATTX teletex access unit

UAU user-to-user information (message), message d'information d'usager à usager (f)

UAWG Universal ADSL Working Group (Compaq, Intel, Microsoft)

UAX unit automatic exchange

UBA unblocking acknowledgement signal, signal d'accusé de réception de déblocage (f)

UBE unsolicited bulk e-mail

UBI Universalisé Bidirectionnalité Interactivité (f) (Videotron *tn*)

ubl unblocking

UBL unblocking message/signal, signal de déblocage (f)

UBM under-bump metallurgy; unsuccessful backward-set-up acknowledgement message

UBR unspecified bit rate (ATM)

UBS user behaviour statistics

UC unité de conversion (f), unidad de conversión (s), conversion facility [CF]; usage context; user-class character

UCA utility communications architecture

UCAID University Corporation for Advanced Internet Development (US)

UCB United Christian Broadcasters

UCCD user-dependent call connection delay

UCCID user-dependent call clear indication delay

UCD unidades de conmutación distantes (s), remote switching units [RSU]; uniform call distributor

UCE unsolicited commercial e-mail

UCF Underground Computing Foundation; Untouchable Cracking Force

UCIC unequipped circuit identification code message

UCL universal communications language; upper confidence limit; upper control limit

UCM unité de contrôle multipoint (f), unidad de control de conferencia multipunto (s), multipoint conference unit [MCU]; universal cable module

UCO user profile check – originating network

U-CODE universal (Pascal) code

UCP update control process

UCR under-colour removal

UCS Unicode conversion support; unidad de conmutación de soporte (s), bearer switchover unit [BSU]; uniform communication standard; universal character set (ISO/IEC); universal communications standards; universal multiple-octet coded character set; Unix SCF (service control function) subsystem; user coordinate system

UCSD University of California San Diego – version of Pascal

UCSD p-system University of California San Diego OS

UCSG Universities and Colleges Software Group

UCT unité centrale de traitement (f), central processor unit; Universal Coordinated Time

UCV user profile check – visited network

UCW Union of Communication Workers (UK)

UD do not know if the party will be there today

UDA unidad de distorsión de atenuación (s), attenuation distortion unit [ADU]; unidad de distorsión de cuantificación (s), quantizing distortion unit [QDU]; universal data access (Microsoft specification); user-defined commands

UDD unconstrained data delay

UDF universal disk/drive format; user-defined functions

UDFS usuario fuera de servicio (s)

UDG user-defined gateway

UDH user profile data – home network

UDI unrestricted digital information (ISDN); user data interface

UDM usage data management

UDMA ultra-direct memory access

UDMPU unidad de datos de protocolo de mensaje de usuario (s), user message protocol data unit

UDP unidad de datos de protocolo (s), protocol data unit [PDU]; Usenet death penalty (post-flaming); user datagram protocol (RFC 768); user data part

UDPA unidad de datos de protocolo de aplicación (s), application protocol data unit [APDU]

UDPAT unidad de datos de protocolo de acceso telemático (s), telematic access protocol data unit [TAPDU]

UDPAU-IE status report user-agent protocol data unit (s)

UDPF unidad de datos de protocolo físico (s), physical protocol data unit [PhPDU]

UDP-GP UDP gestión de pruebas (s), test management PDU [TM-PDU]

UDPM unidad de datos del protocolo de mensajes (s), message protocol data unit [MPDU]

UDPMS service message protocol data unit (s)

UDPO operation protocol data unit (s)

UDPP unidad de datos de protocolo de presentación (s), presentation protocol data unit [PPDU]

UDPP AAC UDPP acuse de alteración de

contexto (s), alter context PPDU [AC PPDU]

UDPP AC UDPP alteración de contexto (s), alter context PPDU [AC PPDU]

UDPP ADC UDPP aceptación conexión presentación (s), presentation data PPDU [TD PPDU]

UDPP ARS UDPP acuse de resincronización (s), re-synchronize PPDU [RS PPDU]

UDPP CP UDPP conexión presentación (s), connect presentation [CP PPDU]

UDPP DA UDPP datos acelerados (s), expedited data PPDU [ED PPDU]

UDPP DC UDPP datos sobre capacidades (s), capability data PPDU [TC PPDU]

UDPP DP UDPP datos de presentación (s), presentation data PPDU [TD PPDU]

UDPP LAP UDPP liberación anormal por el proveedor (s), abnormal release provider PPDU [ARP PPDU]

UDPP LAU UDPP liberación anormal por el usuario (s), abnormal release user PPDU [ARU PPDU]

UDPP RCP UDPP rechazo conexión presentación (s), connect presentation reject PPDU [CPR PPDU]

UDPP RS UDPP resincronización (s), re-synchronize PPDU [RS PPDU]

UDPR unidad de datos de protocolo de red (s), network protocol data unit [NPDU]

UDPS unidad de datos de protocolo de sesión (s), session protocol data unit [SPDU]

UDPT unidad de datos de protocolo de transporte (s), transfer protocol data unit [TPDU]

UDPT AA UDPT de acuse de recibo de datos acelerados (s), expedited acknowledgement TPDU [EA TPDU]

UDPT AR UDPT de acuse de recibo de datos (s), data acknowledgement TPDU [AK TPDU]

UDPT CC UDPT de confirmación de conexión (s), confirm connection TPDU [CC TPDU]

UDPT CD UDPT de confirmación de desconexión (s), disconnect confirm TPDU [DD TPDU]

UDPT DA UDPT de datos acelerados (s), expedited data TPDU [ED TPDU]

UDPT DT UDPT de datos (s), data TPDU [DT TPDU]

UDPT ER UDPT de error (s), error TPDU [ER TPDU]

UDPT PC UDPT de petición de conexión (s), connection request TPDU [CR TPDU]

UDPT PD UDPT de petición de desconexión (s), disconnect request TPDU [DD TPDU]

UDPT RCH UDPT de rechazo (s), reject TPDU (RJ TPDU]

UDS service-data unit (s) [SDU]

UDSED unidad de datos del servicio de enlace de datos (s), data link service data unit [DLSDU]

UDSL ultra-high data-rate digital subscriber line; universal data/digital subscriber line/loop

UDSP unidad de datos de servicio de presentación (s), presentation service data unit [PSDU]

UDSP unidad de datos de servicio de red (s), network service data unit [NSDU]

UDSR network service data units (f) [NSDU]

UDSRA unidad de datos del servicio de red, datos acelerados (s), expedited data network service data unit [ENSDU]

UDSS unidad de datos de servicio de sesión (s), session service data unit [SSDU]

UDSSA unidad de datos de servicio de sesión, datos acelerados (s), expedited data session service data unit [XSSDU]

UDSSN unidad de datos de servicio de sesión, datos normales (s), normal data session service data unit [NSSDU]

UDSST unidad de datos de servicio de

sesión, datos tipificados (s), typed data session service data unit [TSSDU]

UDST unité de données du service de transport (f), unidad de datos del servicio de transporte (s), transport service data unit [TSDU]

UDT uniform data transfer; Unit DaTa; unitdata; universal data-flow and telecommunications; unstructured data transfer; usability design target

UDTS Unit DaTa service; universal data transfer service

UDTX discontinuous transmission up-link

UDUB user-determined user busy

UDV user profile data visited network

UDVM universal data voice multiplexer

UE ulterior estudio (s), further study [FS]; user element

UEC Université d'Eté de la Communication (d'Hourtin) (f)

UEE unidad de señalización del estado de enlace (s), link status signal unit [LSSU]

Ue-g Übertragung gehend (g) (line unit outgoing)

UEL upper explosive limit

UEM unidad de enlace multipunto (s), multipoint junction unit [MJU]

UER Union européenne de radio-télévision (f) (European Broadcasting Union); unité d'enregistrement et retransmission (f), store-and-forward unit [SFU]; unité d'exploitation réseau (f)

UERMS unité d'enregistrement et retransmission du service maritime par satellite (f), maritime satellite store and forward unit [MSSFU]

UES usuario en servicio (s)

UF ultra-filtration; unframed

UFC user-part flow control messages

UFI upstream failure indication

UFL upper flammable limit

UFO UH frequency follow-on (US DoD satellite)

UFS UNIX file system; usuario fuera de servicio (s)

UG underground; user group

UGrl Untergruppen-Vermittslungsleitung (g) (sub-group switching centre line)

UgrVSt Untergruppen-Vermittlungsstelle (g) (sub-group switching centre)

UGW Umsteuergruppenwähler (g) (route-switching group selector)

UHF ultra-high frequency (300-3000 MHz)

UHP ultra-high purity

UHV ultra-high vacuum

UI unité d'interfonctionnement (f), unidad de interfuncionamiento (s), interworking unit [IWU]; unnumbered information, información no numerada (s); unit interval; Unix International; user interface

UIC user identification code

UID user identification/identifier

UIDL unique ID listing (POP3)

UIF unité d'interfonctionnement (f), unidad de interfuncionamiento (s), interworking unit function [IWU]; United Ignorance Front; user information functionality

UIL user interface language

UIM user identity module

UIMS user interface management system

UIN universal Internet number

UIS unidad inicial de señalización (s), initial signal unit [ISU]; universal information services (AT&T ISDN concept)

UISG Utility Industry Standards Group

UIT Union Internationale des Télécommunications (ITU - Genève) (f), Unión Internacional de Telecomunicaciones [ITU]

UITLM unidad de interfuncionamiento telemático (s), telematic(s) interworking unit [TIU]

UITS unacknowledged information transfer service

UIT-T Union Internationale des Télécommunications – Téléphonie (f)

UIV unidad de interfaz videotex (s), videotex interface unit [VIU]

UJT unijunction transistor

UKB universal keyboard

UKCCD United Kingdom Council for Computing Development

UKCOD UK Citizens' Online Democracy

UKERNA UK Education and Research Network

UKPO UK Post Office

UKUUG UK Unix Users' Group

UL underlaid; Underwriters' Laboratories Inc (US product tests); unité logique (f), logical unit [LU]; unordered list; upload; uplink; utility lead

ULA uncommitted logic array

ULCC University of London Computing Centre

U-links links joining sections of communications channels

ULL usable scanning line-length

ULN universal link negotiation

ULPA ultra-low particulate air

ULS user location server

ULSI ultra-large-scale integration, intégration extrêmement grande (f)

ULTRA Underground Legion of Terroristic Research Activists

Ultrix DEC version of Unix OS to run VAX processors *tn*

UM unified messaging; usage metering; user mobile

UMA upper memory area

U-matic composite format using cassette tape *tn*

UMB unidad de medida básica (s), basic unit measurement [BMU]; upper memory block

UMC unassigned multiplex (MUX) channel

UMCO usage metering call originating

UMCT usage metering call terminating

UMD unscrambled mode delimiter

UMDL University of Michigan Digital Library Project

UME unidad de medida en escala (s), scaled measurement unit [SMU]; UNI management entity (protocol)

UM-HO usage metering hand-over

U-MIC UTP medium interface connector (token ring multi-access unit)

UML unified modelling language

UMLU usage metering location update

UM-MOC usage metering mobile originating call

UMPDU user message protocol data unit

UMS universal special service order

UMTK computer-aided translation unit (Universiti Sains Malaysia)

UMTS universal mobile telecommunication system (EU)

UMUR usage metering user registration

UMXU user message transfer unit

UN unassigned; unbalanced normal; universal number; unknown

UNA universal night answering

UNB counter of unreasonable (BSN) backward sequence numbers

unbal unbalanced

UNC universal/(uniform) naming convention

UNCL universal naming code locator

UNCOL universal computer-oriented language

UND échange unidirectionnel (f), comunicación unidireccional (s), one-way communication [OWC]

UNE unbundled network element

UNESCO UN Educational Scientific and Cultural Organization

UNF counter of unreasonable (FIB) forward indicator bits

UNGI Un Nouveau Guide Internet (f)

UNI Ente Nazionale Italiano di Unificazione (i) (member of ISO); user-network interface

UNICCAP universal cable circuit analysis program

UNICOM universal integrated communication system

Unicom China United Telecommunications

unicos Unix for Cray computers

Unics uniplexed information and computing service

Unihi wide-screen hi-definition broadcast standard tape *tn*

UNII unlicensed national information infrastructure

Uninet value-added network (United Telecom Communications Inc)

UNINETT Universities' Network (Norway)

UNIPEDE Union internationale des producteurs et distributeurs d'énergie électrique (f), International union of producers and distributors of electrical energy

Unisys Universal Systems Corp (from merger of Sperry and Burroughs)

UNIVAC Universal Automatic Computer

UNIX 32-bit multi-user operating system

UNMA unified network management architecture (AT&T)

UNN unallocated-number signal, signal de numéro inutilisé (f)

UODR usuario ocupado determinado por la red (s), network-determined user busy [NDUB]

UODU usuario ocupado determinado por el usuario (s), user-determined user busy [UDUB]

UP uniprocessor; unnumbered poll; user part

UPA UltraSPARC port architecture (Sun Microsystems *tn*)

UPAF Union postale africaine (f), Unión Postal Africana (s)

UPC universal product Code (*v* bar code); usage parameter control; user profile check

UPCMI uniform PCM (pulse code modulation) interface (13-bit)

UPD up to date; up-to-date voice-activity detection; user profile data

UPG upgrade

UPH units per hour

UPI United Phreakers Incorporated

UPL user programme language

U-plane user plane

UPM user profile management (IBM)

UPQH user profile query handler

UPS uninterruptible power supply; United Pirate Syndicate; Unix platform subsystem; upper-performance score

UPSim user part simulator

UPT universal personal telecommunications (service)

UPTD user-dependent data packet transfer delay

UPU Universal Postal Union, Union Postale Universelle (f), Unión Postal Universal (s); user part unavailable signal

UR user registration

URATE user transmission rate

URC uniform resource citation; universal reference clock

URG urgent flag (TCP header)

URI universal radio interface; uniform resource identifier; universal resource identifier (RFC 1630); universal resource locator (web site address); localisateur uniforme de ressources (f) [URL]

URL/URI uniform resource locator/universal resource identifier

URM usage reference model

URN Underground Rip-off Network; uniform resource name

URS update report system

URSI Union Radio – Scientifique Internationale (f), International Scientific Radio Union

URTNA Union des radiodiffusions et télévisions nationales d'Afrique (f), Union of National Radio And Television Organizations of Africa

US unidad de señalización (s), signal unit [SU]; unit separator

USA United Software Association; United States Alliance (Amiga Group)

USAM US Association for Computing Machinery

USAModSim US Army ModSim compiler

USART universal synchronous-asynchronous receiver-transmitter

USAT ultra-small aperture terminal

USB universal serial bus; upper sideband (modulation)

USC unidad de señalización de central (s), exchange signalling unit [ESU]; uniprocessor system controller (Sun Microsystems *tn*); user service centre

USC/ISI University of Southern California – Information Sciences Institute

USCP un solo canal por portadora (s), single channel per carrier [SCPC]

USCS unidad de señalización de control de sistema (s), system control signal unit [SCU]

USDC United States' Digital Communications

USENET world-wide e-mail/teleconferencing/BBS network of UNIX systems

USENIX UNIX and advanced computing systems professional association

USERID user identification

USI unidad de señalización de instrumento (s), instrument signalling unit [ISU]; user service interface

USICA US International Communication Agency

USID identificateur de service d'usager (f), user service identifier; unidad de señalización de sincronización (s), synchronization signal unit [SYU]

USITA US Independent Telephone Association

USL Unix System Laboratories

USM unidad de señalización de mensaje (s), message signal unit [MSU]; user session manager

USNO US Naval Observatory

USO universal service obligation; universal service order

USOA uniform system of accounts

USOC uniform service order code; universal service ordering code

USOP ultra-small outline package

USP unshielded twisted pair; usage sensitive pricing

USPAU-MI IP-message user agent protocol data unit [IM-UAPDU]

USR unidad de señalización de relleno (s), fill-in signal unit [FISU]; (list of) users' names/group names (Microsoft Mail); user-to-user information message; US Robotics Corporation

USRT universal synchronous receiver-transmitter

USS unidad subsiguiente de señalización (s), subsequent signal unit [SSU]

USSA User Supported Software Association (UK)

USSB United States Satellite Broadcasting

USSD unstructured supplementary service data

USTA United States Telephone Association

USTSA US Telecommunications Suppliers' Association

Usuario SP usuario del servicio de presentación (s), presentation service user [PS-user]

Usuario ST usuario del servicio de transporte (s), transport service user [TS-user]

USV unidad de servicio videotex (s)

USWST US World Service Teletext

UT unité de taxation (f), metered charging unit; upper tester

UTAM Unlicensed PCS Ad Hoc Committee for 2 GHz Microwave

UTB unité de taxation de base (f), basic metered charging unit

UTC coordinated universal time, Temps Universel Coordonné (f) [CUT], Tiempo Universal Coordinado (s) [CUT]; universal time coordinates

UTE Union Technique de l'Electricité (Paris); universal test equipment

UTF-8 UCS transformation format 8 (Unicode/ISO 10646)

UTI universal text interchange

UTM unité de transfert de messages (f), unidad de transferencia de mensajes (s), message transfer unit [MXU]

UTMC United Technologies Microelectronics Center

UTMS unité de transfert de messages de
service (f), unidad de transferencia de
mensajes de servicio (s), service message
transfer unit [SMXU]

UTMU unité de transfert de messages
d'usager (f), unidad de transferencia de
mensajes de usuario (s), user message
transfer unit [UMXU]

UTOPIA universal test and operations
interface for ATM

UTP unshielded twisted pair (cable)

UTPOS user terminal position

UTRA UMTS terrestrial radio access

UTRAN universal terrestrial radio access
network

UTRC United Technologies Research
Centre

UTS Unix TC-user subsystem; update
transaction system

UTSi ultra-thin silicon (C-MOS
technology)

utt unattenuated

UTV uncompensated temperature
variation

UU user-to-user; Unix-to-Unix

UUCF Unix-to-Unix copy facility

UUCICO Unix-to-Unix copy incoming –
copy outgoing

UUCP Unix-to-Unix copy program (mail
network); Unix-to-Unix copy protocol

UUE UU-encoded files (news groups on
Internet)

UUENCODE Unix-to-Unix encoding
(binary as ASCII)

UUI user-to-user information (AT&T)

UUNet formerly Pipex

UUS user-to-user signalling
(supplementary service)

UUT unit under test

UV ultraviolet; unidad de volumen (s),
unité (américaine) de volume (f), volume
unit [VU]

UVV Usenet Volunteer Votetakers

UWB ultra-wideband radar

UWE upper window edge, limite

supérieure de fenêtre (f), borde superior
de ventana (s)

UXU Underground Experts United

V

V normalized frequency; tiempo de ventana (s), window time [W]; value only

V&V verification and validation

V(A) variable de estado de acuse de recibo (s), acknowledge state variable

V(M) variable de estado de recuperación (s), recovery state variable

V(R) variable de estado en recepción (s), receive state variable

V(S) variable de estado en emisión (s), send state variable

V.pcm pulse code modulation (modem standard)

V/D voice/data

v/m volts per meter

V+TU voice-plus-teleprinter unit

V1 CCITT letter relating to telephone analogue circuits

V20/V30 processor chips (NEC *tn*)

VA value-added; virtual address; volt-ampere

VAB voice answer back

VAC vacant; vacuum; value-added carrier; volts alternating current

VAD voice activity detection/detector

VADIS voice and data integrated system

VADS value-added data service

VADSL very-high-rate asymmetric digital subscriber line

VAFC Vesa advanced feature connector

VAfJ Visual Age for Java (IBM *tn*)

VAI video-assisted instruction

VAL validation execution mode (man-machine comms); variable assembly language (for industrial robots)

VAM virtual access method

VAN value-added network

VAP value-added process; videotex access point

VAPC vector adaptive predictive coding

VAr volt-ampere reactive

VAR reactive volt-amperes; value-added reseller; variant/variable

varactor variable reactor

VARC variable axis rotor control system (wind power)

VAS value-added services, Mehrwertdienst Dienst (g); VISSR atmospheric sounder

VASE variable angle spectroscopic ellipsometry

VASP virtual analogue switching points

VAST variable-array storage technology

VATS Value-Added Telecommunications Service (Advantis *tn*)

VAVP virtueller analoger Vermittlungsstellepunkt (g) (virtual analogue switching point)

VAX operating system and the computer it ran (DEC *tn*); virtual address extension (DEC *tn*)

VAX/VM virtual address extension/virtual memory system (DEC)

VAX/VMS OS for VAX/virtual memory system *tn*

VAXBI VAX back-plane interconnect

VAXen two or more VAX machines

VAXft VAX fault tolerant

VB variable block; Visual Basic (Microsoft *tn*); voice bank

VBA Visual Basic for Applications (Microsoft Office *tn*)

VBA-DO VBA dialog object (Microsoft Office *tn*)

VBE VESA BIOS extensions

VBE/AI VESA BIOS extension/audio interface

VBI vertical blanking interval/interface

VBN Vorläufer Breitband Netz (g) (German broadband network)

vBNS Very High-Speed Backbone Network Service (MCI *tn*)

Vbox VideoBox (Sony *tn*)

VBR variable bit rate

VBR-NRT variable bit rate – non-real-time

VBR-RT variable bit rate - real-time

VBS voice broadcast service

VBV velocidad binaria variable (s) [VBR]

VBX Visual Basic extension

VC virtual calls; virtual channel (ATM); virtual circuit; voice channel; voice circuit

VC/CMS virtual machine/conversational monitor system OS

VCA voice-connecting arrangement

Vcc volts courant continu (f) [vdc]

VCC virtual channel connection; virtual circuit connection

VCCE virtual channel connection endpoint

VCD virtual communications driver

VCG voice channel group

VCI virtual channel identifier (ATM); virtual circuit identifier; virtual connection identifier

VCL Virtual Channel Link (UNI 3.0); Visual Component Library (Borland *tn*)

VCM DRAM virtual channel memory DRAM

VCMUI Version Conflict Manager UI (Microsoft Windows 98)

VC-n virtual container-n

VCO voltage-controlled oscillator; voice-controlled oscillator

VCOE Virginia Center of Excellence

VCOS Visual Caching OS (AT&T)

VCPI virtual control program interface

VCR video-cassette recorder

VCS version control system; virtual circuit switch; Virtual Connectivity Services (Mercury *tn*)

VCSI very large-scale integration

VCXO voltage-controlled crystal oscillator

VD virtual destination

VDA verificación del acondicionamiento (s), training check [TCF]

vdc volts direct current

VDD virtual device driver

VDDM virtual device driver manager

VDE Verband Deutscher Elektrotechniker (Germany) (g); video display editor; visual development environment

VDFG variable diode function generator

VDI virtual device interface

VDL Vienna definition language (IBM)

VDM video display metafile; Vienna Development Method (IBM); virtual device metafile; virtual DOS machine

VDMAD virtual direct memory access device (Microsoft)

VDo Verbindungsdose (g) (link socket)

VDP video display processor

VDPN virtual private dial-up networking

VDR video disk recorder; voice digitization rate; voltage-dependent resistor

VDS vapour distribution system; video distribution system

VDSK very high data rate DSL

VDSL very high-speed digital subscriber line

VDT video/virtual display terminal; visual display terminal

VDU video/visual display unit

VE value engineering; Verarbeitungseinheit (g) (processing unit); Vermittlungseinheit (g) (switching unit)

VEC visual/virtual embedded copyright

VEG virtual environment generator (NASA)

VEI Vocabulaire Electrotechnique International (f), Vocabulario Electrotécnico Internacional (s), International Electrotechnical Vocabulary [IEV]

VEMM virtual expanded memory manager

VEMMI videotex enhanced man-machine interface

VER verification; verify; verified; verifying operator

VERA virtual entity of relevant acronyms

VERONICA very easy rodent-oriented net-wide index to computerized archives

VERR verify read access

vert vertical

VERW verify write access

VESA Video Electronics Standards
Association
VESP video-enhanced service providers
VF variance factor; virtual floppy; voice
frequency
vf band voice-frequency band
VFAT virtual FAT (Microsoft Windows)
VFC vector function chainer; version
first class; video feature connector
VFCT voice-frequency carrier telegraph
VFD vacuum fluorescent display
VFDF voice-frequency distribution frame
VFEI virtual factory equipment interface
VFFT voice-frequency facility terminal
VFL voice-frequency line
VFMS virtual file management system
VFO variable frequency oscillator
VFR visual flight rules
VFT voice-frequency telegraphy
VFTE voice-frequency telegraph
equipment
VFTG voice frequency telegraph
VFU vertical format unit
VFW Video for Windows (Microsoft *tn*)
VG voice-grade
VGA video graphics array
VGCS voice group call service
VGPO velocity gate pull-off
vgrep visual grep (*ie* Unix search)
VGU closed user group validation check
message (f) [CVM]
VGX Variational Graphics eXtended
(SDRC *tn*)
VH/VCS IBM OS *tn*
VH 1 Video Hits 1
VHD very high density
VHDL very high-speed description
language; VHSIC hardware description
language
VHE virtual home
entertainment/environment
VHF very high frequency
VHN Vickers hardness number
VHORG vertically and horizontally
organized grid
VHP very high productivity
VHRR very high resolution radiometer

VHS variable hard sphere; Video Home
System (JVC *tn*); virtual host storage
VHS.C VHS compact (JVC *tn*)
VHSIC very high-speed integrated circuit
VHSOL very high speed optical loop
vi visual interface (Unix screen editor)
VI indicación de datos válidos (s); vector
identifier; voie interurbaine (f), trunk
channel [TC]
VIA VAX information architecture;
versatile interface adaptor
VIAS voice intelligibility analysis set
VIBGYOR visible colour spectrum
(violet to red)
VICAR video image communication and
retrieval
VICDS vehicle integrated
communications and information
distribution systems (UK DoD)
VICS Voluntary Inter-industry
Communications Standards for EDI
vid video
VIDEC video digitally-enhanced
compression (Connectix *tn*)
Video CD compact disk with full-motion
video
VIDEOTEX interactive TV and
telephone computer-controlled system
VIDF vertical side of intermediate
distribution frame
VIDS VICAR interactive display
subsystem
VIE virtual information environment
VIEWDATA BT version of Videotex
(now Prestel *tn*)
VIF VHDL interface format; virtual
interface; virtual interrupt flag
VIM VAX interface manager; vendor
independent messaging (Lotus); vi
improved; video interface module
Vin voltage input
VINCE vendor-independent network
control entity
VINES Virtual Networking System
(Banyan)
VIO video input-output
VIP variable/versatile information

processor/processing; VINES Internet Protocol; virtual interrupt pending; visual information projection

ViP visual programming (Lotus)

VIPER computer implemented in silicon on sapphire

VIR versión internacional de referencia (s), international reference version [IRV]; vertical interference reference signal

VIROS VIRtual memory Operating System (early DEC name for TENEX)

VIS video information system (Tandy); visual instruction set (Sun); voice information service

VISCA video system control architecture (Sony *tn*)

VISCorp Visual Information Service Corp

VisiCalc spreadsheet for Apple format (VisiCorp *tn*)

VISS video/VHS index search system

VISSR visible and infrared spin scan radiometer

VISTAnet Vermont Internet Services and Technology Access Network

Visual Basic visual programming language (Microsoft *tn*)

Visual C visual programming implementation of C language (Microsoft *tn*)

VITC vertical interval time-code (used on video tracks)

VITS vertical insertion test signal

VIU voice interface unit

VL vector length; Verlängerungsleitung (g) (attenuator)

VL bus VESA local bus (33 MHz/32-bit data path)

VLA very large array

VLAN virtual LAN

VLB VESA local bus

VLBI very long base interferometer

VLBR very low bit-rate

VLD variable-length decoder

VLDB very large database

VLE vapour levitation epitaxy

VLF vertical laminar flow; very low frequency

VLFMF very low frequency magnetic field

VLIW very long instruction word (chip architecture)

VLKB very large knowledge base

VLL video lead locator

VLM virtual loadable module (Novell)

VLR visitor location register (GSM-radio)

VLSI very large-scale integration, circuit à très haute intégration (f), Höchstintegration (g); höchstintegrierter Schaltkreis (g)

VLSIPS very large scale immobilized polymer synthesis

VLSM variable length subnet masks

VLT variable list table; Varley loop test (finds position of cable faults)

VM virtual machine (IBM)

VM/CMS virtual machine/conversational monitor system (GEISCO)

VM/PC operating system on IBM XT/370 and AT/370

VM/SP virtual machine/system product

VMA virtual memory address

VMAC virtual media access control

VMB virtual machine boot

VMC variable message cycle

VMCMS virtual machine-conversational monitor system

VMDF vertical side of main distribution frame

VME Versa micro-module extension; virtual machine environment (ICL OS for mainframes); virtual manufacturing enterprise; virtual memory environment

VME bus Versa Module Europa (*originally* Motorola Versabus: IEEE 1014)

VMEL valeur mesurée d'effet local (f), measured sidetone value

VMGG vector de movimiento global de grupo de bloques (s), group of blocks global motion vector [GGMV]

VMIC VME Microsystems International Corp
VML vector mark-up language
VMM virtual machine/memory manager
V-model software life-cycle model
V-MOS vertical-groove metal-oxide semiconductor
VMP virtual modem protocol
VMR voltmeter reverse
VMRS voice message relay system
VMS virtual memory system (VAX operating system DEC *tn*); voice-mail system; voice message service/system
VMSC virtual mobile (services) switching centre
VMSP virtual machine-system product (IBM)
VMT virtual memory technique; virtual method table (Delphi)
VMU Visual Memory Unit (Sega Dreamcast *tn*)
VMW vertically-modulated well
VMX VME bus extension
V$_n$ modem specifications
VN verify number if no answer; version numérique
VNA virtual national architecture
VNBC variación del número de bits de código (s)
VNL via net loss
VNLF via net loss factor
VNN vacant national number
VNOD video nearly on demand
VOA Voice of America
VOC volatile organic compound
VOCODER voice-operated recorder
VoD Video-on-Demand (via ISDN/ADSL)
VODAS voice-over-data access station
vodas voice-operated device anti-sing
Vods vehicle owners' description search (UK police)
VOGAD voice-operated gain-adjusting device
VoIP voice over IP
Vol Vollvermittlungsleitung (g) (full-facility switching centre line)

VOL volume header label
VOLCAS voice-operated loss control and echo/singing suppression circuit
VollVSt Vollvermittlungsstelle (g) (full-facility switching centre)
VOM volt-ohmmeter/milliammeter
VON voice over the Net
VOR VHF omnidirectional radio range
VOS video on sound
vox voice-operated relay circuit; voice-operated transmit/telegraph
voxel 3D equivalent of pixel used in volumetric models
VoxML voice mark-up language
VP valeur de paramètre (f), valor de parámetro (s), parameter value [PV]; virtual path (ATM)
VPB retroceso de la posición de línea (s), line position backward
VPC virtual path connection
VPCE virtual path connection endpoint
VPCI virtual path connection identifier
VPD vapour phase desorption/decomposition; virtual printer device; vital product data (IBM)
VPD-ICPMS vapour phase decomposition – inductively-coupled plasma mass spectroscopy
VPDN virtual private data network
VPDS virtual private data service (MCI)
VPE video port extensions (Microsoft); visual/(virtual) programming environment
VPF validity period format
VPI virtual path identifier
VPL virtual path link (UNI 3.0); visual programming language
VPLC voice packet line card
VPLMN visited public land mobile network
VPM video port manager; virtual private/protocol machine; virtual private network/number
VPR posición de línea relativa (s), line position relative
VPS voice processing system

VPT virtual path terminator (UNI 3.0); voice plus telegraph

VPU voice processing unit

VQ vector quantization/quantizer

VQL voice quantizing level

Vr Verstärker (g) (amplifier – repeater)

VR virtual reality

VRAM variable-rate adaptive multiplexing; video RAM (device)

VRC verificación por redundancia cíclica (s), cyclic redundancy check [CRC]; vertical redundancy check (parity)

vri varistor

VRL virtual reality language

VRM voltage regulator module

VRML virtual reality (3D) modelling/mark-up language

VRML97 virtual reality modelling language (ISO/IEC 14772-1)

VROOMM virtual real-time object-oriented memory manager (Borland)

VRS voice response system

VRT voltage regulation technology (Intel)

vs versus

VS Vermittlungssatz (g) (switching unit); virtual scheduling; virtual storage (Wang Laboratories)

VS/VD virtual source/virtual destination

VSA VME to SCSI adaptor

VSAM virtual sequential access method (IBM)

VSAT very small aperture terminal (satellite dish/ground station)

VSB vestigial sideband (transmission)

VSC videotex service centre

VSE Verkehrssimulationseinrichtung (g) (traffic simulation equipment); virtual storage extended (IBM *tn*)

VSELP vector sample/sum excited linear prediction (Motorola)

V-series CCITT recommendations for data transmission on telephone networks

VSF virtual scanning frequency; Virtual Software Factory (Systematica *tn*); voice store-and-forward

VSI virtual socket interface

VSIO virtual serial input-output

VSLE very small local exchange

VSM vestigial sideband modulation; virtual shared memory; virtual storage management; visual system management (IBM)

VSNL Videsh Sanchar Nigraim Ltd, India (ISP)

VSO very small outline

VSOS virtual storage OS

VSP variable soft sphere; Very Simple Prolog+; virtual switching point

VSPC virtual storage personal computing

VSS variable soft sphere; voice-switching system (US)

VSSP voice-switch signalling point

VSt Vermittlungsstelle (g) (switching centre)

VST volume-sensitive tariff

vswr voltage standing wave ratio

VSYNC vertical sync

VT vacuum tube; vertical tab; verrouillage de trame (f), frame alignment [FA]; vertical tabulation, tabulación vertical (s); video telephony; virtual terminal/tributary

VT-52 video terminal control codes (DEC)

VTAM virtual telecommunications access method (IBM)

VTE virtual terminal environment

VTFL variable to fixed-length (code)

VTMS voice-text messaging system

VTOA virtual trunking over ATM; voice and telephony over ATM

VTOC volume table of contents

VTOM Virtual Tool Object Model

VTP vertical thermal processor; virtual terminal protocol (ISO)

VTPR vertical temperature profile radiometer

VTR videotape recorder

VTS video teleconferencing system

VTVL variable-to-variable length code

vtvm vacuum-tube voltmeter

VTW Voters' Telecommunications Watch

VTX Videotex

VU voltímetro vocal (s) [speech voltmeter]; volume unit

VUE Visual User Environment (Hewlett Packard)

VUI video user interface

VUIT visual user interface tool (DEC)

VUP VAX unit of processing/performance

VWB Visual Workbench (Microsoft)

VxD virtual device/(extended) driver (Microsoft Windows)

VXI VME bus extension for instrumentation

W

W watts; width

W/B wire bonding

w/i within

w/o without

W3/W³ World Wide Web

W3A World Wide Web applets

W3C WWW consortium

w4 what works with what

W4WG Windows for Workgroups
(Microsoft *tn*)

WA wire-armoured

WAAS wide-area augmentation system

WABI Windows (3.1) application binary
interface (Sun *tn*)

WAC wide-area coverage

WACK wait acknowledgement

WACS wireless access communications
system

WADS wide-area data service

WAFL Warwick (University) functional
language

WAI Web accessibility initiative; Web
application interface

WAIS Washington Association of Internet
Service Providers; wide-area information
server (database protocol)

WAITS wide-area information transfer
system

WAL2 second order Walsh function

WAN wide-area network

wand barcode reader

WAOSS wide-angle opto-electronic
stereo scanner

WAP wide-area paging; wireless
application/access protocol

WARC World Administrative Radio
Conference

Warez 'pirated' software website

WAS Wahlaufnahmesatz (g) (dial input
circuit)

WASP Warriors Against Software
Protection

WASS Western and Southern States (US)

WATFIV enhanced successor to
WATFOR

WATFOR Fortran compiler (University
of Waterloo Fortran)

WATM wireless ATM

WATS website activity tracking statistics;
wide area telecommunications
(telephone) service

WATSON UK police charting system
software; wireless activities transmitted
seamlessly over the network

watt-hr watt-hour

WAV waveform audio

WAWS Washington (DC) area wideband
system (US)

WaZOO Warp-zillion Opus to Opus

Wb SI unit of magnetic flux

WB weak base; wet bulb; White Book:
Standard for Video CDs

WB/BB wideband/broadband

WB-95 WideBand 95

WBEM Web-based enterprise
management

WBFH wideband frequency hopping

WBFM wideband frequency modulation

W-bit wait-bit

WBS wideband system, work breakdown
structure

WBSARC World Broadcasting-Satellite
Administrative Radio Conference

WBSC wideband CDMA BSC

WBSEM Wire Bonder Specific
Equipment Model

WBTS wideband CDMA BTS

WBVTR wideband video tape recorder

WC Western Cedar (telephone pole);
wire chief (test clerk); world coordinates

WCC Worldwide Communications
Corporation (US)

WCDMA wideband CDMA

WCS wireless communication service;
world coordinate system; writable
control store

WCTP wire chief test panel

WD watchdog (timer); Western Digital
(US corporation)

WDC World Data Center

W-DCS wideband digital cross-connect system
WDF World Domination Force
WDL Windows driver library (Microsoft)
WDM wave difference method (digital subscriber loop timing recovery); wavelength division multiplexer/multiplexing; Windows driver model (Microsoft *tn*)
WDP wireless datagram protocol
WDR West Deutscher Rundfunk (g)
WDRAM Windows Dynamic RAM (Microsoft *tn*)
WDS wavelength dispersive spectrometry of X-rays
WDT A/B watchdog timer answerback
WDX wavelength dispersive X-ray
WDXA wavelength-dispersive X-ray analysis
Web World Wide Web
Web NFS Web network file system (Sun Microsystems *tn*)
WebDAV Web distributed authoring and versioning protocol
WEC wafer environment control, World Energy Council
WECo Western Electric Co
WEFAX weather facsimile
WEFT Web embedding font tools (Microsoft *tn*)
WELL Whole Earth 'Lectronic Link (early West-coast BBS)
WES Western Electronics Switching; wind electric system
WFA work force and administration
WFA/C WFA control
WFA/DI WFA despatch in
WFA/DO WFA despatch out
WFD waveform distortion
wff well-formed formula
WFG waveform generator
WFL Workflow language (Burroughs)
WFMC Workflow Management Coalition
WFP well-formed path (Microsoft Windows)
WFT wafer fabrication template
WfW Word for Windows (Microsoft *tn*)

WFWg Windows for Workgroups (Microsoft *tn*)
WGS work group system
WH ready with called party; we have; Western Hemlock (telephone pole)
WHAM waveform hold and modify (Microsoft)
Whirlwind first real-time computer (MIT)
whr watt-hour
WHSR White House Situation Room
WI wavelet image (Summus *tn*)
WIA World Interworking Alliance
WIB within-batch
WIBNI wouldn't it be nice if
WID wireless information device
WII worldwide information infrastructure
WIMP windows – icons – mouse – and pull-down menus (Xerox); windows – icons – mouse/menus – pointer/pointing device
Wimpe windows – icons – mouse – pointer environment
WIN wireless in-building network; Wissenschaftsnet (g) (German science network); WWMCCS *qv* inter-computer network
WIN Forum Wireless In-building Network Forum
Win31 Windows 3.1x (Microsoft *tn*)
wince derogatory pronunciation of Win CE/Windows CE (Microsoft *tn*)
Windows 95 Microsoft OS *tn*
Windows 9x Windows 95 or 98 (Microsoft *tn*)
Windows DNA Windows Distributed Internet Applications (Microsoft *tn*)
Windows NT Windows New Technology OS (Microsoft *tn*)
Windows SC Windows Smart Card (Microsoft *tn*)
WINDX Windows' Development Exchange
Wine Wine is not an emulator *tn*
WINE Windows emulator (*ie* by non-Windows systems)

WinHEC Windows Hardware Engineering Conference

WINS Windows Internet naming service (Microsoft *tn*); Warehouse Industry National Standards Guidelines

winsock Windows sockets (API for TCP/IP interchanges)

Wintel Windows and Intel combination

Winword Word for Windows (Microsoft *tn*)

WIP work in progress

WIPO World Intellectual Property Organization

WIS water-induced shift; WATS information system; Web Information Systems

WISC writable instruction set computer

WISE WordPerfect information system environment (Corel *tn*)

WISP wireless ISP

WITS Washington integrated telecommunications system (US); web interface for telescience (NASA/Java applet)

WIW within wafer

WIWNU within wafer non-uniformity

WL Western Larch (utility pole)

WLAN wireless LAN

WLBI wafer-level burn-in

WLL wireless (in) local loop

WLT wafer-level test

WM wireless manager

WMC World Meteorological Center

WMI Windows Management Instrumentation (Microsoft *tn*)

WML WAP mark-up language; wireless mark-up language

WMP Windows media player (Microsoft *tn*)

WMS World Magnetic Survey

WMSC wideband mobile switching centre

WMux/Demux wavelength multiplexer/demultiplexer

WN wrong number

WNP waste neutralization plant

WNS Wahlnachsendesatz (g) (post-dialling unit)

WO work order

WOM write-only memory

WOMBAT waste of money brains and time

WOOL Window object-oriented language

WORD work order record and details

WordStar word processor (MicroPro International Corp *tn*)

WorldTel WorldTel Limited

WORM write once, read many (disk); a generic, network-borne virus

WOSA Windows open services/systems architecture (Microsoft *tn*)

WoW Window on the World (virtual reality)

Woz [Steve] Wozniak, co-founder of Apple Macintosh

WP Western pine (telephone pole); WordPerfect (Corel *tn*); word processing/processor; write protect

WPA wrong password attempts (counter)

WPBX wireless PBX

WPC wafer process chamber; World Power Conference

WPE World Pirate Echo

WPH wafers (manufactured) per hour

WPHD write-protect hard disk

WPP Web presence providers (Microsoft Front Page)

WPS Windows printing system (Microsoft *tn*); workplace shell (IBM)

WPVM Windows parallel virtual machine (Microsoft *tn*)

WRAM Window RAM (Microsoft *tn*)

WRB Web request broker (Oracle Corp)

WRED weighted random early discard

wrg wrong

WRK Windows resource kit (Microsoft *tn*)

WRMA write many read always

WRN World Radio Network

WRNC wideband radio national controller

WRS working transmission reference system

WRT with respect/regard to

WRTC working reference telephone circuit

WRU who are you' character

WS Wahlsatz (g) (dialling unit – circuit); wire send; WordStar; workstation

WSAPI Web site application program interface

WSD World Systems Division (Comsat)

WSE Wannier-Stark effect

WSFN which stands for nothing

WSG wired shelf group

WSH Windows scripting host (Microsoft *tn*)

WSI wafer-scale integration

WSIG wanted type of signalling

WSN Wahlsatz für Nummernschalterwahl (g) (dialling unit for rotary dials)

WSOD Workspace on Demand (IBM)

WSP wireless session protocol

WSS wide-screen signalling

WST world systems teletext

WT will talk; wireless transceiver

WTA Wireless Telegraphy Act (UK); wireless telephony application

WTC wafer transfer chamber

WTD World Telecommunications Development

WTDM wavelength time-division multiplex

WTF who/why/what the fuck?

WTM White Trash Magazine; wired terminal mobility

WTOR write to operator with reply

WTS Windows terminal server (previously Hydra; Microsoft *tn*)

WTSC World Telecommunications Standardization Conference

WTW wafer-to wafer

WTWNU wafer-to wafer non-uniformity

Wü Wechselstromübertragung (g) (AC signalling line unit)

WU Western Union (Telegraph Co)

WUGNET Windows Users' Group Network

WUI Web user interface; Western Union International

wv working voltage

WVDC working voltage direct current

W-VHS wide VHS

WVR water vapour regained

WW wall-to-wall; wire-wound; Weenie Warriors

WWDSA world-wide digital system architecture

WWMCCS Worldwide Military Command and Control System

WWW World Weather Watch; World Wide Web

WWW7 WWW Conference 7 (Brisbane, 1997)

WWWW World Wide Web Worm

WYGIWIG what you get is what is given

WYSIAYG what you see is *all* you get

WYSINWYG what you see is not what you get

WYSIWYG what you see is what you get

WYSIWYN what you see is what you need

X

X any figure 0-9 used in software version nomenclature; CCITT notation for data communications; cross; extension; symbol for inductive reactance; ten; transmission; exchange; cross-/crossed

X terminal graphical terminal for X Window system (protocol)

X Window System network-based graphics windowing system (MIT *tn*)

X.25 CCITT standard for data communications

X.121 addressing format in X.25 base networks

X.25 connection-oriented network facility

X2B hexadecimal to binary (REXX)

X2C hexadecimal to character (REXX)

X2D hexadecimal to decimal (REXX)

x86 Intel series of CPUs

X.400 electronic mail system messaging standard

XA cross-arm; exchange access; extended architecture/attribute

XANES X-ray adsorption near-edge structure spectroscopy

XAPIA X.400 Application Program Interface Association

Xapi-J XML API in Java

XB crossbar

X-band microwave band 8-12 GHz

Xbar cross-bar

XBL extension bell

XBT cross-bar tandem

XCHG exchange

xcmd external command

XCS ten call seconds

xcvr transceiver

XD crossed; ex-directory

XDE Xerox development environment

XDF extended density format (IBM); extended distance feature; Xrm decrease factor

XDR external data representation (RFC 1014)

xDSL one of the digital subscriber line technologies

XDTI Xerox Digital Textile Imaging service *tn*

Xemacs Lucid Emacs

Xenix UNIX-derived 16-bit multi-user multitasking OS running a PC

XEQ execute; control word

Xerox PARC Xerox Palo Alto Research Centre *tn*

Xetra exchange electronic trading

XFA XML forms architecture

XFA-SOM XFA scripting object model

XFC transfer charge

XFCN external function

Xfer transfer

xfmr transformer

XG extension of GM (extra instruments/digital effects) *tn*

XGA extended graphics array (IBM graphics standard); matriz de gráficos ampliada (s)

XHL extensible hyperlinking language

XID exchange identifier/identification

XIE X image extension

XIF Xerox image file

X-Indik Erdschlußindikator (g) (earth connection indicator)

XIOS extended input/output system

XIP execute in place

XLAT translate

XLL XML linking language

XLM Excel macro language (Microsoft)

XLS excimer laser system; extended light scatterer

XMA eXtended memory adaptor

XMI XML metadata interchange

xmit transmit (*also* XMIT)

XML eXtensible Mark-up Language

XMM extended memory manager (LIM/AST *qv*)

Xmodem file transmission protocol for modems

Xmodem 1K version of Xmodem transmission protocol sending 1k blocks

XMP X/Open management protocol

XMS extended memory specification; extended multiprocessor system

xmsn transmission

xmt transmit

xmtd transmitted

xmtg transmitting

xmtr transmitter

XNS Xerox Network Systems (Xerox *tn*)

XNSI Xerox Network Systems Institute

XNSIG Xerox Network Systems Implementors' Group

XO crystal oscillator

x-off transmitter off

XON cross-office highway

x-on transmitter on

XON/XOFF stop-resume method of flow control transmission protocol

X-OPEN European organization of OEMs for common standards in data processing

XOR exclusive OR (logical operator)

XOS cross-office slot

XOW express order wire

XPD cross-polar discrimination

XPG X-Open portability guide

XPRM Xerox print resources manager

XPS X-ray photo-electron spectroscopy; (Rank) Xerox Publishing System

XPT cross-point

XRCE Xerox Research Centre, Europe (Grenoble)

XRD X-ray diffraction

xref cross reference

XRF X-ray fluorescence spectrometry; extended recovery facility (IBM)

Xrm available bit rate service parameter

XRT extensions for RealTime

X-series series of CCITT recommendations

XSI X-Open system interface

XSL extensible stylesheet language

XSLT extensible SLT

XSMD extended storage module drive interface

XSS existing source system

XSSI extended server side includes

XT cross-talk; name of original IBM PC *tn*

xtal crystal

XTC external transmit check

XTCLK external transmit clock

XTI X-OPEN transport interface

XTP express/Xpress transfer protocol

XUDT eXtended unit DaTa

XUDTS eXtended Unit DaTa service

XUI X-user interface

XUM Xerox University Microfilm *tn*

XVGA extended VGA

XVR exchange voltage regulator

XVT extensible virtual toolkit

XWI Centrum voor Wiskunde en Informatica (Holland) (nl)

X-Windows standard set of API commands

Xxx1 addressed signalling system, Système de signalisation (f), Sistema de señalización direccionado (s), Adressiertes Signalisierungs-System (g)

Xxx2 telematics applications

X-Y display vector display

Y

Y yotta, 10^{24}; yttrium (used in superconducting alloys; as yttrium oxide (Y_2O_3) in monitors/radar; symbol for admittance (in ohms)

Y2K year 2000 problem (new millennium date roll-over)

Y2K CAN Y2K Community Action Network (UK)

YAA yet another assembler

YABA yet another bloody acronym

YACC yet another compiler-compiler (UNIX meta-compiler)

YADE yet another DSSSL engine

YAFIYGI you asked for it you got it

YAG yttrium aluminium garnet (used to focus laser light-beam)

Yagi form of radar antenna; end-fire array

YAHOO yet another hierarchically-organized/(officious) oracle

YAM Youth (Youngsters) against McAfee

YAST yet another system tool (Linux)

YBCO yttrium barium and copper oxide (used to make superconductors)

YBTS yellow book transport service

YCPS Yale Center for Parallel Supercomputing

YEL yellow alarm

YIG yttrium-iron-garnet

YIQ Y (luminance) Intensity Q (chrominance) (NTSC colour signals)

YLE YLEisradio (Finnish broadcaster)

YMMV your mileage may vary

Ymodem enhanced version of Xmodem

YMS Young Micro Systems

Y-network three-branch star network; T-network

Yoyo un système très instable (f), a very unstable system

YP Yellow Pages

YPS Yellow Pages service

Y-signal controls chrominance of TV signal

YUV PAL colour signals: luminance and colour difference

YWIA you're welcome in advance

Z

Z formal notation for specification of computing systems; symbol for impedance (in ohms); Zulu time zone (GMT)

Z_0 characteristic impedance

Z3 electromechanical programmed computer

z39.50 data transmission protocol for WAIS

Z80 early 8-bit processor developed by Zilog *tn*

ZAA group area exchange

ZAI zero administrative initiative

ZAK zero administration kit

ZAPP zero assignment parallel processor

ZAS zentrales Ansagegerät (g) (central voice message equipment)

ZAW zero administration for Windows (Microsoft)

ZB Zentralbatteriebetrieb (g) (central-battery working)

ZBR zone-bit recording (IBM)

ZBTSI zero-byte time-slot interchange

z-buffer method for hidden-surface removal

ZC zone code

Z-CAV zoned constant angular velocity

ZCS zero code suppression

ZD zero defects; Ziff-Davis (magazine publisher)

ZDF Zweites Deutscher Fernsehen (g)

ZDL zero delay lock-out

ZDS Zenith Data Systems

ZEN Zero effort networks (Novell *tn*)

Zenon French listening satellite

ZF Zwischenfrequenz (g) (intermediate frequency)

ZFH Zweitfernhörer (g) (second earpiece)

ZFK zentraler Fernsteuerkanal (g) (channel for centralized remote control)

ZG Zusatzgerät (g) (ancillary equipment)

ZGn zone geographique n (f), geographical area n (category)

ZGS Nr. 7 Zeichengabesystem Nr.7 (g)

ZGW Zentralgruppenwähler (g) (central group selector)

Zif socket zero-insertion-force socket

ZIG Zeichenimpulsgabe (g) (signalling pulse-sender)

zip generic term for file compression

ZIP zero in-line package; zig-zag in-line package; Zone Information Protocol (AppleTalk) *tn*

ZIS zone information socket

ZIT zone information table

Zl Zentralvermittlungsleitung (g) (central switching centre line)

Zmodem enhanced Xmodem (corrects broken connections)

zoo anti-virus researchers' laboratory

ZPV zoomed port video (Toshiba)

Zr zirconium (used in chip alloys)

ZSL zero slot LAN

ZST Zentralsteuerwerk (g) (central control unit)

ZTS Zweierteilnehmersatz (g) (two-party subscriber line unit)

ZulB Zulassungsbedingungen (g) (conditions for approval)

ZVSt Zentralvermittlungsstelle (g) (central switching centre)

ZVt Zwischenverteiler (g) (intermediate distribution frame)

ZX81 Clive Sinclair's first PC

ZZF Zentralamt für Zulassungen im Fernmeldetechnisches (g) (central authority for telecommunications approval)

ZZK Zentraler Zeichenkanal (g) (common signalling channel); Zentral-Zeichengabe Kanal (g) (common signalling channel)

zzz Trans-European network in field of telecoms (EU)

zzz(rt) insert with red marking; Kapsel mit roter Markierung (g)

Post-Alphabet

-/b negative on leg b, Ader b negative (g)

-/g gehend (g), outgoing ()/G

-/k kommend, incoming (I/C)

0-D zero dimensional

0TLP zero transmission level [reference] point

16-QAM quadrature-amplitude modulation with 16 phase-amplitude states

286 Intel 80286 processor

2B1Q line code (2 binary encoded into 1 quaternary)

2PPAPM two-pulse amplitude and pulse modulation modem

2-w two-wire circuit

386 MAX Qualitas 386 Expanded Memory Manager (386MAX.SYS)

386 Intel 80386 processor

3BNET AT&T network

3Com US corp: Computer, Communication, Compatibility

3Dfx 3D graphics card (Diamond multimedia)

3IN24 3-in-24 stress pattern

3P three pole

3PDT three-pole, double throw

3PST three-pole, single throw

3PTY three-party service

3WC 3-way calling

486 Intel 80486 processor

4B3T Siemens line-coding method (4 binary into 3 ternary digits)

4GL fourth-generation language

4P four pole

4PDT four-pole, double throw

4-w four-wire circuit

4W four-wire circuit

4WTS four-wire terminating set

586 Intel 80586 processor (Pentium)

68000 8MHz CPU from Motorola for Apple Macintosh

686 Intel 80686 processor (Pentium Pro)

69LC040 CPU from Motorola for Apple Macintosh without FPU

7D seven-digit number

8080 Intel CPU 2MHz (1974)

8086 Intel CPU 5MHz (1978)

8088 Intel CPU 8MHz (1979)

8421 weighted code

8B6T Bellcore line coding method (8 binary encoded into 6 ternary digits)

8B6T Bellcore line-coding method

8-PSK phase-shift keying with eight carrier phase states

(rt) Kapsel mit roter Markierung (g), insert with red marking

λ symbol for wavelength

μA micro-ampere

μm micron

μsec microsecond

Ω symbol for ohm

Annex I

Data Formats and File Extensions – three-letter acronyms expanded

.#24	24-pin matrix printer LocoScript printer file	.acl	Corel Draw 6 Keyboard accelerator
.#st	standard mode printer definitions (LocoScript)	.aca	Project Manager Workbench
.$db	*tn* dBASE IV temporary file	.acc	DR-DOS program
.$ed	Microsoft C Editor temporary file	.acm	Microsoft Windows audio compression file
.$vm	Microsoft Windows 3.x virtual temporary file manager	.act	Actor source code file; FoxPro Foxdoc action diagrams
.~de	Borland C++ 4.5 project backup file	.acv	audio file compression/decompression driver
.~mn	Norton Commander menu backup file	.ad	AfterDark screen saver
.000	*tn* DoubleSpace compressed hard disk data	.ada	Ada source code file
.096	Ventura Publisher 96 dpi display font	.adb	Ada package body
		.adi	AutoCAD graphics
.1st	'Read me first' (software author's text file)	.adl	QEMM adapter description library
.286	Microsoft Windows standard mode driver	.adn	Lotus 1-2-3 add-in
		.ads	Ada package specification
.2GR	Microsoft Windows standard mode grabber	.afl	Lotus 1-2-3 font file
		.afm	Type 1 font metric ASCII data
.386	Microsoft Windows enhanced mode driver	.agb	Microsoft AutoRoute Express GB route
.3fx	CorelCHART effects file	.ai	Adobe Illustrator vector graphic
.3GR	Microsoft Windows enhanced mode grabber	.ais	Xerox array of intensity samples
		.all	arts and letters library; WordPerfect general printer information
.3t4	converter: binary to ASCII	.alt	WordPerfect menu file
.4th	Forth source code file	.ani	animated cursor
.8m	PageMaker printer font - extended character set	.ann	Microsoft Windows 3 help annotations
.a	Ada source code file; Unix library file	.ans	Ansi character animation graphics; Ascii text ANSI character set
.a3w	MacroMedia Authorware Windows 3.5	.apc	Lotus 1-2-3 printer driver
.abk	CorelDRAW automatic backup file	.app	application object file (dBASE); DR-DOS program/executable; FoxPro application; MacroMedia Authorware
.abr	Adobe PhotoShop Brush		

.apr	Lotus Approach file	.bkx	WordPerfect timed backup file
.apx	Borland C++ app-expert database	.bmk	Microsoft Windows 3 help
.arc	compressed format archive		bookmark
.arj	archive: compressed file format	.bmp	bitmap image format
	by Jung	.bnk	Sim City game file
.ark	QUARK compressed file format	.bpt	CorelDRAW bitmap file
.asc	Ascii text file	.brs	batch run script
.asd	Lotus 1-2-3 screen driver; Word	.brx	browse index on multimedia CD-
	autosave file		ROM
.asf	Advanced Streaming format file;	.btm	Norton Utilities batch file
	Lotus 1-2-3 screen font;	.bvx	WordPerfect overflow file
	Microsoft Active Streaming	.bwr	Beware (buglist) (Kermit)
	Format	.c	C source code file; compressed
.asi	Borland C++ assembler		Unix file archive
.asp	ProComm Plus Aspect source	.c++	C++ source code file
	code	.c00	Ventura Publisher print
.at2	Aldus Persuasion 2 template	.ca	Telnet cache data
.atm	Adobe Type Manager data	.cab	Microsoft cabinet (compressed)
.au	audio/sound clip		file
.avi	Microsoft Windows audio video	.cal	Microsoft Schedule+ calendar
	interleaved format		file; Microsoft Windows 3
.b&w	Atari black and white graphics		calendar
.bad	Oracle bad file	.cap	Telix session capture file;
.bak	Microsoft Word backup file		Ventura Publisher caption
.bar	dBASE bar menu object file	.cat	dBASE IV catalogue
.bas	Basic source code	.cbl	Cobol source code file
.bat	DOS batch file	.cbt	computer-based training
.bbs	Bulletin Board System	.cc	C++ source code
	information file	.cch	CorelChart chart
.bch	dBASE batch process file	.cdk	Atari Calamus document
.bcp	Borland C++ makefile	.cdr	CorelDRAW vector graphics
.bcw	Borland C++ 4.5 environment	.cdt	CorelDraw data
	setting	.cfb	Compton multimedia
.bdf	Bitmap Distribution Format font		encyclopedia
	file (X11)	.cfg	configuration file
.bdr	Microsoft Publisher border	.cfl	CorelFLO chart
.bf2	Bradford 2 font	.cfn	Atari font data
.bfc	Microsoft Windows Briefcase	.cga	Ventura Publisher CGA display
	document		font
.bfm	Unix font metrics	.cgi	common gateway interface script
.bgi	Borland graphics interface device	.cgm	Computer Graphics Metafile
	driver	.chk	WordPerfect temporary
.bif	binary image format	.chl	configuration history log;
.bin	binary file; Macintosh binary file		RealPlayer channel file
	extension	.chm	compiled HTML help
.bk!	WordPerfect document backup	.chp	Ventura Publisher chapter file
.bkp	Microsoft Write backup	.chr	Turbo Pascal character set

.cht	dBASE ChartMaster interface file	**.cvw**	CodeView colour file
.cif	Ventura Publisher chapter information	**.cxx**	C++ source code file
		.dat	ASCII data file; temporary Internet movie clip (MPEG)
.cim	Sim City file	**.db**	dBFast configuration file; Paradox database
.cit	Intergraph scanned image		
.ckb	Borland C++ keyboard mapping	**.db$**	dBASE temporary file
.cl	Common LISP source code	**.dbc**	Visual FoxPro database
.clp	Quattro Pro clip art; Microsoft Windows clipboard file	**.dbf**	Lotus approach database
		.dbg	Microsoft C/C++ debugging information
.cls	C++ class definition file		
.cmd	dBASE command; OS/2 batch file	**.dbk**	dBASE IV database backup
		.dbo	dBASE IV compiled program
.cmp	CorelDRAW header	**.dbs**	SQL Windows database
.cmv	CorelDraw 4.0 animation	**.dbt**	Clipper database text
.cmx	Corel clipart file; Corel exchange format	**.dca**	IBM document content architecture file
.cnf	program configuration	**.dcf**	Lotus driver configuration file
.cnv	Microsoft Word data conversion file; WordPerfect temporary file	**.dcs**	Micrografx Photo Magic 4.0 image; QuarkXPress bitmap graphic
.cob	Cobol source code file		
.cod	dBASE template source file; FORTRAN program compiled code	**.dct**	dictionary
		.dcx	Zsoft multi-page Paintbrush
		.dd	Diskdoubler compressed (Apple Macintosh) file
.com	DOS executable command file; Internet domain suffix for commercial bodies		
		.ddb	bitmap graphics
		.ddp	OS/2 device driver profile
.cpi	DOS code page information	**.deb**	DOS debug script
.cpk	Copernic 99 file	**.def**	default
.cpl	Microsoft Windows control panel applet	**.dem**	demonstration
		.dep	Visual Basic dependency file
.cpp	C++ source code file	**.der**	Internet Security certificate
.cpt	dBASE encrypted memo	**.dev**	device driver
.crd	Microsoft Windows 3 cardfile	**.dfv**	Microsoft Word printing form
.crp	dBASE IV encrypted database	**.dgs**	diagnostic
.crs	WordPerfect file conversion	**.dib**	device-independent bitmap
.crt	Internet Security certificate	**.dic**	dictionary
.css	cascading style sheet datafile	**.dif**	Microsoft Excel data interchange format; VisiCalc database
.csv	comma separated value text file		
.ctl	Aldus control file; Visual Basic control	**.dir**	Macromedia Director movie; ProComm Plus dialing directory; VAX directory
.ctx	PGP RSA System cipher text file		
.cur	Microsoft Windows cursor	**.dis**	CorelDraw thesaurus; VAX Mail distribution file
.cut	Dr Halo picture file		
.cvp	WinFax cover page	**.dld**	Lotus 1-2-3 data
.cvs	Canvas image	**.dlg**	digital line graph file extension;
.cvt	dBASE IV converted backup file		

	Microsoft Windows SDK dialog resource script
.dll	Microsoft Windows dynamic link library
.dls	Norton Disklock setup
.dmp	memory dump
.doc	Microsoft Word document
.dos	DOS specific text information
.dot	CorelDRAW line-type definition; Microsoft Word template
.doz	file description out of zip (*aka* VENDINFO)
.dpr	Borland Delphi project file
.drs	WordPerfect display resource
.drt	Micrografx Windows Draw 4.0 template
.drv	device driver
.drw	Micrografx Drawing 4.0 drawing
.ds4	Micrografx Designer 4 vector graphics
.dsc	Oracle discard file
.dsk	Turbo Pascal project desktop
.dsp	Dr Halo graphics display driver
.dsr	WordPerfect driver resource
.dsw	Borland C++ desktop settings information
.dtp	Timeworks Publisher document
.dun	Dial-Up networking export file
.dvc	Lotus 1-2-3 data
.dvi	TeX device independent document
.dvp	AutoCAD device parameter; Desqview program information
.dw2	DesignCAD drawing
.dwg	AutoCAD drawing format
.dxf	AutoCAD vector drawing interchange format
.dxn	Fujitsu dexNET fax
.edt	VAX editor default settings
.edu	Internet domain suffix for educational institutions
.eeb	WordPerfect button bar for Equation Editor
.ega	Ventura Publisher EGA display font
.el	Emacs Elisp source code file
.elc	compiled ELISP code
.emf	enhanced metafile graphics
.enc	Encore music file; Lotus 1-2-3 UUencoded file
.end	CorelDRAW arrow-head definition file
.eng	Sprint dictionary engine
.env	WOPR Microsoft Word enveloper macro; WordPerfect environment file
.eps	Adobe Illustrator Encapsulated PostScript graphic; Micrografx Photo Magic 4.0 image
.eqn	WordPerfect equation
.err	error log/message
.esh	extended shell batch file
.evt	event log
.ex3	Harvard Graphics 3 device driver
.exc	Rexx source code file
.exe	executable program
.exm	MS DOS executable
.ext	Norton Commander Extension file
.f	Fortran source code file
.f0x	Dos text font - height x pixels
.f77	Fortran 77 source code file
.fac	Face graphics
.faq	frequently asked questions file
.fax	fax raster graphics
.fd	Fortran declaration file
.fdf	Adobe Acrobat forms document
.feb	WordPerfect figure editor button bar
.ff	Agfa outline font
.fH3	Aldus Freehand 3
.fhx	Aldus FreeHand x vector graphics
.fi	Fortran interface file
.fif	fractal image file
.fil	dBASE list object file; WordPerfect overlay
.fit	Microsoft Windows NT file index table
.fix	patch file
.flc	Autodesk animation ('flicks')
.fli	Autodesk animation ('flicks')
.flm	AutoCAD film roll

.flt	FileMaker Pro graphics filter file; Micrografx Publisher filter file	**.glb**	Microsoft Mail global system file
.fm	FileMaker Pro spreadsheet	**.gly**	Microsoft Word glossary
.fm1	Lotus 1-2-3 spreadsheet	**.gov**	US government Internet domain suffix
.fm3	Harvard Graphics 3 device driver; Lotus 1-2-3 v3 spreadsheet	**.gph**	Lotus 1-2-3 graph
		.gr2	Microsoft Windows 3 screen driver
.fmb	WordPerfect file manager button bar	**.grb**	MS-DOS shell monitor
.fmf	IBM font/icon file	**.grp**	Microsoft program group; Microsoft Windows 3 group file
.fmk	Fortran makefile	**.gz**	GZIP GNU zip archive
.fmm	FileMaker Pro	**.gzl**	Go!Zilla file list
.fmt	dBASE IV format file	**.h**	header file
.fmv	FrameMaker file extension (vector/raster format)	**.h++**	C++ header file
		.hdl	ProComm Plus download file list
.fn3	Harvard Graphics 3 font file	**.hdr**	ProComm Plus message header
.fnt	font file	**.hdw**	Harvard Draw vector image
.fo1	Borland Turbo C font file	**.hdx**	AutoCAD help index
.fon	ProComm Plus calls log; Microsoft Windows 3 font file; Telix dialing directory file	**.hel**	Microsoft Hellbender game file
		.hex	hex dump
		.hgl	Hewlett Packard graphics language
.for	Fortran source code		
.fot	font file	**.hh**	C++ header file
.fox	FoxBase	**.hhp**	ProComm Plus help
.fp	Claris FileMaker Pro; FoxPro configuration file	**.hlb**	VAX help library
		.hlp	Microsoft WinHelp
.fpc	FoxPro catalog	**.hmm**	ProComm Plus alternate mail menu
.fpx	Kodak flashPix		
.frm	order form Ascii text; Visual Basic form	**.hpf**	Hewlett Packard LaserJet font
		.hpg	Hewlett Packard GL plotter vector graphic
.frs	WordPerfect screen font resource		
.frt	FoxPro report	**.hpi**	GEM font information
.fsx	Lotus 1-2-3 data file; Microsoft embedded object form	**.hpj**	Microsoft help compiler
		.hpm	ProComm Plus Alternate menu for privileged users
.ftm	Micrografx font file		
.fw	FrameWork database	**.hpp**	C++ header file
.fw2	Framework 2 database	**.hqx**	BINHEX compressed Macintosh ASCII archive
.fxp	FoxPro compiled format		
.fxs	WinFax transmit graphic	**.hrf**	Hitachi raster format graphic
.gdf	GEOS dictionary file	**.hrm**	ProComm Plus alternate menu for limited/normal users
.gem	Digital Research graphics environment manager; Ventura GEM drawing		
		.hst	ProComm Plus history file
		.ht	HyperTerminal
.gen	dBASE Generator compiled template; Ventura Publisher generated text	**.htm**	document with HTML coding
		.html	hypertext markup language
		.htr	Microsoft Windows NT password-change request file
.gfb	compressed GIF		

.htx	hypertext file
.hxm	ProComm Plus alternate protocol selection
.hxx	C++ header file
.hy1	Ventura Publisher hyphenation algorithm
.hyd	WordPerfect hyphenation dictionary
.i	Borland C++ intermediate file
.icl	icon library
.ico	Microsoft Windows icon
.ide	Borland C++ project
.idw	IntelliDraw vector graphic
.idx	index
.iff	Amiga (bitmap) interchange file format
.ifs	OS/2 system file
.im8	Sun raster graphics
.img	Ventura Publisher bitmap image
.in$	Installation file (HP NewWave)
.in3	Harvard Graphics 3 input device driver
.inc	include file
.ind	dBASE index
.inf	generic information file; generic install/installation script (ASCII)
.ini	generic initialization file
.ink	CorelDRAW Pantone reference file
.ins	install/installation script (ASCII)
.int	Borland interface unit
.inx	Foxbase index
.ipl	CorelDRAW Pantone spot reference palette
.iqy	Microsoft Excel web query file
.irs	WordPerfect resource
.isd	Isbister Time and Chaos
.ism	Isbister Time and Chaos
.ist	Isbister Time and Chaos
.iwa	IBM Writing Assistant
.jav	Java source code
.jbf	Paint Shop Pro browser file
.jff	JPEG file interchange format
.jor	SQL journal file
.jou	VAX editor journal backup
.jpg	Joint Photographic Expert Group image
.jtf	Hayes JT fax; JPEG tagged interchange format image
.kar	karaoke Midi file
.kb	Borland C+ keyboard script
.kbd	ProComm Plus keyboard (mapping)
.kcl	Kyoto Common Lisp source code
.kdc	Kodak digital camera file
.key	Iolo Macro Magic keyboard macro
.l	Lisp source code file
.lab	MS Excel mailing label
.lbg	dBASE IV label generator
.lbl	dBASE IV label
.lbo	dBASE IV compiled label
.lbr	Lotus 1-2-3 display driver
.lbx	FoxPro label
.lcf	Norton linker control file
.lck	Paradox Lockfile
.lcn	WordPerfect Lection
.ld	Telix long-distance codes
.ldb	MS Access data
.lex	lexicon
.lgo	Microsoft Windows startup logo
.lha	LHA compressed archive
.lib	library file
.lif	HP logical interchange format
.lis	VAX listing
.lj	Hewlett Packard LaserJet printer text file
.ll3	LapLink III document file
.lnk	Microsoft Windows link (ie shortcut)
.lod	loading file
.log	(Microsoft Windows) log file
.lrf	Microsoft C++ linker response file
.lrs	WordPerfect language resource
.lsf	Advanced Streaming format file
.lsp	Lisp source code file
.lst	list file
.lsx	Advanced Streaming Redirector file
.ltm	Lotus form
.lwd	LotusWorks text file
.lwp	Lotus Word Pro 97 document
.m3	Modula 3 source code file

.m3d	3D animation macro	.mke	Microsoft Windows SDK makefile
.ma3	Harvard Graphics 3 macro		
.mac	Macintosh MacPaint bitmap image; macro	.mlb	Symphony macro library file
		.mmf	MS Mail message file
.mai	VAX mail	.mnd	AutoCAD menu source
.mailrc	Unix mail configuration file	.mng	Paint Shop Pro Animation Shop animation
.mak	Microsoft Visual Basic Project file		
		.mnt	FoxPro menu memo
.man	manual	.mnu	AutoCAD menu; Norton Commander menu
.map	Micrografx Picture Publisher format data		
		.mnx	AutoCAD compiled menu; FoxPro menu
.mar	VAX macro assembly		
.mas	Lotus Freelance Graphics SmartMaster	.mny	MS Money account book
		.mod	Commodore Amiga sample music file; Microsoft Windows kernel module
.mbk	dBASE IV multiple index file backup		
.mbs	Micrografx brush stroke	.motif	Microangelo motif
.mbx	mailbox	.mov	AutoCAD movie
.mcf	Mathcad font	.mp2	MPEG audio file (Xing)
.mcp	Mathcad printer driver	.mp3	audio compression format
.mcr	Iolo Macro Magic macro	.mpc	MS Project calendar file
.mcw	MacWrite II text file	.mpeg	MPEG movie clip
.mda	MS Access data	.mpg	MPEG animation
.mdb	MS Access database	.mpm	WordPerfect mathplan macro
.mdm	TELIX modem definition	.mpp	MS Project project file
.mdx	dBASE IV multiple index file	.mpr	FoxPro generated program; Lotus Approach SmartMaster
.me	READ.ME text file		
.med	WordPerfect macro editor delete save	.mpv	MS Project view file
		.mpv2	MPEG movie clip
.mem	dBASE IV memory variable save; WordPerfect macro editor macro	.mpx	FoxPro compiled menu
		.mrb	Microsoft C/C++ multiple resolution image
.meq	WordPerfect macro editor print queue		
		.mrs	WordPerfect macro resource file
.mes	generic message; WordPerfect macro editor work space	.msc	Microsoft C makefile
		.msg	message
.met	WordPerfect macro editor top overflow	.msp	Microsoft Paint bitmap
		.mst	Microsoft Windows SDK setup script
.meu	DOS shell menu group		
.mex	WordPerfect macro editor expound file	.msw	Microsoft Word text file
		.mtm	Multitracker Module music
.mgf	Micrografx font	.mu	Quattro Pro menu
.mic	Microsoft Image Composer	.mus	MusicTime sound file
.mid	MIDI file	.mvb	Microsoft Multimedia Viewer
.miff	ImageMagick file format	.mvf	AutoCAD stop-frame file movie clip
.mil	US armed services Internet domain suffix		
		.mvi	AutoCAD movie command file
		.mwp	Lotus Word Pro 97 SmartMaster

.mys	Myst saved game file	**.pak**	Quake game data
.ncd	Norton Commander change directory file	**.pal**	CorelDRAW colour palette; Micrografx palette; Paintbrush palette file
.ndx	dBASE IV index file		
.neo	Atari Neochrome raster image	**.pan**	CorelDRAW printer-specific file
.net	Internet domain suffix for network operators; network configuration/information file	**.part**	Go!Zilla partial (incomplete) download
		.pas	Pascal source code file
.ng	Norton Online Guide documentation database	**.pat**	AutoCAD patterns; CorelDRAW vector pattern file
.nlm	Netware loadable module	**.pb**	WinFax Pro phonebook
.npi	source for dBASE Application Generator interpreter	**.pbi**	Microsoft Source Profiler binary input
.nt	Microsoft Windows NT startup files	**.pbm**	planar bitmap graphic; portable bitmap (UNIX picture)
.nuf	ProComm Plus message for new users	**.pbo**	Microsoft Source Profiler profiler binary output
.nxt	NeXT sound format	**.pc**	IBM PC-specific information
.o	UNIX object file	**.pc3**	Harvard Graphics 3 custom palette
.oab	Microsoft Outlook Address Book		
.ob	IBM LinkWay Object cut-paste	**.pcc**	PC Paintbrush picture vector graphic
.obd	Microsoft Office Binder		
.obj	Intel object module code	**.pcd**	Kodak PhotoCD image
.obr	Borland C++ object browser	**.pcf**	Microsoft Source Profiler command file
.ocx	OLE custom control		
.odl	Visual C++ type library source	**.pch**	Microsoft C/C++ precompiled header; patch file
.olb	VAX object library		
.old	backup file/renamed	**.pck**	Turbo Pascal pickfile
.oli	Olivetti text file	**.pcl**	Hewlett Packard printer control language
.opt	QEMM optimize support file		
.or3	Lotus Organizer 97	**.pct**	Macintosh black and white bitmap image
.ora	Oracle parameter file		
.org	Internet domain suffix for non-commercial organizations; Lotus Organizer calendar file	**.pcw**	PC Write text file
		.pcx	PC Paintbrush bitmap image
		.pdc	Mijenix PowerDesk configuration
.otl	Z-Soft Type Foundry outline font description	**.pdd**	Adobe PhotoDeluxe image
.otx	Olivetti Olitext Plus	**.pdf**	Adobe Acrobat portable document format
.out	output file		
.ov1	overlay file	**.pdl**	Borland C++ project description language
.ozm	Sharp Organizer memo bank		
.ozp	Sharp Organizer telephone bank	**.pdv**	Paintbrush printer driver
.p	Pascal source code file	**.pdw**	Professional Draw document
.p16	ProTracker Studio 16 channel music	**.pem**	WordPerfect program editor macro
.pab	Microsoft Exchange address book	**.peq**	WordPerfect program editor print queue

.pfa	PostScript font type 3	**.pro**	DOS graphics profile; Prolog source code file
.pfb	encrypted Type 1 font		
.pfm	Postscript font metrics	**.prs**	dBASE IV procedure; WordPerfect printer resource (*ie* fonts)
.pgl	UNIX portable greymap		
.pgm	portable greymap		
.pgp	Pretty Good Privacy	**.prx**	FoxPro compiled program
.ph	Perl header file	**.ps**	ASCII Postscript file
.phr	LocoScript phrases	**.psd**	RGB/indexed PhotoShop image file
.pic	Lotus pictor/PC Paint picture file		
.pict	Apple Macintosh image format	**.psm**	Turbo Pascal symbol table
.pif	picture interchange format; (Microsoft Windows) program information file	**.pt3**	Harvard Graphics 3 device (printer) driver; PageMaker 3 template
.pit	PackIt - Apple Macintosh compressed file	**.pub**	Microsoft Publisher page template; Pretty Good Privacy public key ring (RSA); Ventura Publisher publication
.pjt	FoxPro project memo		
.pkg	Next installer script		
.pkt	Fidonet packet	**.pw**	Professional Write text file
.pl	Harvard Graphics palette; Perl script; Prolog source code file	**.pwd**	Microsoft Pocket Word document
		.pwl	Microsoft Windows password file
.pl3	Harvard Graphics 3 chart palette		
.plb	FoxPro library	**.pwp**	Professional WritePlus text file
.pln	WordPerfect spreadsheet	**.px**	Paradox primary database index
.plt	Lotus Pro Organizer layout	**.qag**	Norton Desktop quick access group
.pm3	PageMaker 3 document		
.pn3	Harvard Graphics 3 printer device driver	**.qbe**	dBASE IV query
		.qbw	QuickBooks spreadsheet
.png	portable network graphics bitmap image	**.qdk**	QEMM startup backup files
		.qic	backup set for Microsoft Backup
.pnt	Apple Macintosh painting	**.qlb**	Microsoft C/C++ quick library
.pop	dBASE pop-up menu object; PopMail messages index	**.qpr**	FoxPro generated query program; OS/2 print queue driver
.ppb	WordPerfect print preview button bar	**.qpx**	FoxPro compiled query program
		.qrk	Quark document
.ppd	PostScript Printer description	**.qry**	dBASE IV Query
.ppm	Portable PixelMap	**.qt**	Apple Quick Time movie
.ppp	Serif PagePlus publication	**.qtp**	Apple QuickTime preferences
.pps	Microsoft PowerPoint Slideshow	**.qtx**	Apple QuickTime extension
.ppt	Microsoft PowerPoint presentation	**.qxd**	QuarkXPress document
		.qxl	QuarkXPress element library
.pr2	Aldus Persuasion 2 presentation	**.ra**	RealAudio music file
.pr3	dBASE IV Postscript printer driver	**.ram**	RealAudio Ramfile
		.ras	Sun raster image
.pre	Freelance Graphics presentation	**.raw**	RAW RGB 24-bit image
.prg	Atari program file	**.rbf**	Rbase datafile
.pri	LocoScript printer definitions	**.rc**	Emacs configuration file;
.prn	Lotus 1-2-3 text file		

	Microsoft C/C++ resource script; Unix run commands file		LocoScript screen font; Microsoft Windows screensaver
.rcg	Netscape newsgroup file	.sct	FoxPro screen memo; Lotus screen capture text
.rec	Microsoft Windows 3 recorded macro file	.scy	security file
.ref	(cross) reference	.sda	Fidonet's Software Distribution Network archive
.reg	registration entry		
.rem	generic remarks	.sdf	system data format (ASCII file)
.rep	Lotus Organizer 97 report	.sdi	Software Distribution Network information
.req	generic request		
.res	Borland C++ compiled resource	.sdl	SmartDraw library
.rev	Geoworks revision file	.sdn	Software Distribution Network compressed archive
.rex	Oracle report definition; Rexx source code file	.sdr	SmartDraw drawing
.rez	resource file	.sdt	SmartDraw template
.rf	Sun raster image	.sea	self-extracting compressed archive (Apple Macintosh)
.rft	IBM DisplayWrite revisable format	.sec	Disney Animation Studio secured animation; PGP secret key ring file
.rh	Borland C++ Resource header		
.ric	Ricoh fax		
.rif	Fractal Design Painter riff bitmap	.sep	printer separator page
.rle	run-length encoded	.set	setup options file
.rm	RealMedia file (Real Player)	.sgi	Silicon Graphics Inc (IRIS *qv*)
.rno	VAX runoff file	.sh	Unix shell script
.rnx	RealPlayer (first run) file	.sh3	Harvard Graphics 3 presentation
.rp	RealPix clip (Real Player)	.shb	CorelSHOW background
.rpt	generic report	.shg	segmented-graphics bitmap
.rs	Amiga resource data file	.shk	SHRINKIT Apple II compressed archive
.rsc	resource file		
.rtf	rich text format	.shm	WordPerfect shell macro
.rtl	runtime library	.shp	AutoCAD shape file; Micrografx Windows Draw Cool Shapes
.rws	Borland C++ resource workshop data file		
.s	UNIX assembly source code file	.shw	CorelSHOW presentation; WordPerfect Presentation slide show
.sav	saved game position		
.sc	Paradox Pal script; signed byte audio file	.shx	AutoCAD shape entities
		.sif	Microsoft Windows NT setup installation information
.sc3	Harvard Graphics 3 screen device driver; renamed dBASE III screen mask file	.sit	Stuffit file extension (Apple Macintosh)
.sci	scalable content interface; system configuration information	.sks	Fractal Design Expression stroke definition
.scm	Lotus ScreenCam movie clip	.slb	AutoCAD slide library
.scp	dial-up networking script	.slc	Telix compiled SALT script
.scr	dBASE IV screen snapshot; DOS Debug source code file;	.sld	AutoCAD slide
		.slk	Microsoft Excel SLK data import format

.sll	sound data file		(device driver/hardware
.slt	Telix SALT script application		configuration)
	language	.tah	Borland C++ Turbo assembler
.sm	Smalltalk source code file		help file
.smm	Lotus Word Pro 97 SmartMaster	.tar	tape archive (Unix compressed
	macro		archive)
.smp	sample sound file	.tb1	Borland Turbo C font file; OS/2
.snd	Macintosh/PC digitized sound		table of values
	clip	.tbm	Visual Basic TextBase custom
.snm	Netscape mail		manager
.sno	Snobol4 source code	.tc	Borland Turbo C configuration
.sol	game solution		file
.som	Paradox sort information;	.tch	Borland Turbo C help file
	Quattro Pro network serial	.tcw	TurboCAD drawing
	number	.td	DOS Turbo Debugger
.spc	WordPerfect temporary file		configuration file
.spg	Sprint glossary	.tdh	Turbo Debugger help file
.spi	Siemens/Philips scanner image	.tdk	Turbo Debugger keystroke
.spl	Microsoft Windows 3 print		recording
	spooling file	.tel	Telnet host file
.spr	FoxPro generated screen	.tem	Borland C++ Turbo Editor macro
	program; sprite		script
.sql	SQL query/report	.tex	Scientific Word text file
.srf	Sun raster file image	.tfm	Intellifont tagged font metric file
.ssm	standard Streaming metafile	.tga	Truevision Targa graphics
.st	Smalltalk source code file		adaptor bitmap
.sto	Pascal stub OBJ file	.tgz	TAR/GNUzip compressed
.str	dBASE structure list object file		archive
.sts	Microsoft C/C++ project status	.ths	WordPerfect/Lotus thesaurus
	information	.tif	tagged image file format (RFC
.sty	style sheet		1314)
.sum	summary	.tlb	VAX text library; Visual C++
.sun	Sun raster graphic		type library
.sup	WordPerfect supplementary	.tmf	WordPerfect tagged font metric
	dictionary	.tmp	Microsoft Windows temporary
.svd	Microsoft Word autosave		file
	document	.toc	table of contents
.swf	Macromedia Shockwave Flash	.tp	Turbo Pascal configuration
	object	.tp3	Harvard Graphics 3 template
.swp	Microsoft Windows swap file	.tph	Turbo Pascal help file
.sy1	Ami Pro Smartpix symbol library	.tpl	Harvard Graphics template;
.sy3	Harvard Graphics 3 symbol file		Turbo Pascal resident units
.sym	Borland C++ precompiled		library
	headers	.tpp	Borland Pascal 7.0 protected
.syn	Microsoft Word synonym file		mode unit
.sys	Microsoft Windows system file	.tpu	Turbo Pascal library file
		.tpw	Turbo Pascal for Windows unit

.tpz	TAR/GNUzip compressed file archive	**.vsd**	Visio diagram
.trm	Microsoft Windows 3 terminal settings	**.vsh**	McAfee VirusShield configuration
.trn	Quattro translation support	**.vss**	Visio SmartShapes file
.trs	Micrografx executable file	**.vst**	Vista Truevision bitmap
.tst	WordPerfect printer test file	**.vxd**	Microsoft Windows virtual device driver
.ttf	TrueType Font	**.w31**	Microsoft Windows 3 start-up file
.tut	tutorial		
.tv	Paradox table view settings	**.wav**	Microsoft Windows wave sound file
.tvf	dBASE table view settings		
.txi	TeX support file	**.wbk**	Microsoft Word document backup file; WordPerfect workbook
.txt	text file		
.tym	PageMaker 4 time stamp		
.uld	ProComm Plus uploaded files information	**.wcd**	WordPerfect macro token list
		.wcm	Corel WordPerfect 7 macro; Microsoft Works data transmission file
.unx	UNIX information text file		
.upd	dBASE update data		
.upo	dBASE compiled update data	**.wdb**	Microsoft Works database
.usp	PageMaker USASCII extended character set printer font	**.wfn**	CorelDRAW font
		.wfx	WinFax data file
.usr	FileMaker Pro 3.0 runtime database; ProComm Plus user database file	**.wht**	Microsoft NetMeeting whiteboard file
		.wid	Ventura Publisher width table
.uu	UUDE/ENCODE compressed ASCII archive	**.wiz**	Microsoft Office wizard
		.wkq	Quattro spreadsheet
.uue	UUENCODE compressed ASCII file archive	**.wks**	XLisp workspace
		.wmc	WordPerfect macro file
.val	Paradox validity checks/referential integrity	**.wmf**	Microsoft Windows metafile vector image format
.van	VistaPro animation file	**.wp**	WordPerfect text file
.vbp	Microsoft Visual Basic project	**.wpd**	WordPerfect demonstration
.vbx	Microsoft Visual Basic eXtension	**.wpf**	WordPerfect form
.vc	VisiCalc spreadsheet	**.wpg**	WordPerfect Graphics vector format
.vcw	Microsoft Visual C++ Visual workbench		
		.wps	Microsoft Works text file
.vcx	VisiCalc Spreadsheet	**.wq!**	Quattro Pro compressed spreadsheet
.vew	Lotus Approach view file		
.vga	VGA display driver	**.wri**	Microsoft Write document
.vid	MS-DOS 5 shell monitor file	**.wrp**	WARP Compressed Amiga archive
.vmc	Acrobat Reader virtual memory configuration		
		.wrs	WordPerfect Windows resource
.vmf	Ventura Publisher font	**.ws**	WordStar Text file
.vms	virtual memory system	**.ws2**	WordStar 2000 text file
.vrm	QuattroPro overlay file	**.wsd**	WordStar document
.vrs	WordPerfect video resource	**.wsp**	Fortran PowerStation Workspace
.vsc	McAfee VirusScan configuration		

.wst	Claris Home Page site definition; WordStar text file
.wwk	WordPerfect keyboard layout
.x	Lex source code file
.x32	Macromedia Shockwave decompression Xtra; RealAudio Director Plug-in
.xar	Corel Xara drawing
.xbm	X11 bitmap image; X-Window bitmap
.xdm	StreamWorks metafile
.xif	Xerox Pagis Pro file format
.xla	Microsoft Excel add-in
.xlb	Microsoft Excel worksheet
.xlc	Microsoft Excel chart
.xlk	Microsoft Excel backup file
.xll	Microsoft Excel dynamic link library
.xlm	Microsoft Excel 4.0 macro
.xls	Microsoft Excel worksheet
.xlt	Microsoft Excel template
.xlw	Microsoft Excel workspace/workbook
.xmi	compressed eXtended MIDI music
.xrf	cross-reference file
.xtb	LocoScript external translation table
.xwd	X-Window system window dump file
.xwp	Xerox Writer text file
.y	YABBA-compressed Amiga archive; Yacc grammar file
.z3	Infocom game module
.zip	PKZip/WinZip compressed archive
.zom	ZOOM-compressed Amiga archive

Annex II

Country codes – top level domains

Code	Country	Code	Country
.ac	Ascension Island	.cg	Congo
.ad	Andorra	.ch	Switzerland
.ae	United Arab Emirates	.ci	Côte d'Ivoire
.af	Afghanistan	.ck	Cook Islands
.ag	Antigua & Barbuda	.cl	Chile
.ai	Anguilla	.cm	Cameroon
.al	Albania	.cn	China
.am	Armenia	.co	Colombia
.an	Netherlands Antilles	.cr	Costa Rica
.ao	Angola	.cs	Czechoslovakia
.aq	Antarctica	.cu	Cuba
.ar	Argentina	.cv	Cape Verde
.as	American Samoa	.cx	Christmas Island
.at	Austria	.cy	Cyprus
.au	Australia	.cz	Czech Republic
.aw	Aruba	.de	Germany
.az	Azerbaijan	.dj	Djibouti
.ba	Bosnia & Herzegovina	.dk	Denmark
.bb	Barbados	.dm	Dominica
.bd	Bangladesh	.do	Dominican Republic
.be	Belgium	.dz	Algeria
.bf	Burkina Faso	.ec	Ecuador
.bg	Bulgaria	.ee	Estonia
.bh	Bahrain	.eg	Egypt
.bi	Burundi	.eh	Western Sahara
.bj	Benin	.er	Eritrea
.bm	Bermuda	.es	Spain
.bn	Brunei Darussalam	.et	Ethiopia
.bo	Bolivia	.fi	Finland
.br	Brazil	.fj	Fiji
.bs	Bahamas	.fk	Falkland Islands
.bt	Bhutan	.fm	Micronesia
.bv	Bouvet Island	.fo	Faroe Islands
.bw	Botswana	.fr	France
.by	Belarus	.fx	France (Metropolitan)
.bz	Belize	.ga	Gabon
.ca	Canada	.gb	United Kingdom
.cc	Cocos Islands	.gd	Grenada
.cd	Zaire	.ge	Georgia
.cf	Central African Republic	.gf	French Guiana

.gg	Guernsey	.la	Lao People's Democratic
.gh	Ghana		Republic
.gi	Gibraltar	.lb	Lebanon
.gl	Greenland	.lc	Saint Lucia
.gm	Gambia	.li	Liechtenstein
.gn	Guinea	.lk	Sri Lanka
.gp	Guadelope	.lr	Liberia
.gq	Equatorial Guinea	.ls	Lesotho
.gr	Greece	.lt	Lithuania
.gs	South Georgia & South	.lu	Luxembourg
	Sandwich Islands	.lv	Latvia
.gt	Guatemala	.ly	Libyan Arab Jamahiriya
.gu	Guam	.ma	Morocco
.gw	Guinea-Bissau	.mc	Monaco
.gy	Guyana	.md	Moldova
.hk	Hong Kong	.mg	Madagascar
.hm	Heard & McDonald Islands	.mh	Marshall Islands
.hn	Honduras	.mk	Macedonia
.hr	Croatia	.ml	Mali
.ht	Haiti	.mm	Myanmar (Burma)
.hu	Hungary	.mn	Mongolia
.id	Indonesia	.mo	Macau
.ie	Ireland	.mp	Northern Mariana Islands
.il	Israel	.mq	Martinique
.im	Isle of Man	.Mr	Mauritania
.in	India	.ms	Montserrat
.io	British Indian Ocean Territory	.mt	Malta
.iq	Iraq	.mu	Mauritius
.ir	Iran	.mv	Maldives
.is	Iceland	.mw	Malawi
.it	Italy	.mx	Mexico
.je	Jersey	.my	Malaysia
.jm	Jamaica	.mz	Mozambique
.jo	Jordan	.na	Namibia
.jp	Japan	.nc	New Caledonia
.ke	Kenya	.ne	Niger
.kg	Kyrgyzstan	.nf	Norfolk Island
.kh	Cambodia	.ng	Nigeria
.ki	Kiribati	.ni	Nicaragua
.km	Comoros	.nl	The Netherlands
.kn	Saint Kitts & Nevis	.no	Norway
.kp	Korea (Democratic People's	.np	Nepal
	Republic of)	.nr	Nauru
.kr	Korea (Republic of)	.nt	Neutral Zone
.kw	Kuwait	.nu	Niue
.ky	Cayman Islands	.nz	New Zealand
.kz	Kazakstan	.om	Oman

.pa	Panama	.to	Tonga
.pe	Peru	.tp	East Timor
.pf	French Polynesia	.tr	Turkey
.pg	Papua New Guinea	.tt	Trinidad & Tobago
.ph	Philippines	.TV	Tuvalu
.pk	Pakistan	.tw	Taiwan
.pl	Poland	.tz	Tanzania
.pm	St Pierre & Miquelon	.ua	Ukraine
.pn	Pitcairn	.ug	Uganda
.pr	Puerto Rico	.UK	United Kingdom
.pt	Portugal	.um	United States Minor Outlying
.pw	Palau		Islands
.py	Paraguay	.us	United States
.qa	Qatar	.uy	Uruguay
.re	Réunion	.uz	Uzbekistan
.ro	Romania	.va	Holy See (Vatican City State)
.ru	Russia	.vc	Saint Vincent & the Grenadines
.rw	Rwanda	.ve	Venezuela
.sa	Saudi Arabia	.vg	Virgin Islands (British)
.sb	Solomon Islands	.vi	Virgin Islands (US)
.sc	Seychelles	.vn	Vietnam
.sd	Sudan	.vu	Vanuatu
.se	Sweden	.wf	Wallis & Futuna Islands
.sg	Singapore	.ws	Samoa
.sh	St Helena	.ye	Yemen
.si	Slovenia	.yt	Mayotte
.sj	Svalbard & Jan Mayen Islands	.yu	Yugoslavia
.sk	Slovakia	.za	South Africa
.sl	Sierra Leone	.zm	Zambia
.sm	San Marino	.zr	Zaire
.sn	Senegal	.zw	Zimbabwe
.so	Somalia		
.sr	Surinam		
.st	Sao Tome and Principe		
.su	USSR (former)		
.sv	El Salvador		
.sy	Syrian Arab Republic		
.sz	Swaziland		
.tc	The Turks & Caicos Islands		
.td	Chad		
.tf	French Southern Territories		
.tg	Togo		
.th	Thailand		
.tj	Tajikistan		
.tk	Tokelau		
.tm	Turkmenistan		
.tn	Tunisia		